The Invincible

Deadly Mosquitoes

India's health and economy enemy # 1

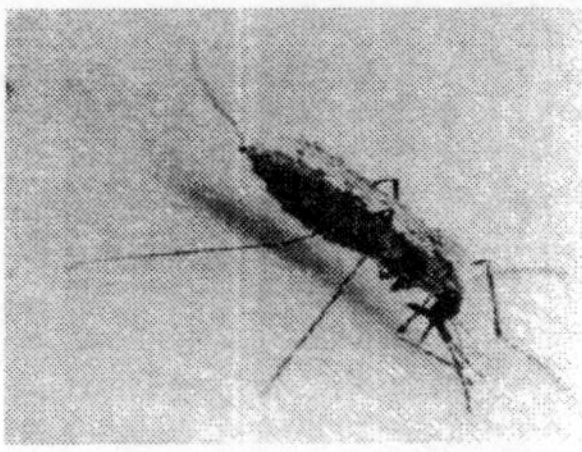
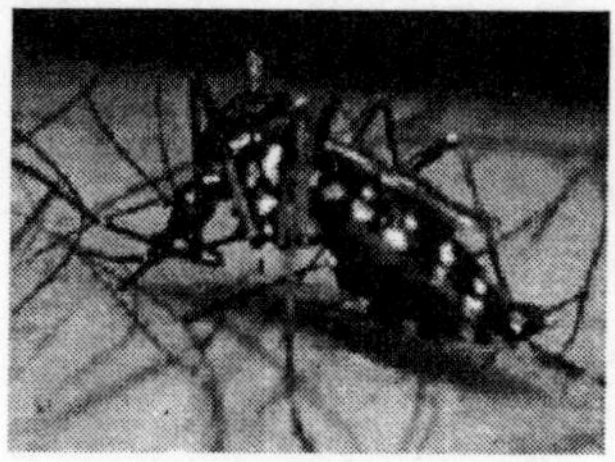

Dr. B.K. Tyagi

MSc, DPhil, FRES, FISCD, FISP, FAZRA

Deputy Director (Senior Grade)

Head, Division of Insecticide and Vector Control

Centre for Research in Medical Entomology (ICMR)

Madurai 625002, India

&

President, Jodhpur Natural History Society

(Formerly *the Officer in-Charge, Desert Medicine Research Centre, Jodhpur; National Representative in-Charge, National Office in India for the Societas Internationalis Odonatologica; Member in-Charge in South Asia, Odonata Specialist Group of the Species Survival Commission, IUCN*)

SCIENTIFIC PUBLISHERS (INDIA)

P.O. BOX 91 JODHPUR

Published by:
Pawan Kumar
Scientific Publishers (India)
5-A, New Pali Road, P.O. Box 91
JODHPUR – 342 001
Tel.: +91-291-2433323 Fax: +91-291-2512580
E-mail: info@scientificpub.com

ISBN: 81-7233-365-X

Lasertype set : Rajesh Ojha
Printed in India

DEDICATED TO

All my predecessors and contemporaries who devoted whole-heartedly their lives for furthering our understanding about the biology, as well as significance in public health and national economy, of this age-old *bete noire*, mosquito - the *female fatale*, and fought against it through all thick and thin, leaving behind a great legacy and stimulus for future generation scientists to continue an unfinished task so that there would be a better world to live in.

PREFACE

"You sleep under a mosquito net,
even one insecticided.
You close the windows,
even screened additionally.
You spray the walls in
and drains outside your home.
You use all kinds of repellents,
and vaporisers, too.
Then you are praying that mosquitoes don't bite."

Such is the fear of this tiny creature, called mosquito, that nobody feels safe from its attacks, including the perils of transmission of deadly and highly debilitating infections such as malaria, dengue, filariasis, yellow fever and several types of encephalitides. India houses all the important diseases but yellow fever and most of encephalitides, although the deadly Japanese encephalitis continues to prevail in several parts of the country. The physical, psychological and monetary losses to the mosquito-borne diseases and their transmitters *per se* are astronomical and nearly impossible to figure out correctly. The annual deaths and morbidities as well as consequential impacts on economies of both the individuals and nations are sufficient as an indicator for the growing impoverishment of the affected countries worldwide. Many countries like United States of America, France, United Kingdom, The Netherlands, which were until early 20th century infested by both mosquitoes and several types of infections these transmitted, particularly malaria, had got themselves rid of most of these maladies due to extensive and intensive control programmes aided adequately by improved economic situations, living conditions and general awareness about the vector-borne infections. Some other countries in the Tropics, such as those in Central America and Southeast Asia, which were always the worst affected for centuries, launched some of the exemplary and massive eradication/control programmes against the culprit – mosquito, and overcame it with thundering successes. Still, close to 100 countries continue to suffer with malaria alone which saps off much of the human energy and national economy in underdeveloped countries. Mosquito, the man's deadliest foe, is small in size but has caused insurmountable problems to the successful survival of human himself. It seeks blood of man and his associates, enfeebling his domination on earth.

Mosquitoes are pure pests benefiting nothing other than the pathogens they carry. Yet, with more than 3,500 species in their family, mosquitoes are among the most varied animals on the planet found in every climate, from the wind-swept frozen Arctic tundra to the humid jungles of the Amazon, and from the inhospitable hot deserts to the deeply holed coal mines! Interestingly, not all mosquitoes, not even all female mosquitoes, drink blood. One species relies solely on the nutrients in water that collects in the leaves of the pitcher plants. Another swoops in on obliging ants, sticks its proboscis down the ants' throats, stimulate the ants to regurgitate, and sucks up the resulting food. Some species of mosquitoes are quite elegant in their appearance and/or behaviour. Mosquitoes present a highly fascinating and varied mating behaviour in different species. Some form huge swarms above the treetops, creating a buzzing, dusky cloud. In one particular rapacious species, the males typically impregnate females as the latter emerge helplessly from their pupal casings; the females fly away, minutes old and impregnated. Remotely though, a few mosquitoes help reduce vector population of certain other cousins, thereby proving their worth as effective biological tools in the man's armory to combat dangerous mosquitoes.

Man (*Homo sapiens*) may boast himself to be the most superior species in the animal evolution, but this is only one side of the coin. Evolutionists may possibly have a different opinion since mosquitoes were born on earth much more earlier (in fact, several million years ago!) than either the present day man or even his distant predecessors, the apes. Mosquitoes, as minute nothings in contrast with large animals have at least two striking evolutionarily acquired reproductive advantages. (i) First, they can mobilize vast numbers in small space and little time. Considering a female mosquito lays between 50 and 500 eggs, with an average of 200 eggs, in her first blood, slightly lower in each subsequent delivery, of which they may be eight or, maximally, ten before she dies. If by general hypothesis, 50% of these are destined to be females, then theoretically there will be formed 20 million females in five generations out of solitary mother insect. In some warm countries, including India where in some parts reproduction goes on throughout the year, one female mosquito could presumably be the ancestor to the progeny comprising twice ten to the twenty-sixth power - quite an inconceivable figure! Providentially, in nature it does not happen, but even so the minimal survivors are capable to replenish the loss and keep the density pretty high to cause concern to human and his associates. On feeding side, a female mosquito's reproductive exuberance demands large quantities of protein for maturation of eggs. She is therefore a voracious blood feeder. An average meal comes to two and half times her unfed weight, meaning 1-4 mg. If all the world's anthrophagic mosquitoes are theoretically considered at once then, to one's astonishment, the amount of blood squeezed by the glutton mosquitoes is good enough to equal over a thousand men's! (ii) Secondly, in going through a large number of generations in a short time, the mosquito acquires a genetic amicability or adaptability to its environment, making the sucker more fitter to survive successfully in any unforeseen event of environmental subtleties. To say least, in the evolution sense, the weak are those easily carried off by disease or by other foes, but the strong always stand longest against adversity.

Mosquitoes, in fact, are both ugly and beautiful at the same time. Surely, mosquitoes should be subjected to serious scientific inquiry for all kinds of damages they inflict on human, but they must also be appreciated for what they can be used to alleviate human suffering. In spite of the fact that over a century has elapsed in having explored the mosquito with microscopic details, this is nevertheless no exaggeration to accept that we still need to do a good homework to actually comprehend the little beast, mosquito!

The birth of this publication coincides with the year of receiving the Nobel Prize – 1902 by Dr. Ronald Ross, in 1904. It is hoped that this book written keeping in mind both the professional medico-entomology students and amateurs alike will serve a good purpose to emulate the legacy of untiring perseverance, unflinching patience and focused dedication in pursuing good quality research as preached through their deeds by the great mosquito men of the late 19th and early 20th centuries. In this reference the golden words of the famous Italian neuro-biologist and Nobel Laureate Dr. Camillo Golgi (1902; see Mazzarello, 1999) are good to follow: **"The more solid scientific knowledge .... are never conquered with flights of fancy, which can only lead to an appearance of progress, but with minute, methodical, daily work, which leading to the solid acquisition of a single fact, creates the unshakeable knowledge of the laws of life".**

Dr. B.K. Tyagi

FOREWORD

The insect world has always spawned mixed reactions among humankind throughout history – some love them and some abhor them; few are indifferent to them. Many insects are beautiful, others are useful. Some are a nuisance, others a menace.

Mosquitoes are not exactly the favourite insects of humankind. The reaction of the average human to a visiting mosquito is to squash it as soon as possible. Or use a bewildering array of killing techniques, ranging from insecticides to electrical currents, to dispose of it. Or, at the very least, drive it away (temporarily, of course) with repellents. Most never see the beauty of these beasts, their wonderful adaptation to what is essentially a piratical life style, their subtle (sometimes brilliant) colouring and scalation that are only fully revealed under a microscope. And not all mosquitoes are harmful – indeed, relatively few of the vast array of mosquito species buzzing around our ears are the deadly disease carriers that we imagine them to be. Most do not even feed on humans, preferring animal blood instead. While their buzzing and singing irritate, their painful bites infuriate – but they are simply going about the business of ensuring that there will be a next generation, something that every self-respecting life form attempts to do.

B.K. Tyagi's new book, *Deadly Mosquitoes*, takes us into the complex world of the mosquito, painting a picture of the bad and the good, the ugly and the beautiful about this little insect. Written in a racy, populist style that will appeal to the general reader, it nevertheless contains much in the way of little-known facts to interest the specialist as well. Tyagi is best known internationally among malaria and medical entomology circles for his long years of research into mosquitoes, malaria and irrigated agriculture in the Thar Desert of Rajasthan, India. In this book, he sheds the traditional staid prose of the serious researcher and gives full rein to a vivid dramatization of the mosquito's life and times. I wish him and this book well.

07 April 2004.

Dr. Felix P. Amerasinghe
International Water Management Institute
Colombo, Sri Lanka

ACKNOWLEDGEMENT

This book is actually a tribute to all those scientists who relentlessly toiled to free humanity from the fears of both the deadly and debilitating diseases, and their transmitters – mosquito! Their legacy of undaunted hard work and application has been my encouragement to write this book.

Foremost, I wish to respectfully express my gratitude to my mother organization - Indian Council of Medical Research, under whose tutelage I have gained limitless opportunities and experiences on mosquito sciences in various parts of the country during past two and a half decades, and the much needed stimulus to write this book.

In accomplishing this task I have been enormously helped by a large number of colleagues and friends both in India and abroad who generously provided published or unpublished material. My grateful thanks are due to all of them, particularly Dr. Felix P. Amerasinghe, IWMI, Sri Lanka, Dr. Bos, WHO, Geneva, Dr. Wim van der Hoek, IWMI, Sri Lanka, Dr. Vijay Veer, DRDE, Gwalior and Prof. D. Mohan, JNV University, Jodhpur. I am also indebted to the Librarian (Shri Prabhakar Reddy), British Council, Hyderabad and the Librarian (Miss Mary E. Gibson), London School of Hygiene and Tropical Medicine, London, UK for their generous manifold help and guidance with respect to bio-bibliographical work on Dr. Ross.

As always in past, my wife Ajita and two daughters Anupama and Akansha were most critical to my day to day work on this publication, for which I am thankful to them.

Last but not the least, I am thankful to Mr. Pawan Kumar, Scientific Publishers (India), for his excellent support and encouragement throughout the exercise of book writing and for timely bringing out the publication.

CONTENTS

CHAPTER - 1

INTRODUCTION :
UGLY, BOLD AND BEAUTIFUL MOSQUITOES

There are over two million catalogued species of insects in the world, with an equal number of taxa awaiting description, but fewer can match the popularity and importance that these tiny creatures, the mosquitoes, carry to the human life! No animal on earth has affected human lives so directly and profusely as is brought about by the mosquito. A mosquito is small in size but is giant of a killer. It is indeed no small enemy of humans.

Mosquitoes are ubiquitous in distribution and are found in all regions of the world, except the regions near the two poles and altitudes beyond 2000 meters. There are some 3500 species of mosquitoes, with a worldwide population of more than 100 trillion!! Most of them exist in tropical climates, but there are mosquitoes in arctic and desert regions, ranging in distribution from sea level to altitudes as high as 10,800 feet (3,600 meters). References to their very being can be traced in the annals of yore of several countries; in India they have been known from very ancient times, having been cited in century-old scriptures such as 'Susruta Samhita', a Sanskrit work dated fifth century B.C. of tremendous medical value, and described as a cause of both nuisance and human disease.

The common belief is that all mosquitoes are 'bad' and there are no 'good' mosquitoes. Although few exceptions exist where some specific mosquitoes are not only quite innocuous but are useful as well, nevertheless, the adage "Mosquitoes are harmful" is largely true since mosquitoes are often incriminated to have been responsible, amongst others, for (i) *felling great leaders* (Alexander The Great was suspected to have fallen to *Plasmodium falciparum* infection transmitted by anopheline mosquitoes while leaving the Indian borders following the famous battle with the King Porus of Punjab. So also, the much dreaded and ruthless ruler, Sultan Muhammed bin Turghluk contracted malaria in 1351 while on a military campaign against rebels, and within a short time succumbed to the disease), (ii) *decimating armies* (Yellow fever was stated to have taken its toll on thousands of U.S. Army recruits during the brief Spanish-American War, and that death from disease among American troops was much more common than death from bullets), and (iii) *deciding the fates of nations* (Panama, which still belonged to Columbia until late nineteenth century, raged a war for independence, following which a bitter civil war ensued from 1900 until 1903, when it finally won its independence (Spielman and D'Antonio, 2001). Mosquitoes are branded objects of deep abhorrence and dislike for all they offer to the mankind – the vexation, the pain, the nuisance and deadly infections, albeit the fact that all mosquitoes are neither equipped to bite nor capable to spread diseases. But, for the whole world they are only nuisance creators and disease transmitters! In India, like most of other fast developing countries, mosquitoes, merely by virtue of their transmitting deadly infections like malaria, dengue, filariasis and Japanese encephalitis, have through several decades been responsible for wasting hard-earned national resources, including human health (Fig. 1.1 & 1.2).

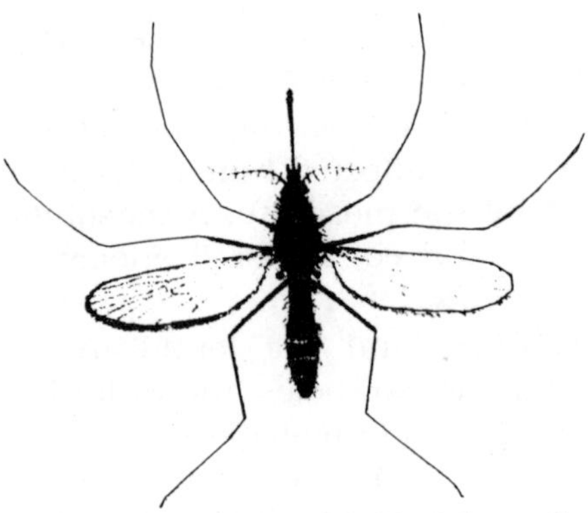

Fig. 1.1. General body plan of a female mosquito

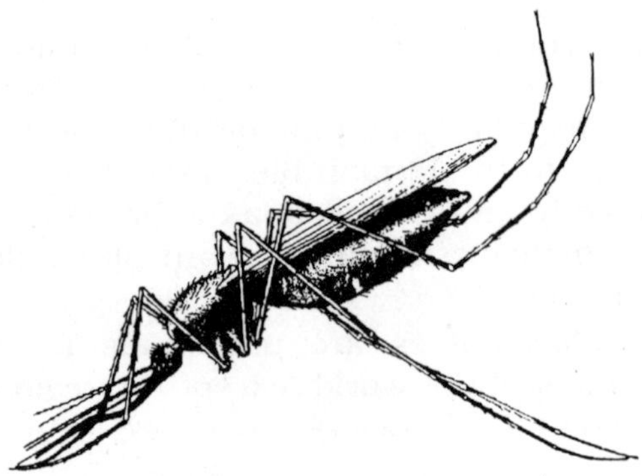

Fig. 1.2. Female anopheline mosquito resting after meal

It is amazing that this tiny and feeble-looking organism has in fact assumed the reputation of the most formidable envy of man in modern times. Hardly measuring the size and weight of a grape seed, a mosquito, notwithstanding all its startling elegance and beauty persona, is ubiquitously abhorred for the nuisance, pain and death it hails every year on millions of humans. And just as nations guard against biological terrorism, they are now being constantly pressed hard to conduct surveillance against naturally occurring agents that are delivered, not by human enemies, but by the haematophagous insects, amongst which blood-feeding mosquitoes figure in the frontline. In the industrialized world, living with mosquitoes now means accepting that an outbreak of disease might occur at any time. Rao (1981), a celebrated Indian medical entomologist, while stressing upon an intimate knowledge of mosquito, of course, *the fenmale fatale* - her strengths and weaknesses, however emphatically warned not to underestimate the miniscule enemy who has so far outwitted humans in their fight against various mosquito-borne deadly and/or debilitating infections.

A RENDEZVOUS WITH THE MOSQUITO

Mosquitoes are insects belonging to the order Diptera, the true flies (Phylum Arthropoda : Class Insecta) (Pittaway, 1991). Like all true flies they have two wings but, unlike other flies, mosquito wings have scales. Female mosquito's mouthparts form a long piercing-sucking proboscis. Males differ from females by having feathery antennae and mouthparts not suitable for piercing skin. A mosquito's principal food is nectar or a similar sugar source; only fewer than 100 species (less than 3% of the entire mosquito fauna worldwide) transmit disease pathogens to humans! India is a home for about 325 species and subspecies of mosquitoes, which is close to 10% of the global faunal wealth (Christophers, 1933; Barraud, 1934; Wattal & Kalra, 1961; Rao, 1981; Das et al., 1990; Nagpal & Sharma, 1995; Tyagi & Chaudhary, 1997; Tyagi, 2002).

Every mosquito is assigned a Latin scientific name under binomial nomenclature system, such as *Anopheles culicifacies* or *Aedes aegypti.* The first word is the "generic" name of a group of closely related mosquitoes, while the second word is invariably the "species" name which denote a group of individuals that are similar in structure and physiology, and capable of interbreeding. Some mosquitoes, like *Anopheles culicifacies*, *An. gambiae* and *An. macullipennis*, have well-characterized sibling species complexes whose individual species though morphologically indistinguishable from other members in the complex (both *sensu stricto* and *sensu lato*) is still a good species as demonstrated by cytogenetical and immunological techniques. Such sibling species complexes of certain vector mosquito species carry tremendous biomedical significance as there are some, and not all, of the species of the complex which actually are capable to transmit the human disease pathogens, thereby helping focus to target a given species for its control.

The mosquito names are used in a descriptive manner so that the name tells something about each particular mosquito. For instance, in the name *Anopheles culicifacies*, the generic word `Anopheles', taken from Greek, which in turn is itself derived from Sanskrit means`hurtful' or `harmful' (An = not; opheles = fruitful), while the specific word 'culicifacies' possibly refers to the Culex-like posture or resting habit, which the mosquito characteristically assumes soon after engorgement on blood. Some mosquitoes have what are called 'common names' as well as 'scientific names', such as *Aedes albopictus*, the 'Asian tiger mosquito', alluding to its possible cradle point – the Asia. Scientific investigators or taxonomists are continuously striving hard to discover new taxa of mosquitoes, and possibly one of the last mosquito species described was *Anopheles grabhamii* from the Florida Keys, United States of America, in 2001 (Darsie et al., 2002).

MOSQUITOES ARE LOATHED BECAUSE THEY BITE, BLOODFEED AND TRANSMIT DISEASE PATHOGENS

The impact of painful biting of mosquitoes is so long-lasting and deeply ingrained on humans' psyche that any amount of arguments to appreciate other much less known, but beneficial, traits of some of their cousins get simply underscored, giving an impression that there is no such thing as a good mosquito. Indeed, mosquito bloodfeeding is the most important characteristic of its life cycle which offers a unique amalgamation of various arcane processes. The complexity of their mechanical and chemical systems for finding human host, finding a nice juicy spot once they land, and then manipulating the body chemistry while they draw their nourishment, deserves respect as a remarkable example of evolution (Fig. 1.3).

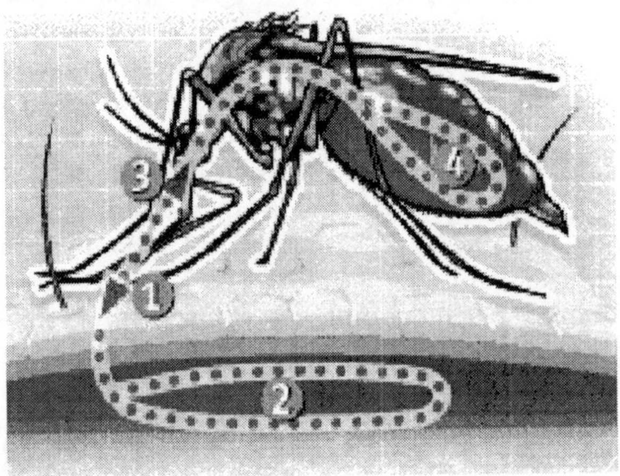

Figure 1.3. A mosquito's vicious cycle, depicting bloodfeeding and how it spreads diseases like malaria: (1) Parasite injected by infected mosquito as it feeds on blood, (2) Parasite develops in body tissues, (3) Parasite infects other feeding mosquitoes, and (4) Parasites live in salivary glands, waiting for next victim.

Mosquitoes have antennae that can detect carbon dioxide, which humans exhale. They can also detect heat, given off by warm-blooded animals. They can even detect lactic acid on man's skin, a byproduct of muscular exertion. Then, once they land, they have chemical receptors that find a likely spot to go drilling for blood by detecting heat and chemicals on the skin. They also have mechanoreceptors that push ever so gently against the skin to detect differences in resistance that indicate arterioles and venules in certain spots.

Mosquito's feeding apparatus then begins to work. It is a complex seven-piece set of devices for doing a variety of jobs. One of those parts is a tube that allows the mosquito to secrete its saliva into the body being bitten. That saliva, remarkable in its action, contains a set of chemicals that act like an anesthetic as well as anticoagulant. On one hand, the little bugger numbs the biting spot on the skin so that humans do not know mosquitoes are probing around on skin, and, on the other, make the blood flowing so as to enable the animal continue to relish on it to her satisfaction! Meanwhile, the other parts begin an intricate pattern of coordinated movements that let her hack into man's flesh. The mandibles, on the edges, are serrated on the tip, like tiny steak knives, to carve their way in. Just below the surface, the whole package of savagely armoured parts bends and takes a 90° turn, to begin the attack.

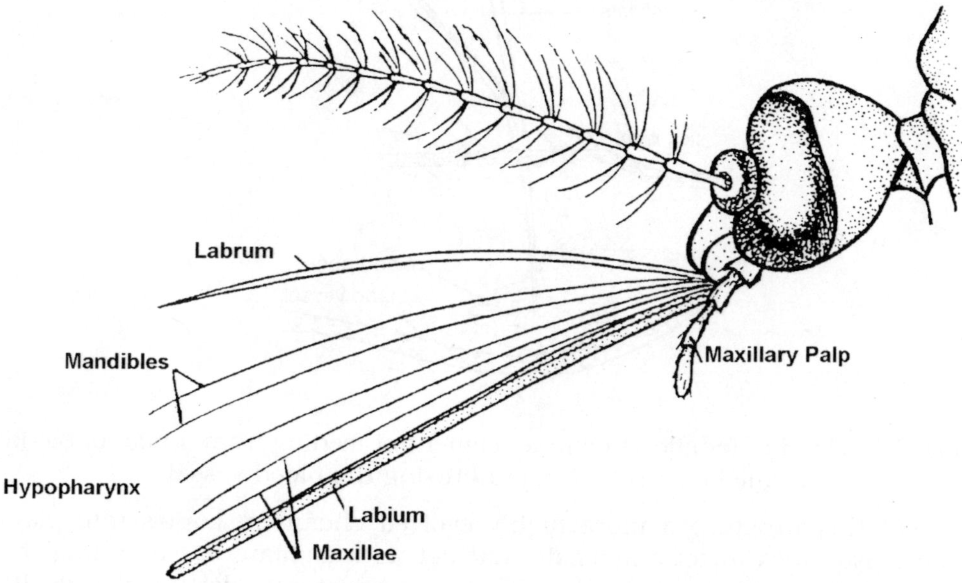

Fig. 1.4. Piercing-sucking mouthparts of a mosquito. Stylets removed from labial sheath (labium)

That set of needlelike parts begins to randomly poke away, anywhere between 12 and 18 times in just several seconds. It's not actually looking for a venule or arteriole just yet. It's indiscriminately poking around to try and poke into anything in the area that may bleed. That tube in the middle of their mouth assembly that does the spitting is actually two straws in one, one inside the other. As they continue to spit out their anesthetic, the other wider tube in the straw is sucking in samples of fluid to see if it has hit paydirt. Mosquitoes are the only known creatures that can spit and suck at the same time. If they don't find blood, they draw their probing mouth parts back a little, not all the way out, and dig into another spot. This process can last up to two minutes, if somehow the human body remains unaware that the bugs are there. If a mosquito does taste blood, it turns on more powerful pumps in its mouth, increasing the suction on its "drinking" straw. Then she draws on the tiny cloud of blood that has formed under the skin. The nearest venule or arteriole, even if it's not the one already wounded, is pulled over to the needlelike tip of the whole apparatus, where it's impaled. That opens up the hole through which the mosquito starts to drink (Fig. 1.5).

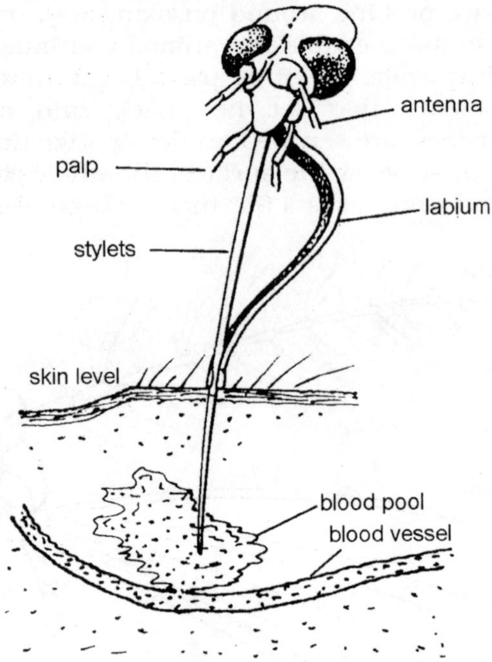

Fig. 1.5. Head of female at commencement of feeding from a pool of blood formed as a result of puncturing the blood vessel.

At this juncture a more highly evolved chemistry comes into play. The mosquitoes inject chemicals that act as vasodilators, expanding the blood vessel to increase the flow. They also inject anti-clotting chemicals, to keep the tiny wound from sealing up before they have had its fill. Mixed in with these chemicals are parasites that the mosquitoes had earlier picked up while feeding on the host. The parasites are lodged in the mosquitoes' salivary glands, waiting for the mosquitoes to bite on their next victim. When the mosquitoes squirt their chemical payload into human body, out come the parasites. This is how mosquitoes spread diseases.

WHY DOES A MOSQUITO BITE ITCH ?

Mosquito's bite often leaves behind an imprint, evanescent though, of a slowly developing itchful wheal. However, if it is a first mosquito bite ever, the roundish wheal fails to appear. The immune system in human body begins thereafter, and develops antibodies to recognize and flag the foreign chemicals. But after just one dose of mosquito spit, the familiar itchy bump left behind by subsequent bites is caused by an allergic reaction to all those chemicals the mosquitoes are squirting into the body. The immune system releases histamines, chemicals that make the cells in the lining of human's blood vessels spread apart. That lets fluid leak out into the skin. That fluid is carrying components of man's immune system that attacks and destroys the foreign chemicals from the mosquito. The extra fluid from the leaking blood vessels causes a bump, which irritates other receptors in the skin, leading to the itching. Children get bigger bumps from mosquito bites than adults because their immature immune systems respond with larger amounts of histamines. More leaking blood vessels equal bigger bumps, which mean more itching and scratching.

However, following repeated mosquito biting, the human immune system also become more stronger and exhibits a different reaction to mosquito saliva. Instead of triggering histamines, another part of the immune system grabs onto the foreign chemicals and drags them away to macrophages, which destroy the chemicals in the mosquito's saliva.

THERE ARE SOME 'GOOD' MOSQUITOES, TOO

Despite their being universally despised for creating nuisance, painful bites, loss of sleep and blood, and, above all, their potential to transmit a large number of fatal and/or debilitating infections, mosquitoes, or at least a few of them, do offer a wonderful blend of colours and sheen, in addition to highly propagandized biological control tool, in the name of *Toxorhynchytes* spp., for the immature populations of certain vector species such as *Aedes aegypti*. As we know all mosquitoes feed on plant nectors, some of these mosquitoes (*Aedes sollisittans*, *Ae. provocans* and *Ae. tritaeniorhynchus*), who are floral feeders or are both extra-floral and floral feeders, turn out to be very efficient pollinators of some selected plant species in some countries. Interestingly, a few mosquito species particularly those belonging to the Sebathinae group of genera such as *Malaya*, whose two species occur in India, have been observed to perform an arcane feeding behaviour on the ants belonging to the genus *Crematogaster*. These mosquitoes imbibe the sugary volume in the oral cavity of ants. The Malaya mosquitoes are totally innocuous to man and other animals as they never bite due to presence of a highly specialized dew-seeking proboscis or mouthparts only.

The complex nature of saliva of mosquitoes has ever been a subject of constant scientific investigations and its anticoagulatory properties are being determined so as to derive a medicinal significance. So, mosquito – the magnificent eternal enemy of humans – is dangerous and unadorable, but there is still something very special in the beast that makes it a highly sought after animal. In recent years scientists have contemplated using fossilized mosquito's stomach blood to unravel mysteries surrounding the Mesozoic Period-inhabited dinosaurs, by way of isolating and decoding DNA from the blood supposedly taken from the giant animals by mosquitoes.

THE "MOSQUITO" SEMANTICS

The mosquito has several names in various different languages. The Spanish called the mosquitoes as "musketas", meaning "little fly", and the native Hispanic Americans called them "zancudos", again a spanish word, which means "long-legged". Although modern literature advocates that the present day 'Mosquito' word, purportedly having its etiology in Spanish or Portuguese language, is apparently of North American origin dating back to about 1583. The famous Indian medical treatise of yore, Susruta Samhita, written some 2500 years ago, has numerous references to the creature. Aristotle, in 300 B.C., too, has named mosquitoes as "empis" in his *Historia Animalium*. In Europe, mosquitoes were called "gnats" by the English, "Les moucherons" or "Les cousins" by the French, while in Germany the name assigned was "Stechmucken" or "Schnacke". In Scandinavian countries, mosquitoes were called by a variety of names including "myg" and "myyga". Greeks nemed mosquitoes as "konopus".

And, what is the correct plural form of the word mosquito? In Spanish it should be "mosquitos", but in English it is "mosquitoes".

ALL ONE WANTS TO KNOW ABOUT MOSQUITOES

Among mosquitoes, it has been established that only the females desire and extract blood. Male mosquitoes drink only sugary fluids such as flower nectar. Both in the wild and in the laboratory, mosquitoes will visit certain flowers and will feed on fruit placed in their cage. Since they (*Toxorhynchites*) vigorously probe the flowers of some plants, and can distinguish between different types of sugars, they play a certain role in the pollination activity of certain plants. The haematophagous females will also drink sugary fluids, but when hungry, if females are given a choice between sugar water and blood, they will always choose blood. If males are offered the same choice, they will always drink the sugar water as their mouthparts are not suited (or designed) for blood-sucking. Since male mosquitoes do not suck blood, they also do not transmit diseases. Like the males of many other insect families, they are important for just one reason – the mating, and then they become superfluous. The female usually needs to mate just once in her life, although she invariably blood feeds following each oviposition which may, under favourable conditions, go up to as many as six or more in her life time. She stores sperm in her body and fertilizes her eggs at the moment when she lays them. Shortly before or after mating, she takes a meal of blood to provide the eggs with protein. When the eggs are mature and ready to be fertilized, the female searches for a suitable place to lay them which is usually water that is full of decaying matter that includes bacteria and minute organisms. Some mosquito species lay their eggs in places that are likely to contain water in the future; such as, rusty coffee cans, old discarded tires, *etc*. In such situations, the eggs lapse into a state called diapause and they will not hatch until this dormant period passes, and the water level, temperature, and oxygen content are just right. The eggs of some species can survive for years in diapause, even in sub-freezing temperatures. After hatching from the eggs, larvae swim about and begin to feed by sweeping the water with two large fan-shaped bundles of chitinous bristles (or brush-like structures) at either side of their heads. The fans, or brushes, create currents in the water that direct food toward the larvae's mouthparts and help stuff the food into their deeply concealed mouths. The larvae are omnivorous; feeding on bacteria, pollens, microscopic plants, and a wide variety of other minute things. After going through four different larval stages, there is a motile but non-feeding stage called pupa which is, like larvae, is aquatic by nature. The pupa finally transforms into the aerial adult whose fair sex, emerging somewhat belatedly following males, alone bites (Fig. 1.6 & 1.7).

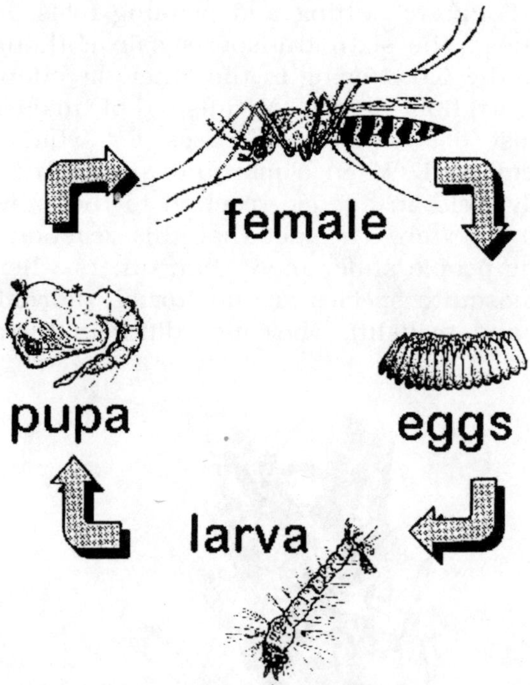

Fig. 1.6. The four different shapes of mosquito's life: egg, larva, pupa and adult.

Haematophagous mosquitoes do not really sting, in the sense of a hypodermic needle piercing the skin. Instead, they saw their way through, using four cutting stylets kept in a protective sheath along with a duct that carries anticoagulant into the wound and a tube that carries blood out. Mosquitoes fly into the wind, picking up scents which lead them to their victims. Just about everything attracts them, including the carbon dioxide we exhale. Mosquitoes will even suck the blood of birds, frogs, turtles, snakes and just about every warm-blooded animal. In fact, some mosquitoes prefer their sources of blood to be other than human. There are many species of mosquitoes that prefer the blood of birds or other animals. In fact, some species have been observed to feed on numerous mammalian groups, on a variety of reptilian species, and even on other kinds of insects.

Anthropophilic mosquitoes are attracted to certain ranges of temperature and humidity, and to carbon dioxide from the exhalation of humans. They seem to prefer dark-colored objects to light. Certain chemical smells may also come into play; for example, components of blood and sweat such as haemoglobin and amino acids. The best explanation of what brings the mosquito to its victim is that she simply flies upwind until she comes upon a potential victim exuding a "host beam" of warm, moist air laden with carbon dioxide. The mosquito proboscis (long tubular mouth parts used for feeding) consists of six different shafts. Four are cutting and piercing tools; a fifth transports blood from the host; the sixth transports saliva, thought to act as an anticoagulant for the blood going in the other direction. The saliva also facilitates transportation of the organisms of malaria, yellow fever, dengue, and most of the other diseases for which mosquitoes are notoriously remembered. When a mosquito punctures ("bites") into the flesh, one usually feels an allergic reaction to the saliva, which causes the swelling and the itch. The fact that this reaction is allergic helps explain why some people suffer more than others when their skins are invaded. Some mosquito species are nocturnal, diurnal, or crepuscular (active at dawn and twilight). They also differ in their preferences for altitudes.

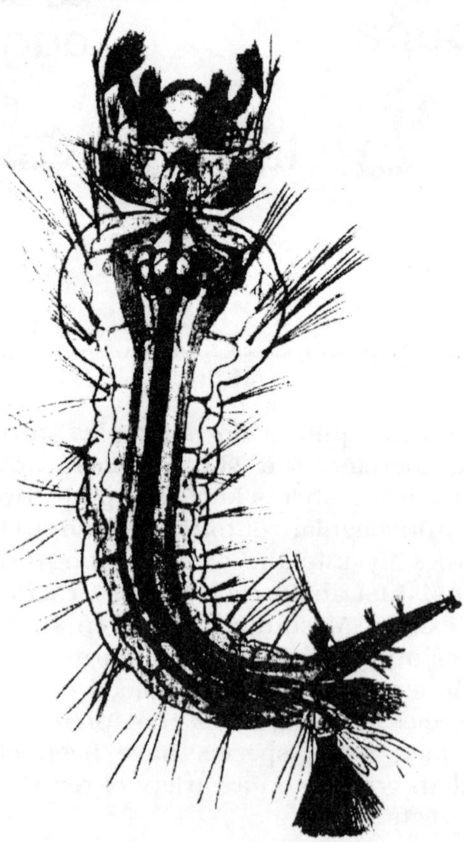

Fig. 1.7.Mosquito larval anatomy.

AMAZING MOSQUITO FACTS

Mosquitoes are wonderful creatures inducing enormous scientific inquisitiveness about their various hard-to-believe traits! Some interesting facts and/or observations are listed below. It may, however, be noted that a few of these appear to have emerged directly from fairytales (for instance, mosquitoes are attracted to a person who has recently feasted on banana; the common belief that mosquitoes prefer children to adults, and blondes to brunettes, and biting activity increases by 500 times when there is a full moon, *etc.*) which need much of a scientific explanation.

1. Mosquitoes don't see very well, but they zoom in like a heat-seeking missile.
2. In the spherical arrangement of their compound eyes, blind spots separate each eye from the next one. As a result, they can not see a human until they are 10 m away. Even then, they have trouble distinguishing a person from any object of similar size and shape, for instance, tree stump, 55-gallon drum, *etc.*
3. When they are 3 m away they use extremely sensitive thermal receptors on the tip of their antennae to locate blood near the surface of the skin. The range of these receptors increases threefold when the humidity is high.
4. Mosquitoes find their host by sight as they are very receptive to movement of any kind. Therefore, if a person waives his arms in trying to keep the mosquitoes at bay then he is actually working against himself because not only movement stimulates a mosquito to find its target, it detects a moving target at 6 m away . Man's movement catches her eyes and she will probe the skin as many as twenty times, like a blind phlebotomist who can taste blood with the end of his needle.

   Mosquitoes do so by detecting infra-red radiation emitted by warm bodies and by chemical signals since they are attracted to carbon dioxide and lactic acid, among other chemicals. Human scent tickles the mosquito's antennae which the mosquito uses as its sensory detectors to chemical stimuli. A mosquito smells human, or a cow or another host at a distance of 20 to 35 meters.
5. Mosquitoes like the scent of estrogen, hence women get bitten by mosquitoes more than men do.
6. Mosquitoes dislike citronella because it irritates their feet.
7. An excellent flyer having wings to beat 500 times a second, a mosquito flies an estimated 1.5 to 2.5 km per hour. Occasionally, however, taking advantage of the wind current, some mosquito species can fly up to 16 kmph, dart between raindrops and even fly backwards. Some species remain close to their breeding site rarely flying farther than 100 m from their breeding place, while others will fly up to 16 km or more in search of food.

8. Although there are different records as to how far mosquitoes can fly, yet certain salt marsh mosquitoes easily migrate 120 to 160 km mostly using the benefit of strength of wind currents. Mosquitoes fly across 240 km in their lifetime.
9. Only female mosquitoes bite. They need the protein that animal and human blood provide to produce eggs. Male mosquitoes feed only on nectar from plants.
10. Female mosquitoes are capable to bite more than once in order to engorge her, although this trait is observed in certain species more than others. Invariably, a female mosquito has to be fed on blood before becoming able to lay the eggs. Immediately after laying the eggs she has an urge for seeking another blood meal. Mosquito can drink 150% as much blood as they weigh (2 to 2.5 milligrams). This is approximately, in case of *Aedes aegypti*, about 5-millionths of a liter. A mosquito retains $1x10^{-3}$ to $1x\ 10^{-4}$ nm of blood in its sucking tube, a comparison of which is given in Table 1.1 (Napoli & McGowan, 1987). The welts (or wheals, but sometimes expressed simply as 'weals') that appear after mosquito leaves are, in fact, not a result of her biting *per se* but actually are caused as a result of an allergic reaction to saliva the mosquito injected under the skin to prevent the blood from clotting.

Table 1.1. Quantity estimates of blood retained or transferred in mosquito compared with that of hypodermic needle and syringe

Agent	Circumstances under which blood retention/ transfer was recorded	Estimates determined when blood was retained or transferred	Amount of blood (in nanometers)	Authority
Mosquito	After engorgement (based on transmission trials)	Transferred	$1x10^{-3}$ to $1x\ 10^{-4}$	Miike, 1987
22-gauge needle	After blood collection	Transferred	$1x10^{3}$	Napoli & McGowan, 1987
22-gauge needle	After intramuscular inoculation	Retained	9	Knaus *et al.*, 1993
Neck of a 20 ml syringe	After blood collection and expulsion	Retained	$1.37x10^{5}$ to $1.42x10^{5}$	Gunn, 1993

11. Mosquitoes, like all insects, are cold-blooded creatures. As a result their body temperatures are the same as their surroundings. In temperate climates, adult mosquitoes become inactive with the onset of cool weather and enter hibernation to live through the winter. In spring, the females emerge from hibernation, find a blood meal and lay their eggs.

12. Though records show the average life span of a female mosquito can range from as low as 3 days to as many as 100 days, in principle a mosquito may normally survive for about 2-3 weeks. Males live shorter than females. Female mosquitoes lay up to 300 eggs at a time, which develop to adulthood, under optimum conditions, in 4 to 7 days. Oddly enough, if for theoretical reasons, a female mosquito is allowed to live unscathed for her full normal lifespan, with all her periodic subsequent progenies together with those of the resultant off-springs born during her lifetime, it will be observed with unbelievable bewilderment that a massive population of 300,000 could result from a single female (*i.e.*, the first procreating mother mosquito) during her life!! This does not happen, though, thanks to the Mother Nature! Even so, considering an unaccountable number of mosquitoes extant in nature, what is lost in a long developmental perigrination is actually far greatly replenished through prodigious and constant breeding – the subsequent hungry adult females from whom lurk around to derive away from the human host his most precious vital fluid, the blood!

CHAPTER - 2

GLOBAL WEATHER CHANGE, MOSQUITOES AND DISEASES: *IMPACT OF RISING TEMPERATURES AND WEATHER EXTREMES*

Vector-borne diseases result from infections transmitted to humans and other animals by blood-feeding arthropods, such as mosquitoes, sand flies, ticks, mites and fleas. The vector-borne pathogens, which include viruses, rickettsiae, bacteria, protozoa, and worm parasites, spend part of their life cycle in a cold-blooded arthropod vector and thus are influenced by environmental changes. The transmission pattern of these diseases (*e.g.*, malaria, dengue *etc.*) is, therefore, effectively regulated by a mix of several factors of which temperature is the key one, followed by CO_2 as a byproduct of the green house effect (Fig. 2.1). Mosquito-borne diseases like Japanese encephalitis and Yellow Fever involving warm-blooded mammals as zoonoses are also adequately affected directly by temperature. Of the thirty odd new diseases that emerged in the last 20 years, many thrive in warmer and wetter weather. Lyme disease is linked to warmer, humid conditions that breed more deer ticks. Malaria kills about 1-2 million people a year worldwide, with about 90% of new cases occurring in Africa and southeast Asia only. Although the disease is now rare in developed countries, however, that could change with global warming. As soon as 50 years from now, malaria could spread to parts of the world that are now too cold to support life cycle of the mosquitoes and their parasites that transmit the disease.

The global warming is blamed to have helped extend the distribution of the serious mosquito vectors like *Aedes aegypti* and *Ae. albopictus* in North America, where they are responsible for several viral infections. There has also been a fourfold increase in malaria in the last five years which is associated with temperature and humidity. The development of the mosquito larvae is faster in warmer climates, resulting in the mosquitoes becoming adults sooner. Also, extrinsic incubation periods of yellow fever and dengue viruses in the mosquito vectors are dependent on temperature. With warmer temperatures, the incubation time required from when the mosquito first encounters an infected host until the mosquito is able to transmit an infection virus may be shortened. Recent years have also witnessed a marked increase in dengue fever, with 320,000 reported cases in the Americas alone.

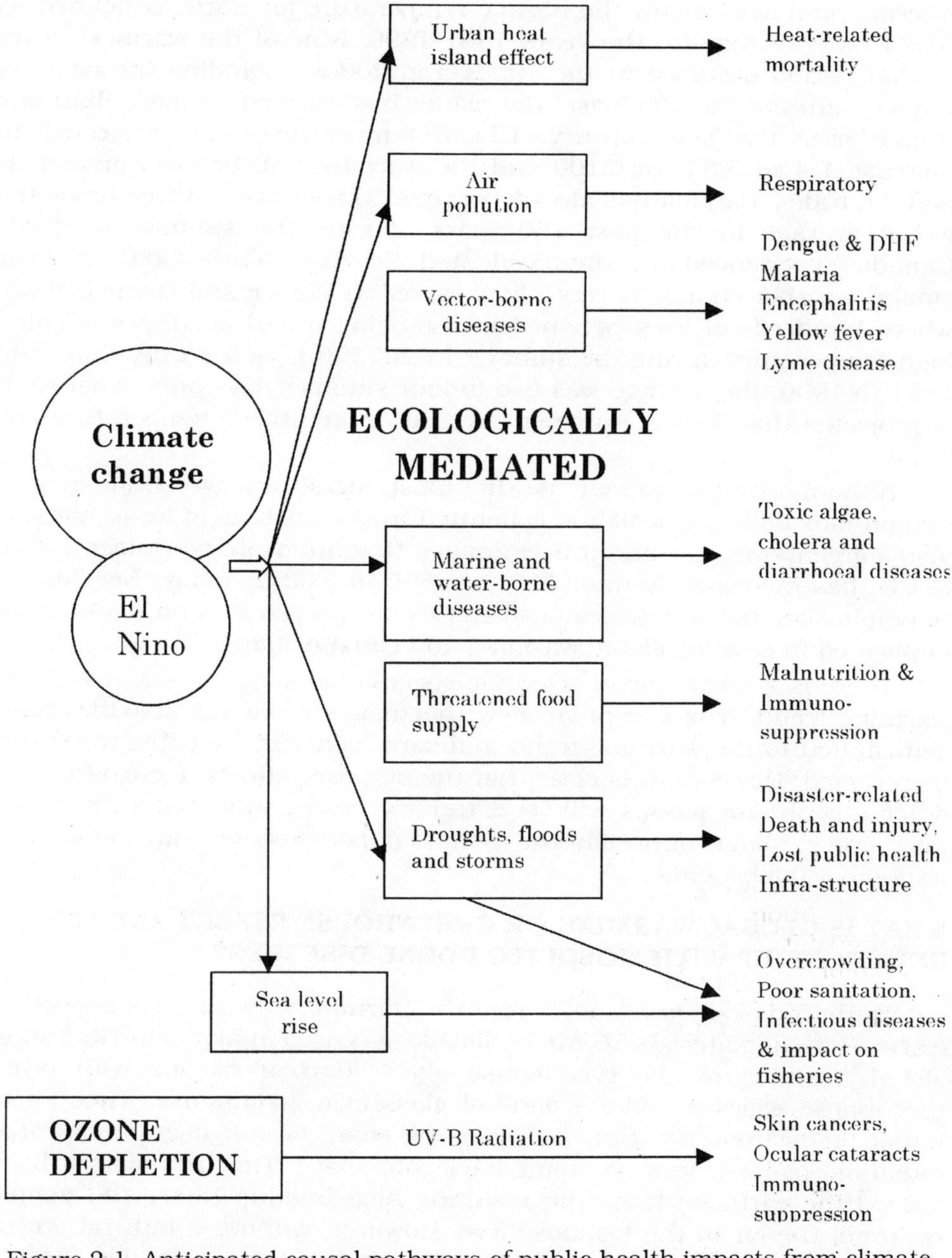

Figure 2.1. Anticipated causal pathways of public health impacts from climate change.

The world is currently experiencing a general warming phase that began in the early part of the 18th century, and temperatures in the northern hemisphere are now similar to those experienced during the Middle Ages. Earth's average temperature has been rising for 150 years, particularly in the past two decades, something most climate experts attribute mainly to greenhouse gas emissions from human activities. By international agreement, the normal temperature for Earth is defined as 61.7°F, the average for the years 1961-1990. Nine of the warmest years in that period occurred in the 1990s and 2000s, including the past six years. Counting the 1997 rise, the planet has warmed by more than one degree over the last century. Global temperatures are projected to increase 1.4 to 5.8°C by 2100 and the warming will be even greater at high latitudes. Canada has already warmed almost two to three times the global average in the past 150 years. During the summer of 2001, Canada experienced an unprecedented 24 days above 30°C. A near similar weather change is recently observed in France and Great Britain, where hundreds of lives perished to a prolonged and an unprecedented high temperature during the summer in 2003! It is noteworthy that from 1961 to 1990, the average was two to four summer days only, whereas it is projected that 50% of summer days may reach those temperatures by 2050.

Although water vapour is the most significant greenhouse gas, carbon dioxide (CO_2), which is generated in the burning of fossil fuels, is also a greenhouse gas and it is estimated that atmospheric concentration of CO_2 has increased from 0.029% in 1890 to 0.037% today. Needless to re-emphasize that the greenhouse gasses are proven beyond doubt to be implicated in causing global warming and climate change.

There is a great deal of scientific speculation as to the causes of this warming trend. It is a popular view that man's economic activities have contributed to the warming trend and some consider that this trend can be reversed. However, it is clear that the proposed efforts to curb CO_2 and other greenhouse gasses will be extremely costly, and that these costs may prove to far outweigh the potential benefits, or may only be a fraction of the benefits.

WHAT IS GLOBAL WARMING OR GREENHOUSE EFFECT AND ITS RELATIONSHIP WITH MOSQUITO-BORNE DISEASES?

Carbon dioxide is released from the burning of fossil fuels and other carbon-based materials. Carbon dioxide gas is a major contributor to global warming or the greenhouse effect. Carbon dioxide with other greenhouse gases act like a pane of glass in a greenhouse. They let in visible light from the Sun but prevent some of the degrade infrared radiation, or heat, from escaping back into space. They reradiate it back toward the earth surface. The resulting heat buildup raises the temperature of the air in the trophosphere. However, without a natural greenhouse effect, Earth would be 33°C cooler than it is now (Reuber, 1997). The global average atmospheric concentration of CO_2 in 1990 was 353 parts per million by volume (ppmv), and was increasing at an average rate of 1.8 ppmv/yr. There is a strong correlation between CO_2 concentrations and polar temperature over this entire period, with CO_2 concentrations of 280 to 300 ppmv during relatively warm periods such as the present (0-10,000 yr BP and 120,000 yr BP) and concentrations of 180-200 ppmv during ice ages (18,000 yrBP and 160,000 yrBP). Current CO_2 concentrations are substantially higher than they have been anytime in the last 220,000 years (Jouzel *et al.*, 1993).

It is estimated that the climatic sensitivity (*i.e.*, the equilibrium surface air temperature rise due to a doubling of carbon dioxide) was between 1.5 and 4.5°C in 1996 according to the Intergovernmental Panel on Climate Change (IPCC). The prediction for future rate of global warming ranges between 0.1 and 0.3°C per decade. United States of America, which has 4% of the world's population, is responsible for 22% of the carbon emissions – the largest share amongst all the countries!

Greenhouse gases make up only a tiny fraction of the atmosphere, but they can have a big impact, and their proportion is rising rapidly as economic development speeds up around the world. Since the beginning of the industrial evolution in middle of the 18th century, levels of CO_2 have jumped 30%, nitrous oxide 15%, and methane 100%. At present, developed countries' emissions account for approximately 60% of global total, but developing countries' emissions, too, are growing rapidly, and by 2020, will account for more than half of the world's emissions.

ENVIRONMENT DEGRADATION AND CLIMATIC VARIABILITY VIS-A-VIS MOSQUITO-BORNE DISEASES: *WATCH OUT FOR MORE DEADLY ACTIONS!*

Environment change is a never ending dynamic phenomenon governed by a mélange of forces and characteristics both within and without the Earth's atmosphere. Amongst several characteristics that have apparently affected dissemination of mosquito-borne diseases (*e.g.*, malaria, dengue, encephalitis *etc.*), besides other infections (such as cholera, leptospirosis *etc.*), are unarguably the discernible increase in atmospheric concentration of carbon dioxide (CO_2) and the resultant upward changes in Earth's surface temperature. It is an acknowledged fact that a source of carbon dioxide, such as the human or animal body, attracts mosquitoes. In this way, it can be safely surmised that emanation of more CO_2 in Earth's atmosphere is causing a rather suitable living condition for these insects. Man is generally credited with this possibly irreversible feat of long-term global warming which has ushered several deadly mosquitoes into regions never visited by these creatures earlier (*e.g.*, invasion of *Aedes albopictus* in the Americas). It is worthwhile to note here that although the normal temperature for Earth, according to international agreement, is defined as 61.7°F, being the average for the years 1961-1990, it has already increased by 1 degree by 1997 over the last century!

There is a great deal of scientific debate concerning the extent of climate change, its potential consequences and man's involvement in that change (McMichael & Kovats, 1999). The way in which climate change is measured is fraught with problems, as there are very few accurate measurements of historical temperatures, rainfall and other climatic data. One thing which is, however, more than clear is that the climate changes, at all levels - local, regional and global, bring about a tremendous impact on the epidemiological features of the mosquito-borne diseases since both the vectors and pathogens are grossly abiotically motivated.

Environment, particularly the climatic variability, plays an inevitably indispensable role in regulating the dynamics of a number of vector-borne diseases (Gubler et al., 2001). Most vector-borne diseases exhibit a distinct seasonal pattern, which clearly suggests that they are weather sensitive. The significance of environment in shaping the biology of these diseases can be amply gauged from the fact that the entire gamut of development, behaviour and survival of arthropod vectors as well as the intensity of the diseases they transmit, is strongly influenced by climatic factors (Fig. 2.2). Temperature (in all seasons), rainfall, and humidity are especially important, but other factors such as wind and duration of daylight can also be significant. In particular, daily maximum and minimum temperatures affect the pathogen's rate of multiplication within the insect, which in turn affects the rate of salivary gland infection and hence the likelihood of successful transmission to the human host. If the development time of the pathogen exceeds the life span of the insect, transmission can not occur. Vector longevity is thus very important and can be shortened by elevated temperatures. The tremendous growth in international travel increases the risk of importation of vector-borne diseases, some of which can be transmitted locally under suitable circumstances at the right time of the year. But, demographic and sociologic factors, as well as land use, agricultural practices, residential patterns, human behaviour and vector preponderance also play a critical role in determining disease incidence. It is, in fact, a complex interplay of all these factors that determine the overall effect of climate on the dynamics of arthropod-borne diseases.

Extreme weather events are also projected to increase, including the boom-and-bust cycle of severe droughts followed by heavy rains that has occurred in North America in recent years. This increased climate variability will probably be even more important than temperature increases in promoting outbreaks of some diseases. In fact, climate affects all the links in the chain of disease transmission: disease-causing organisms such as viruses and parasites, vectors such as mosquitoes, ticks and rodents that transmit the diseases, hosts, such as birds, that provide a reservoir for disease organisms, and predators that keep disease-carriers under control. Weather extremes are playing a significant role in the emergence and spread of infectious diseases and are accelerating the potential for more nasty surprises. The wildfire advance of West Nile across North America is possibly only a harbinger of deadly surprises to come. While the impact of new diseases will vary depending on the resilience of health programmes in affected countries, severe acute respiratory syndrome (SARS) has shown that even in developed countries human and financial resources can be pushed to the wall by unanticipated outbreaks.

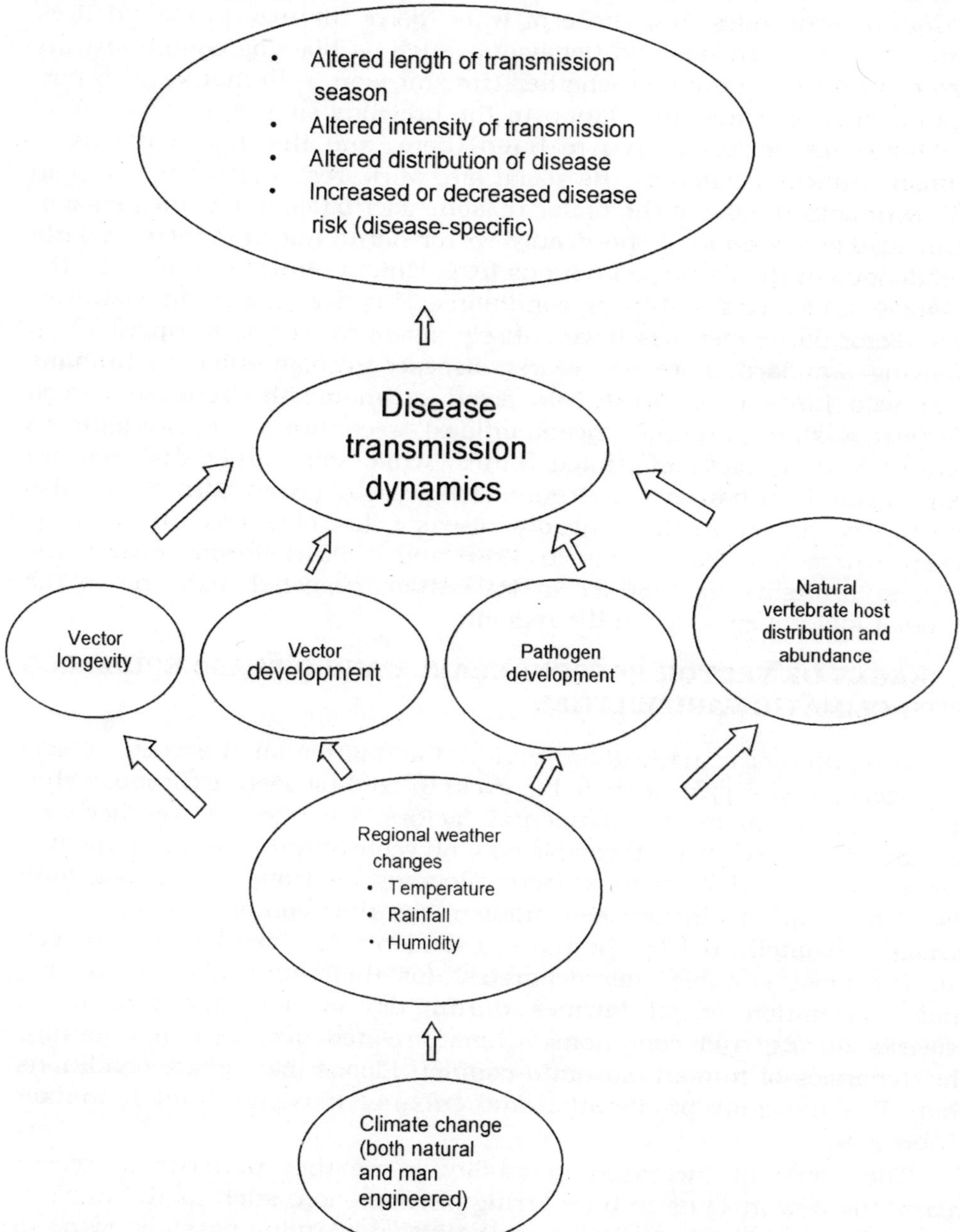

Fig. 2.2. Interactive factors underlying prevalence of potential mosquito-borne diseases.

Weather and climate are, of course, not the only factors that influence the outbreak and spread of diseases; social and economic factors also are important. Surveillance programs and testing and control measures - such as monitoring for dead birds and eliminating mosquito breeding areas - can make a difference. However, it is certain that climate change provides opportunities for diseases, particularly the mosquito-borne ones and those of water-borne nature, to extend their range. Human activities and behaviour, such as housing conditions, are also crucial to determining whether transmission to humans will occur. The location of homes in relation to the breeding sites, the structure of buildings, the material used to build them, and the daily patterns of human behaviour, such as the social life, work, rest and recreation, can all be important. One of the major reasons ascribable to low incidence or near absence of some of the deadly vector-borne infections like malaria and dengue in the developed nations (*e.g.*, United States) is said to be the changes in lifestyles and living conditions. This does not mean, however, that vector-borne diseases have entirely disappeared due to upgradation in living standard in recent years; diseases transmissible to humans from wild birds (*e.g.*, West Nile fever) or mammals (zoonoses) (*e.g.*, Western/Eastern Equine encephalitides) recontinue to circulate in nature in many parts of United States, while some other diseases are reappearing in urban areas. Drought followed by heavy rainfall can also encourage the outbreak of deadly diseases (as observed in the Thar desert where the heavy rains in 1988 and 1990, following one of the severest droughts in 1986-87 in Rajasthan, triggered also one of the severest malaria episodes in the region).

LINKAGES OF VECTOR PREPONDERANCE AND DISEASE EPIDEMICS WITH CLIMATIC VARIABILITIES

Mosquito-borne pathogens, or for that matter all the vector-borne pathogens, spend part of their life cycle in cold-blooded arthropods that are subject to many environmental factors. Changes in weather and climate that can affect transmission of vector-borne diseases include temperature, rainfall, wind, extreme flooding or drought, and sea level rise. The relationship between ambient weather conditions and vector ecology is complicated by the natural tendency for insect vector to seek out the most suitable "microclimates" for their survival, *e.g.*, resting under vegetation or pit latrines during dry or hot conditions or in culverts during cold conditions. Climate-related disasters may change the dynamics of human-mosquito contact. Floods may create conditions that allow mosquito proliferation and enhance mosquito-human contact (Table 2.1).

The apparent increased instability of weather patterns in recent years, the new insights into recurring phenomenon, such as the *El Nino*, that influences world weather, and the global warming possibly owing to the greenhouse-gas accumulation in the troposphere, have re-emphasized to draw fresh attention on the actual and potential health consequences. The inter-relationship between climate, weather and health (in the present context it is to be understood as the state of well-being under the impact of mosquito-borne infections) is quite intricate. It is now realized the world over that climatic changes across time-scales influence ecological systems through direct and indirect events, in turn affecting disease conditions.

Climatic changes in time scales are of different kinds and intensities. The *El Nino* seasonal variations occupies a special place in the global weather formation, even though the ENSO (El Nino Southern Oscillations) is one of several natural variabilities. Its influence on global climate is second only to that of seasons, and it is predictable. An *El Nino* event is a major change in the global climate system associated with epidemic warming of the upper ocean layer in the eastern tropical Pacific Ocean, lasting for three or four months. The *El Nino* events are linked with rainfall and temperature extremes, and are a major cause of inter-annual climate variability. During the 1997 *El Nino* event a postulation was cast for increased risk of serious drought in India, Australia, Brazil and east Africa, on one hand, and heightened the danger of torrential rains and flood in American deserts from Peru to California, and of forest fire in Indonesia. Studies have already proved a strong correlation between outbreaks of malaria and the ENSO in the Thar Desert, in north western India (Bouma & van der Kaay, 1994, 1995, 1996; Bouma *et al.*, 1997; Tyagi, 1999). Even for certain other highly infectious diseases, like cholera, a definite understanding has now been developed over its epidemic occurrence and the rise in the local sea surface temperatures and growth of algal blooms. In similar fashion, therefore, knowledge is being strengthened on ascertaining perceptible correlation for eruptions of other equally devastating diseases like cryptosporidiosis, leptospirosis, giardiasis and Legionnaire's disease to the *El Nino* (warm phase) and *La Nina* (cold phase) situations, since the transmission of these diseases is affected by the quality and distribution mechanisms of drinking water, which in turn, is affected by rainfall patterns triggered by the *El Nino* patterns. In fine, therefore, it has become more than crystal clear that all the vector-borne diseases have a very strong correlation with climatic variabilities.

Table 2.1. Impact of weather and climate on mosquito-borne infections

Environmental factor	Vector/pathogen/ vertebrate host	Impact
Temperature	Vector	Survival can increase or decrease depending on species. Some vectors have higher survival at higher latitudes & altitudes with higher temperatures. Changes in the susceptibility of vectors to some pathogens, e.g., higher temperatures reduce the size of some vectors but reduce the activity of others.

		Changes in the rate of vector population growth.
		Changes in feeding rate and host contact (which may alter the survival rate).
		Changes in the seasonality of populations.
	Pathogen	Decreased extrinsic incubation period of pathogen in vector at higher temperatures.
		Changes in the transmission season.
		Changes in distribution.
		Decreased viral replication.
Precipitation	Vector	Increased rain may increase larval habitat and vector population size by creating a new habitat.
		Excess rain or snowpack can eliminate habitat by flooding, thus decreasing the vector population size.
		Low rainfall can create habitat by causing rivers to dry into pools (e.g., dry season malaria).
		Decreased rain can increase container-breeding mosquitoes by forcing increased water storage.
		Epic rainfall events can synchronize vector host-seeking and virus transmission.
		Increased humidity increases vector survival; decreased humidity decreases vector survival.
	Pathogen	Few direct effects but some data on humidity affects on malarial parasite development in the anopheline mosquito host.
	Vertebrate host	Increased rain can increase vegetation, food availability, and population size.
		Increased rain can also cause flooding and decrease population size but increase contact with humans
Increased sea level	Vector	Altering estuary flow and changing salt marshes may affect drastically mosquito species by decreasing or eliminating selected mosquito breeding sites (e.g., reduced habitat for *Culiseta melanura*).

The possible pathway through which climate affects health, particularly the onset and/or intensities of vector-borne infections, vary widely (Fig. 2.2). However, this highly complex web of events cannot be comprehended within a single-discipline framework and, perhaps to understand relationship of climatic variability with that of any weather-based discipline as well as dynamics of any vector-borne disease, the cross-pathways of inter-disciplinary or inter-events cohesiveness have to be essentially traced out and modeled up. One major point of discourse, notwithstanding, is that of all those health impacts working under climatic variability, the climate *per se* is most certainly not the only factor responsible for bringing about any catastrophic disease pandemic. In fact to build up a strong relationship between climate and health, it is inevitable to also consider a multitude of social, political and cultural factors. Economically and politically unstable populations are most vulnerable under extremes of climatic variabilities due to their inability to adapt or respond adequately and rapidly to counteract the devastating outcomes such as those during malaria and dengue epidemics, to example a few of the most virulent vector-borne diseases associated with high morbidity and mortality. According to a recent study (Hales et al., 2002) based on projected population and climate change worldwide, nearly 5-6 billion people (50-60% of the projected world population) will be at risk of dengue fever transmission compared with 3-5 billion people (35% of the population) if climate change did not happen. Based on a model developed by them, they have suggested that excessive resource consumption in rich countries should be addressed to avoid this increase in the proliferation of dengue fever.

Even though there is no panacea available to completely revert the consequences of the climatic variability such as the onset of vector-borne diseases like malaria, dengue, Japanese encephalitis and filariasis, all of which prevail in India, nevertheless a complete scientific understanding of weather events and development of appropriate early warning tools may certainly still do a good help and save several lives. However, the use of climate forecast information in public health arena not only requires greater scientific understanding of climate-health symptoms, but also a paradigmatic shift in thinking from response to preparedness and mitigation. Therefore, to offer useful insight into causality and ultimately to integrate climate tools into public health policy and decision making, research must be strongly interdisciplinary amongst at least, so to say, climatology, disease epidemiology, medical entomology, sociology and geography, but not to forget in the least the administration and political support for the cause in reference. On any level, be it regional (like that of the Thar Desert), national or international, a concerted research effort involving various different agencies, organizations and the private sectors is a must.

CLIMATIC SENSITIVITY AND MOSQUITO-BORNE DISEASES

On reflection one can count down as many as 30 of the new diseases which thrive in warmer and wetter weather. Invariably all mosquito-borne diseases are characterized to prosper in a 'hot and humid' climate of the tropics and subtropics.

With the passing of time, in recent decades, both the vector mosquitoes and the diseases transmitted by them have witnessed a remarkable dissemination from one continent to another, owing largely to a greater degree of human mobility, his pets and belongings as well as his means of transport, such as aeroplanes, ships, trains and other automobiles. As classical examples, *Anopheles gambiae*, the most dangerous and infamous vector of malaria in the world, was inadvertently transported to South America from Africa; the dangerous Asian Tiger mosquito, *Aedes albopictus*, expanded its distribution with man's aid from Africa to the USA; and apparently *Anopheles stephensi*, another highly important malaria vector from Indian point of view, had spread from its cradle in India, the rural Thar Desert (where, too, it seems to have traveled over from the Middle East), to other peninsular areas in the country, particularly the urban metropolises.

With regard to the mosquito-borne diseases, a dynamic shift has been witnessed over past several years. Great Britain and United States of America, from where malaria had long been eradicated, have both been recently reporting indigenous cases of malaria. In India, most of the arthropod-borne infections have been found increasing their geographic ranges; malaria, particularly *P. falciparum*-dominated malaria, has formed new foci and the Thar Desert region is the most recent classical example if its exacerbation (Tyagi, 2002), while dengue including dengue haemorrhagic fever (DHF) and dengue shock syndrome (DSS), too, has spread to hitherto nearly virgin lands of the Kerala State in south India and the Thar Desert in north-western Rajasthan. It is well documented now that global rise of temperature has induced a serious vector mosquito like *Aedes aegypti* increase its altitudinal distribution range by as much as 1000 ft in the lower Himalayas in India.

Some important mosquito-borne diseases like malaria, dengue and certain encephalitides are best demonstrable to have a working interaction with the environment.

1. Dengue fever

There are clear indications that an increase in ambient temperature in rather cooler areas, such as the lower outer scarps of Himalayas, has resulted in elevating the altitudinal distribution of the dengue vector, *Aedes aegypti*. Similarly, denudation of forest cover, comprising mainly rubber plantation supporting preponderance of another important dengue vector, *Ae. albopictus*, in the Western Ghat region in Kerala State, coupled with the consequent increase in the average annual temperature, has obviously resulted in the vector's spatial and temporal distribution over entire state in a short period of less than two decades. As to the dengue virus, there is a proof of its proliferation in *Ae. aegypti* mosquitoes in the laboratory when temperature was increased (Watts *et al.*, 1987). Appropriate models developed to prove this hypothesis suggest that relatively small increases in temperature in temperate regions, given viral introduction into a susceptible human population, should increase the potential for epidemics (Jetten and Focks, 1997). However, epidemic potential depends on the interaction of many parameters including virus, host and environmental factors. To predict actual risk of human cases it is important to incorporate site-specific human, viral and environmental determinants of transmission (Fig. 2.3).

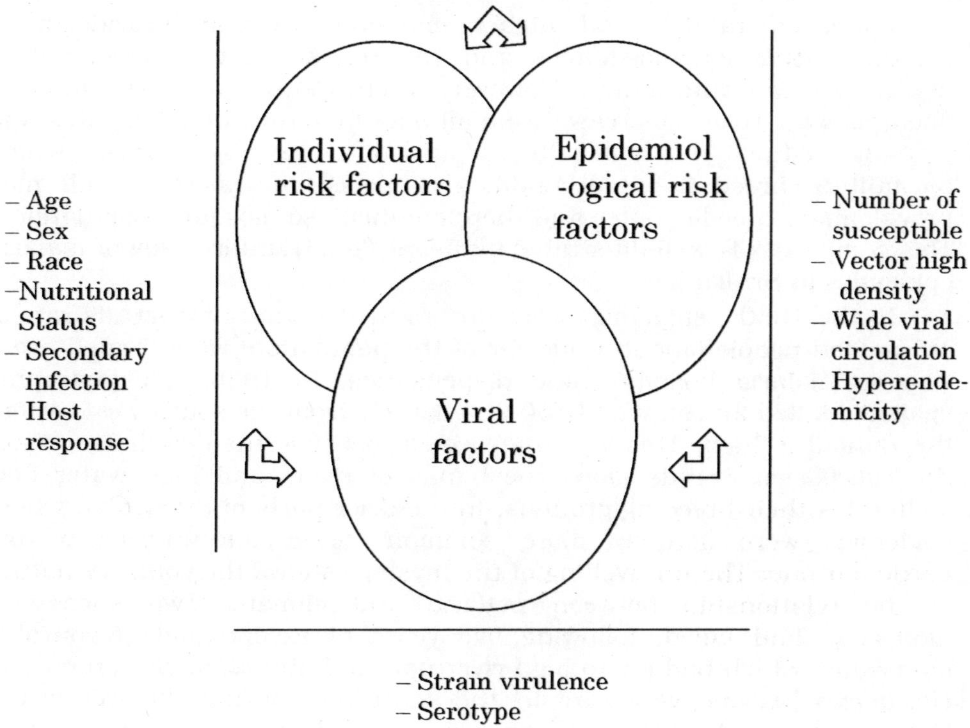

Figure 2.3. A graphical representation of various risk factors involved in DHF/DSS.

2. Malaria

In recent years both the climate change and the global warming have been under strict scrutiny (McMichael and Kovats, 1999). Public health, particularly infectious diseases, has not escaped the global warming debate and it is a widely held view that in future global warming and climate change will more deleteriously affect infectious disease like malaria. A simple internet search using the words 'malaria', 'global warming' and 'climate change' produces about 1000 results. While these studies may be interesting and of some scientific value, they ignore the far more important fact that man's activities have a far greater impact on the spread of malaria than climate (and certainly far more than the small part that man's emissions may be playing in changing the climate).

Climate's role in the spread of malaria may seem somewhat obvious. The spread of the disease relies upon sufficient numbers of anopheline mosquitoes and a large enough parasite pool amongst humans and animals. The behaviour, development and population of mosquitoes are strongly influenced by climate. Temperature, rainfall, humidity all affect mosquito populations, as well as other climatic factors, such as wind and sunlight. The relationship between these climatic factors as they affect mosquitoes, however, is highly complex and varies from country to country and between the different *Anopheles* species.

Increased rainfall and higher temperatures may provide more breeding pools for mosquitoes and can quicken the development of mosquito larvae into adults. However, where mosquitoes breed in large open pools or rivers, increased rainfall may flush out breeding, sites and actually reduce the mosquito populations. In other areas where mosquitoes breed in small puddles or pools, increased rainfall may provide more breeding sites and therefore increase mosquito populations. This complexity is well-illustrated by Reiter (2001) in his study of malaria epidemics in Sri Lanka.

In the 1930s, epidemics were rare in Sri Lanka, but a steady rate of 1.5 million people (about a quarter of the population) were treated every year in malaria hospitals and dispensaries. In 1935, a catastrophic epidemic killed an estimated 100,000, largely from the south west, where the rainfall is high. This was unexpected, because the dominant vector, *An. culicifacies*, breeds along the banks of rivers and high water flow obliterates their breeding grounds. In the drier parts of Sri Lanka, where epidemics were actually more common, fewer succumbed to this particular one. The unravelling of the mystery shows the contrary nature of the relationship between malaria and climate. Two successive monsoons had failed, following five years of exceptionally favourable monsoons, which had led to high river flow and abundant rice crops. *An. culicifacies* became very rare in the southwest during the wet years. However, when the breeding grounds returned as rivers and irrigation ditches dried out after the monsoon failures, numbers increased enormously. In the intervening years, the population in the wet region was barely exposed to malaria, and the group immunity was low, making it extremely vulnerable to epidemic. By contrast, in the drier parts, where *An. culicifacies* continued its normal breeding patterns, immunity remained constant and that population was better protected against epidemic.

The Thar Desert in western India is a glaring example as to how drastic climatic changes brought about in the region through mostly man's own activities such as the massive Indira Gandhi Nahar Pariyojana (IGNP), world's largest canal-based irrigation system in a desert ecosystem, could usher the deadly *Plasmodium falciparum*-dominated malaria in the region, in addition to the fears of plague and cutaneous leishmaniasis (Tyagi, 2001, 2002).

There are around 3,500 species and subspecies of mosquitoes and they are found throughout the world, except in those areas that are permanently frozen. The diversified success of mosquitoes is a proof to their remarkable ability to develop strategies to cope with different climatic conditions. Mosquitoes are known to survive extreme cold and heat; for example, studies in Memphis, Tennessee revealed that mosquitoes were able to survive extremely cold winters with temperatures below -10°C for more than 9 consecutive days. The mosquitoes managed to survive these temperatures by remaining in an underground storm water drain (Reiter, 2001). In Sudan, *An. gambiae* is able to survive where the outdoor summer temperatures can exceed 55°C. Mosquitoes cope with these temperatures by remaining indoors and hiding in thatch buildings during the day and only emerging to feed after midnight. Clearly, while climatic factors play an important role in the transmission of malaria, their impact is highly complex and difficult to model. Mosquitoes are highly sensitive to rising temperatures, which accelerate their metabolic rate, biting rates and feeding frequency. Temperature also affects their reproduction and growth. In USA, at temperatures below 23°C, a female of *Anopheles* mosquito, one of three Ontario-native species that can transmit malaria, takes 72 hours to lay eggs; above 23°C, she takes 48 hours; at 20°C, the eggs hatch in 2.5 days; at 33°C, they hatch in 33 hours. To reach adulthood the eggs take 20 days at 20°C, while at 31°C they take just seven days.

Temperature is the most important single factor that affects both the vector bionomics and the disease epidemiology of malaria (Fig. 2.4). Most vector species can not survive temperatures above 40°C and below 8°C. Pathogens inside mosquitoes are also sensitive to temperature. At 18°C, one species of the malaria parasite plasmodium takes 30 days to develop, but at 25°C it takes just 10 days. While the temperature threshold for the development of *P. falciparum* in its anopheline hosts is 18°C, for *P. vivax* it is 15°C. Since *Anopheles* mosquitoes live only a few weeks, higher temperatures increase the odds the parasite will mature and become infective while the mosquito is alive and biting. And, because warming accelerates mosquito's breeding cycle, there are more generations of vectors. Moreover, warmer nights and warmer winters resulting from climate change can extend the geographic range of mosquitoes, increasing their ability to spread disease. One study estimated that an increase of several degrees in the global temperature by 2100 could multiply the vectorial capacity of mosquitoes by 100 times in temperate regions. The incidence and prevalence of malaria may be limited by altitude (Taylor and Mutambu, 1986), a good proxy for temperature, but malaria transmission depends on many factors besides temperature. In the Thar Desert of western India, which until recently remained virtually a *terra incognito* for high intensity malaria, particularly *P. falciparum*- dominated one, malaria epidemics were associated with the catastrophic transformation in the xeric physiography brought about by extensive canalized irrigation (Tyagi, 2002). Although malaria is seasonal, there is substantial interannual heterogeneity of malarial incidence around the globe (Swaroop, 1949; Fontenille et al., 1997). Extremes of rainfall (both drought and floods) associated with El Nino events are linked to variability in malarial incidence in different regions (Bouma and van der Kaay, 1994; Bouma et al., 1997). Even though, there are clearly two different schools of thoughts on whether or not malaria endemicity will change in already high-risk areas, it is possible that small increases in minimum temperature in cooler regions may disproportionately increase malaria transmission, provided there are no socio-economic barriers to transmission.

In temperate countries, which were either never affected by malaria or where the disease once occurred but subsequently eradicated, malaria may well be there knocking on the door if climate conditions continue to warm. This situation may be exacerbated by the ever increasing inflow of infected travelers and/or vectors, particularly the drug resistant parasites, from the endemic countries.

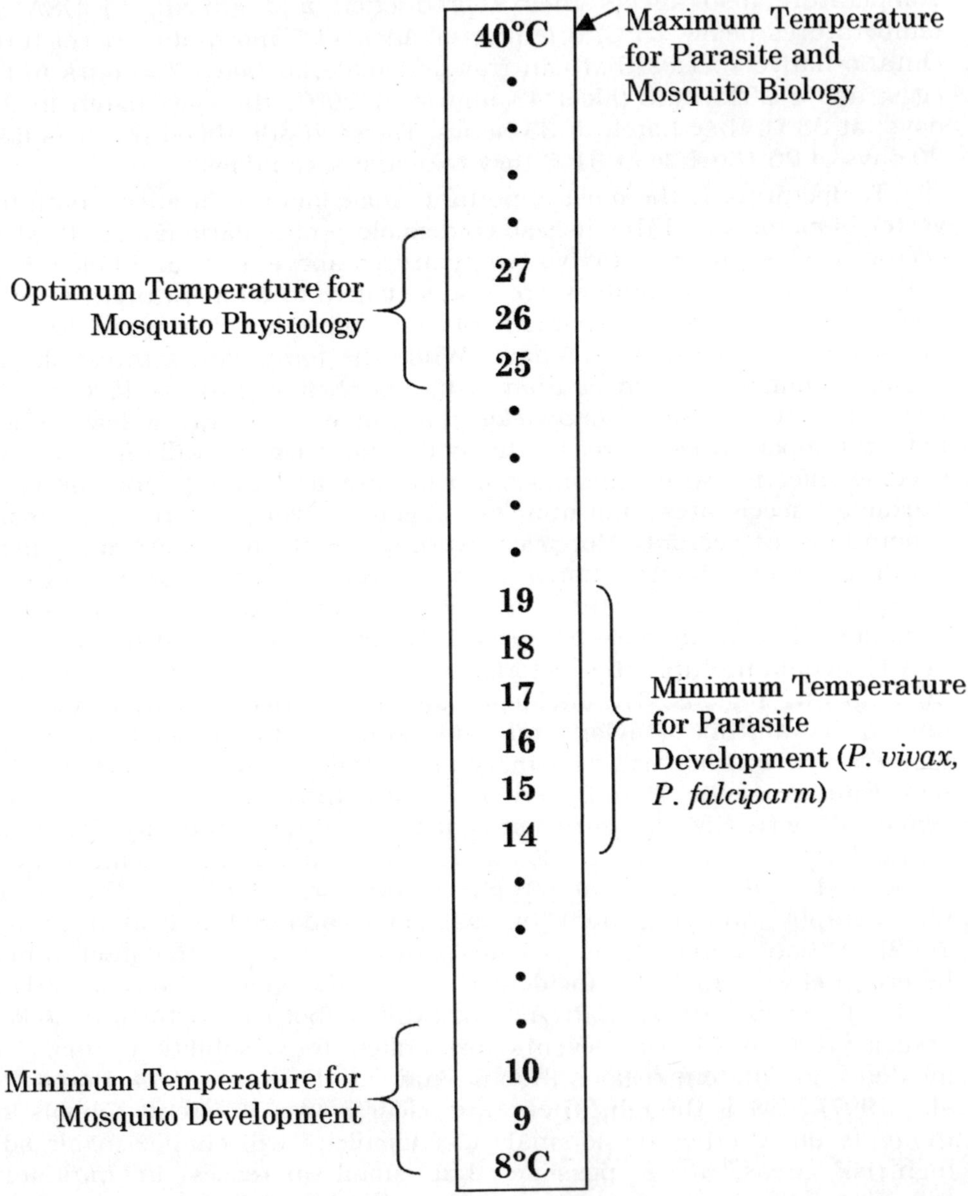

Figure 2.4. Effect of variability in temperature on the development of a mosquito and the parasites in the body of the vector.

By far the greatest impact upon the spread and incidence of malaria are man's economic activities and efforts to control the disease. While DDT finally eradicated malaria from Europe, the incidence of the disease had been declining since the late 19th century. As Europe became wealthy and more developed, so the incidence of malaria declined. As the area of land under agricultural production increased, so the number of mosquito breeding pools declined. Improvements in agricultural technology meant that fewer manual labourers were required, which resulted in increased urban populations and reduced rural populations. By separating people and parasite pools from the malaria vector, the probability that the disease would spread was reduced. As households began to own more livestock, such as pigs and cattle, mosquitoes had a greater number of potential animals upon which to feed, this in turn reduces the probability that the malaria parasite will be transmitted. In addition, improved housing both separated farmers from their livestock and provided better protection from mosquito ingress.

Most of the reductions in the incidence of malaria occurred in parts of Europe before Dr. Ronald Ross confirmed the link between the *Anopheles* mosquito and the transmission of malaria. Indeed, targeted malaria control efforts were only implemented in the early 1900s, when the disease was already retreating in Europe. While the incidence of malaria was declining during the late 19th and early 20th centuries, the disease continued to devastate large areas of Poland and the Soviet Union. While these areas were all subjected to similar, or even cooler, climatic conditions, the incidence of malaria was vastly different. Malaria epidemics were experienced in many parts of the Soviet Union. Outbreaks were not confined to the warmer climes around the Black Sea, but were experienced as far north as Archangel in the Arctic Circle. While increased economic development, better housing, nutrition and sanitation led to decreases in most infectious disease in Europe, social upheaval and economic disruption in the Soviet Union and Poland entrenched and increased the incidence of malaria. South Africa recently experienced a malaria epidemic, mostly concentrated in the northern Kwa-Zulu Natal province. The epidemic began after the South African Department of Health removed DDT from the vector control programme. The major vector, *An. arabiensis* is resistant to the synthetic pyrethroid alternatives that were introduced. In addition, *An. funestus*, one of the most efficient malaria vectors, returned to South Africa after an absence of about 40 years.

The epidemic also occurred during a period of above-average rainfall in Kwa-Zulu Natal and indeed the whole of southern Africa. During 2000, DDT was reintroduced and because of parasite resistance to the drug combination sulphadoxine pyramithamine, a new artemisin-based therapy was introduced. The result of these interventions was an 80% reduction in cases, even though the rainfall in 2000 and 2001 was lower than that for 2000. The South African Department of Health identified DDT as the most important factor in the reduction of malaria cases.

The success of malaria control in South Africa was clearly far more dependent on vector control and medical interventions than on rainfall. Malaria, as are other infectious diseases in developing countries, is a major determinant of poverty. Policies, such as those proposed under the Kyoto Protocol that aim to limit CO_2 emissions and economic growth, will not materially affect the spread of infectious diseases but will hamper the abilities of poor countries to cope with health crises.

While in United States, increased frequency of local transmission as a result of increased imported disease due to migrant workers and immigrant populations living in substandard housing has been observed in the past decade, other non-endemic countries have also reported malaria recently. After 4 years of being malaria free, transmission of *P. vivax* malaria by local vectors has been reported in Italy. In Britain, indigenous cases have been recently detected, while in France, about 5,000 imported cases are reported every year, with seldom local transmission of the disease. However, a major resurgence of epidemic malaria has occurred in the southern republics of the former Soviet Union because of the breakdown of public health services. In India, cross-border infiltration of refugees from across the neigbouring countries like Myanmar and Bangladesh have helped resurge the malarial disease with possibly more drug-resistant cases sneaked in the country.

3. West Nile Virus

The surprise outbreak of West Nile in New York in 1999 is an eye opener. The primary vector, the *Culex pipiens* mosquito, lives mainly in cities, laying eggs in stagnant, polluted water in sewers and elsewhere.

The mild winter of 1998-99 was followed by the hottest, driest spring and summer New York City had experienced in a century. It is likely that drought killed off mosquito predators and also caused birds to congregate around sparse water sources where mosquitoes bred. During a drought, the pools become richer in rotting vegetation that the mosquito needs to survive. The heat wave allowed the virus to mature rapidly inside mosquitoes, which quickly infected many birds, creating a large reservoir of the virus. As some birds died and others flew off in search of water, the mosquitoes increasingly turned to humans for blood. Then came heavy rains that created new breeding sites, causing the population of infected mosquitoes to explode.

In 2002, the second warmest year on record, conditions were ripe for further spread. Many parts of North America experienced a very warm winter and a spring and summer of drought followed by heavy rains. West Nile made "a furious dash" westward, spreading to 44 states and five provinces. In early August, the number of U.S. cases suddenly tripled from the previous week and was expected to exceed last year's totals. The virus has spread rapidly to western states not previously affected. It now infects 230 animal species, including more than 130 species of birds.

The winter of 2002-03 was colder than last year, but dead infected birds still turned up in Ontario more than a month earlier than last year and the first Canadian cases have also been reported then.

The pattern of warm winters followed by hot, dry summers may encourage the spread of other vector-borne diseases, such as encephalitis, transmitted by mosquitoes, and Lyme disease and Rocky Mountain spotted fever, spread by ticks. Warming temperatures may create conditions suitable for the re-introduction of malaria in North America if they allow parasites sufficient time to mature inside mosquitoes.

4. Japanese encephalitis and other encephalitides

Vectors of Japanese encephalitis, belonging mainly to *Culex vishnui* group (e.g., *Cx. tritaniorhynchus, Cx. pseudovishnui* etc.), have a strong relationship to rainfall and particular agricultural practices pertaining to paddy cultivation, as observed in several south-east Asian countries including India. Warming of climate as well as precipitation both have been associated with outbreaks of SLE, transmitted by *Cx. tarsalis* and *Cx. nigripalpus*, in North America, where in certain parts it appears to be associated with the sequence of warm, wet winters, cold springs and hot dry summers. Mere 3-5°C increase in average temperature may cause a northern shift in the distribution of both WEE and SLE viruses and a decreased range of WEE in southern California (Reisen *et al.*, 1993).

CHANGE IN ECOLOGY THROUGH CANAL-BASED IRRIGATED AGRICULTURE AND EXACERBATION OF MOSQUITO-BORNE DISEASES IN THE THAR DESERT: *A MODEL PAR EXCELLENCE*

The famous Thar Desert had been the abode for the once existing historic Harappa civilization in Kalibanga along the extinct holi river Saraswati, traversing across the Great Indian Thar Desert, together with Yamuna and Ghaggar rivers. At present only the relic Ghaggar marks its presence during monsoon, while Yamuna has in course of time changed its direction, now flowing far away from Rajasthan's north-eastern border into the neighbouring State of Uttar Pradesh. *Inter alia* several different kinds of changes took place in the Great Indian Thar Desert region, characteristically experiencing a more or less set frequency of decadal droughts leading to emigration of local folks across the border to other States with high disease endemicities (*e.g.*, Punjab, Gujarat, Madhya Pradesh and Uttar Pradesh etc.), intermittent floods in the lower reaches of the Luni river basin particularly Barmer district and adjoining districts in the Jodhpur zone leading invariably to the eruption of the *Plasmodium falciparum*-dominated malaria, and the excessive recurrence of the sand-storms resulting into expansion of the desert. Yet, the greatest discernible change the Thar Desert seemed to have undergone with a human interruption is the provision of extensive canalization - probably one of the world's largest canal systems of its type in the desert ecosystem. Though, originally developed to boost the future prospects of agriculture in the virgin deserts and animal keeping through planned and sustained irrigation, nevertheless, due to multiple reasons associated with canal-water management, particularly those relating to formation of stagnant water bodies, the canal waters inadvertently became also the favourable breeding sites for a large number of vector mosquitoes (Tyagi and Chaudhary, 1997) (Fig. 2.5).

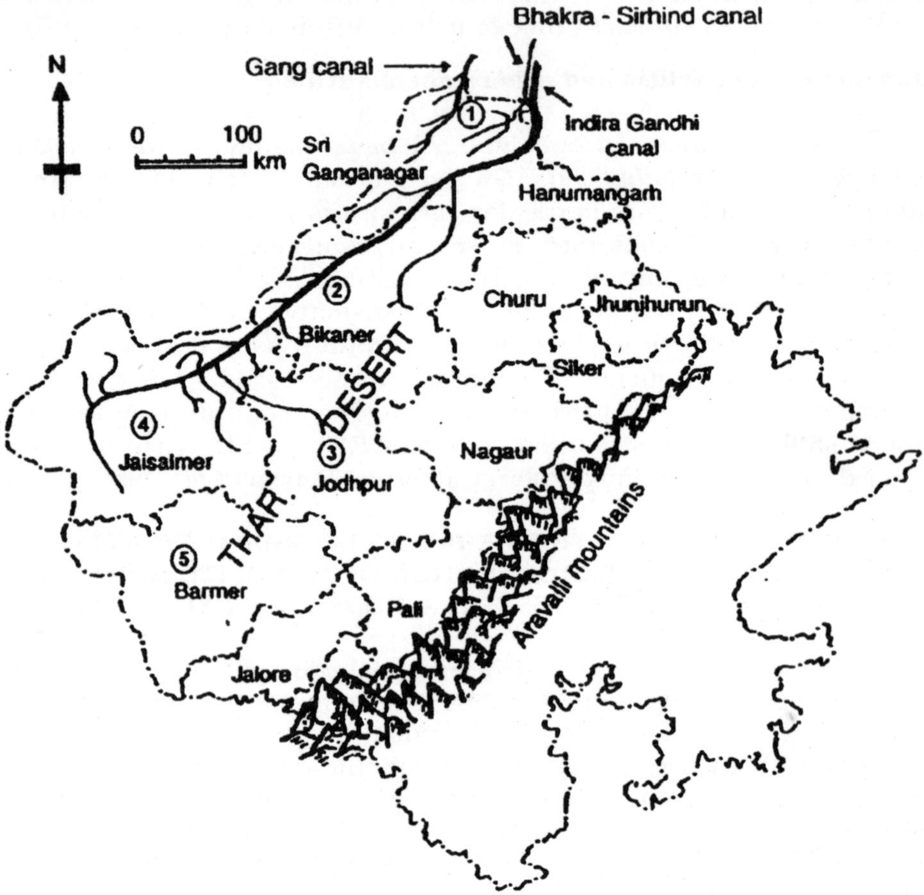

Fig. 2.5. Twelve desert districts int he Thar Desert. Those districts depicted by encircled numerals have experienced repeated *P. falciparum* dominated malaria epidemics; all (except Barmer) falling under canal based irrigation, particularly IGNP.

Desert physiography and the impact of three major canal systems in the Thar desert

The geographic limits of the Thar region are generally understood to be between the irrigated lands of the Indus and Sutlej and the western edge of the Aravalli Hills (i.e., approximately between 24°-30°N and 68°-

78°E). In Rajasthan, the entire desert region is organized in 12 districts (Fig. 2.5), spread over nearly 75,000 km^2 (62% of the total land of the State) and harbouring a population of 12.8 million people (39% of the population of the State). The seasonal Luni river divides the desert into two parts; in the north a vast stretch of sand dunes or levees and in the south an expansive basin rimmed by flat- topped hills of sandstone and limestone.

Three major canal systems exist in the Thar Desert region today. The first was the Gang canal project which became fully operative in 1928, with its 1251 km distribution system in Sri Ganganagar district only. Along with the Bikaner canal feeder, it provides water to about 65% of the 300,000 ha of cultivable land. The Bhakra-Sirhind canal project, with its 1219 km long distribution system, was started in 1955 to utilize a further 300,000 ha to cultivable land. The most important of all the canal systems is the Indira Gandhi Nahar Pariyojana project (IGNP) which is one of the world's largest projects of its kind in the desert. It started in 1958 and irrigation of part of the area began in 1961. The project, when completed, will include a 649-km long main canal (of which 445 km is in the Thar Desert) and over 9000 km of distributaries. Although only two of the three stages envisaged are completed by now, the main canal including feeder and 5500 km of distributaries is already irrigating 950,000 ha of land (approximately 50% of the irrigated area that is planned) (Fig. 2.6). This Herculean task, bringing Himalayan waters to the Thar Desert, aims at transforming vast stretches of western desert terrain into verdure. As a result, the annual agriculture production from the newly irrigated land now exceeds Rs. 6500 million. While several irrigation-intensive crops such as sugarcane, mustard, groundnut, etc., have been established in some of the canalized desert districts, the production of rice in Sri Ganganagar district, the most irrigated among all desert districts, has topped the list in the whole State. Rice cultivation has often been associated with malaria transmission, although a recent study has found a negative correlation between rice cultivation and the annual parasite incidence (API) in most States and Union Territories on a nationwide survey.

At many places along the main Indira Gandhi canal, in particular, the problem of seepage water and water-logging have created serious mosquitogenic conditions. Besides the surplus irrigation water that accumulates on irrigated land, vast areas along the course of the Indira Gandhi main canal have been engulfed by seepage water from the irrigation canals on a perennial basis. An estimated 8600 ha in the Nalibed area, Sri Ganganagar/Hanumangarh district, is permanently inundated under the IGNP Stage I by the Ghaggar river floodwater while 1000 ha is converted into marshy land due to waterlogging and excessive seepage from the canal in Stage II. Supply of canal water to the desert region over several decades has caused a rise of the water table and water-holding potential of the soil, and a plentiful growth of vegetation, particularly hydrophytic weeds. Such extensive breeding grounds have proved to be favourite sites for breeding by several different vector mosquitoes many of which were even not found in the interior of the Thar Desert. The changed physiography of the Thar Desert has also brought about an impact on another important vector-borne infection, cutaneous leishmaniasis, in respect of both the bionomics of sand fly vectors and the disease epidemiology. Interestingly, Tyagi & Hiriyan (pers. comm.) have recently demonstrated spread of dengue vector, *Aedes aegypti*, in the interior of the typical Thar Desert environs, attributable to replacement of time-tested 'tanka' (the underground water reservoirs) by the conduit-supplied canal water, and as a consequence the dengue vector found it convenient to breed in the discarded 'tanka' and propagate in the desert which was struck by dengue epidemic in as latest as 2002.

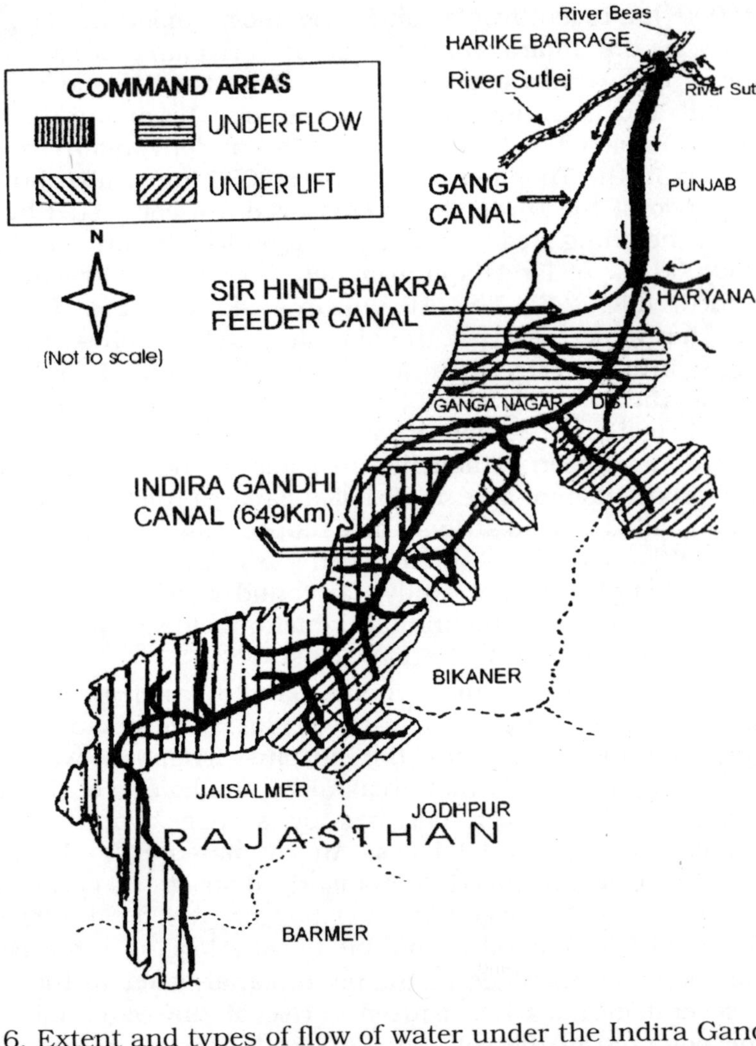

Fig. 2.6. Extent and types of flow of water under the Indira Gandhi Nahar Pariyojana (IGNP) in the Thar Desert, Western Rajasthan.

Changes in rainfall, air temperature and relative humidity in the Thar desert's IGNP command area

A recent compilation of percentage departure of monsoon season rainfall (June to September) for the period 1901–2000 over west Rajasthan by Dr. S.M. Kulshrestha (*pers. comm.*) depicts how significantly the drift has been over past 100 years (*cf.* Tyagi, 2002). An increase in the mean decadal rainfall has been found in the district, which had been subjected to canal-based irrigation for a considerable period.

The Gang canal and IG canal projects aimed to utilize 7.59 MAF of water from Ravi-Beas duo waters to irrigate an 1.143 mill-ha area. In the Ganganagar district, the Bikaner district and the Jaisalmer district, 708, 775 ha, 37,022 ha and 84 ha of land, respectively, has been put under irrigation. The impact of irrigation in Rajasthan is seen from the vegetational and micro-climatic changes that have occurred in the region. Sri Ganganagar receives an annual rainfall of 243 mm. The annual rainfall of Sri Ganganagar showed an increasing trend, at the rate of 1029 mm/year, during the period 1926-1993. The mean annual rainfall of Bikaner is 286 mm with a coefficient of variation of 48%. The region experienced droughts once every 3 years (38% chance). The rainfall (annual) at Bikaner showed an overall decreasing trend at a rate of 0.147 mm/year during 1926-1993, which is forcibly attributable to consecutive droughts which occurred after 1961. Jaisalmer receives a mean annual rainfall of 185 mm with a coefficient of variation of 64%. The rainfall has until recently been erratic and poorly distributed with droughts occurring once every 4 years (24% chance). However, the overall trend in annual rainfall at Jaisalmer which remained the same during 1901-1993 seems to be shifting towards an escalation in recent years.

The ambient temperature, too, has shown decreasing trend being 0.039°C/year at Sri Ganganagar, 0.023°C/year at Bikaner, and 0.009°C/year at Jaisalmer. With regard to relative humidity in the Thar desert, it was recorded to be 5-10% higher humidity measurements in cropped surfaces at 1-m height in the irrigated villages compared to the unirrigated cropped surfaces. This highlights the fact that the amount of canal water available in the desert region can influence atmospheric humidity.

These observations sufficiently indicate towards a shift in the abiotic characteristics of the Thar Desert under the IGNP command area as compared to that in the typical desertic environs.

Exacerbated disease scenario in the Thar Desert under the changing environmental imapct

(1) ***Malaria***: Rajasthan State has been experiencing scourges of malaria for decades, but the disease intensity varied a great deal and mostly confined to north-east and south-western parts of Rajasthan, that is the

Thar Desert. In fact there existed patches of hyper-endemicity which for known reasons had been experiencing incessant episodes of focal and/or widespread epidemics. Yet, the overall impression of the State has been that of a non-problematic area with few local outbreaks here and there. Based on malaria endemicity Rajasthan is classified as given hereunder;

i. Zone or Stratum II, signifying moderately refractory areas with high epidemic potential (*i.e.*, eastern plains and southern forested and mountainous districts of Rajasthan State), and

ii. Zone III, signifying non-refractory areas with moderate to high epidemic potential (*i.e.*, western and northern districts of Rajasthan State, including whole of the Thar Desert).

It is interesting to note that in last nearly two decades the physiography of the Thar Desert, in particular, has undergone such a major change that the zoning of the region needs a re-characterization of its epidemiological potential. Under prevailing grim circumstances, the malaria transmission dynamics in Rajasthan State particularly the Thar Desert region requires a serious thinking to device appropriate control methodologies. In general, the malariometric data (1961-94) show that the overall number of positive cases in the State had increased 27 times, from 8494 cases in 1961 to 229,772 in 1994, while the targeted population increased only 2.5 times during the same period. It is also quite evident that while the slide positivity rate (SPR) had significantly increased from 1.43 in 1961 to 5.07 in 1994 (3.5–fold increase), the slide *P. falciparum* rate (SFR) had also increased as many folds only, from 0.55 to 1.95. However, the increase in either the SPR or the SFR was not steady and, therefore, no definite trend can be noticed. The maximum SPR was recorded in 1976, with several epidemics in non-desert parts of the State. From the mid-eighties onwards, however, the trend had become clearer and both the SPR and SFR had steadily increased. In fact, the highest ever SFR had been recorded in 1994 – the year when the State as well as the Thar Desert region experienced the maximum deaths due to malaria in their entire history. A cursory review of the malariometric data reveals it all. Since early 1970s, many districts in the Thar Desert region have had repeatedly several epidemics, and this trend is a cause of major concern in areas where physiographic changes are continuing to happen under the snail-paced work progression of the IGNP in Phase II and III to create new canal-based irrigational avenues.

Until about two decades from now the Thar Desert was hardly considered an important habitat for the malarial disease or its vectors. This was due essentially to very low prevalence of malaria and without any epidemic in the region. In the wake of such knowledge about malaria, no significant anopheline survey was ever made in the interior parts of the Thar Desert, and the fewer studies then directed on the subject have been from around the townships. Also, medical entomologists got more attracted to the forested and mountainous districts of the Udaipur zone in southern Rajasthan, being highly malarious. It is not surprising , therefore, that in the early days nearly all the major faunal surveys in the State were organized in the Udaipur zone districts only.

Although, the first ever account of Indian anopheline was presented on a 'wall map' in 1912 by the Central Malaria Bureau, at Kasauli (subsequently transformed first into malaria Institute of India and, finally, the National Institute of Communicable Diseases, Delhi), it became possible only in the late twenties when for the first time four anopheline species from the then Rajaputana State were put on record. Later in his *magnum opus* on 'Family Culicidae: Tribe Anophelini' in the classic 'Fauna of British India' series, now largely out of date, Christophers (1933) described in detail the taxonomy and bionomics of these four species, *viz.*, *An. stephensi*, *An. culicifacies*, *An. annularis* and *An. subpictus*, which were stated to be occurring in at least a few desert districts as well. The total number of taxa in the State was first raised to seven and subsequently to 16 (Puri, 1955), without any change in the strength of the desert fauna. According to Puri (1960), most of the 16 anopheline species were restricted to the two southern forested districts, Sirohi and Udaipur, contributing 56% of the total fauna from the State, while the Thar Desert had a meager share of seven (44%) species from five of the then eleven desert districts. Wattal & Kalra (1961) offered a comprehensive taxonomic key, also State-wise, for the adults of anopheline mosquitoes known to be occurring in India till then. Interestingly, though Rao (1981) extensively dealt with the Indian anophelines, but no mention was made to either the Rajasthan State or the Thar Desert while discussing the regional distribution of the taxa. Two more species, *An. nigerrimus* (new to Sri Ganganagar) and *An. splendidus* (new to Rajasthan State), collected out of a total of seven species in Sri Ganganagar district, were added to the list of anopheles from the Thar Desert raising the total to nine (52.9%), with only two additions to the faunal strength of the state, including *An. d'thali*, new to Rajasthan, and elevating the fauna from the Thar to 10 (55.6%), and in the State to 18 (34.6% of the total anophelines recorded from India). Subsequently, *An. barbirostris* from Bikaner and *An. fluviatilis* from Jaisalmer have also been recoded making the latest inventory shooting to 12 (66.6% of the total taxa enlisted from Rajasthan. It is comprehensible from the distribution of taxa in each district, as given in Table 2.2, that there stands a good possibility of discovering more new records from the Thar Desert, particularly the unexplored and/or less explored north-eastern and southern desert districts, provided long term surveys are undertaken, though at the moment the Thar Desert remains under represented from districts' point of view.

Table 2.2. District-wise distribution of anophelines (1= An. culicifacies, 2= An. stephensi, 3= An. subpictus, 4= An. vagus, 5= An. annularis, 6= An. nigerrimus, 7= An. splendidus, 8= An. barbirostris, 9= An. d'thali, 10= An. fluviatilis, 11= An. pulcherrimus)

Thar Desert Districts	Species										
	1	2	3	4	5	6	7	8	9	10	11
Sri Ganganagar	+	+	+	+	+	+	+	–	–	–	+
Hanumangarh	+	+	+	+	+	+	+	–	–	–	–
Churu	ns	ns	ns	ns	ns	ns	ns	ns	ns	ns	ns
Jhunjhunu	ns	ns	ns	ns	ns	ns	ns	ns	ns	ns	ns
Sikar	ns	ns	ns	ns	ns	ns	ns	ns	ns	ns	ns
Bikaner	+	+	+	+	+	+	+	+	–	–	–
Nagaur	ns	ns	ns	ns	ns	ns	ns	ns	ns	ns	ns
Jodhpur	+	+	+	–	+	–	–	–	–	–	–
Jaisalmer	+	+	+	+	+	+	+	–	+	+	
Barmer	+	+	+	–	–	–	–	–	–	–	–
Pali	ns	ns	ns	ns	ns	ns	ns	ns	ns	ns	ns
Jalor	ns	ns	ns	ns	ns	ns	ns	ns	ns	ns	ns

Species complexes in vectors of the Thar Desert

As far as the Thar Desert is concerned, sibling species have been confirmed to be occurring in *An. culicifacies*, *An. subpictus* and *An. fluviatilis* (Table 2.3). *Anopheles stephensi* does not possess any species complex.

Anopheles culicifacies complex comprising 5 species (A,B,C,D and E) is the most important group since the malaria flare up in the Thar Desert seem to be influenced by the irrigation of the desert lands and it is confirmed that species A, which is a vector, occurs prodigiously in the canal-irrigated areas (Subbarao *et al.*, 1999). From the point of view of distribution of sibling species in different vector species complexes, the Rajasthan State is stratified in Division I and II. The northeastern districts of the Thar Desert like Sri Ganganagar, Hanumangarh, Churu, Jhununu and a part of Bikaner and Sikar are classified into Division I which is harbouring species A. The remaining districts in the Thar Desert are grouped inti Division II where species B and C are most common. Therefore, it is suggested that areas with species A and C need be thoroughly monitored for various behavioural characteristics to help design the control strategies more purposefully.

Table 2.3. Malaria vectors with species complexes in the Thar Desert

Species	No. of sibling species	Nomenclature assigned to sibling species	Sibling species found in India	Sibling species found in the Thar Desert
An. culicifacies	5	A,B, C,D and E	all	A,B and C
An. fluviatialis	3	S,T and U	all	Not known yet
An subpictus	4	A,B, C and D	all	Not known yet
An. annularis	2	A and B	all	Not known yet

As regards *An. annularis, An. subpictus* and *An. fluviatilis* still nothing is determined about the sibling species prevailing in the Thar Desert, and this presents a new pathway for future research.

Emergence of *Plasmodium falciparum*-dominated malaria in the Thar Desert

The correlation of environmental changes brought about in the Thar Desert by extensive canalization and irrigational activities with malaria prevalence of varying intensities in different physiographic situations is so clearly discernible and demonstrable that the Thar Desert practically offers a inafalliable model to understand the causes underlying resurgence of malaria in a desert ecosystem and control of vector mosquitoes with proper water management.

Decades ago, particularly before the canalized irrigation had been ushered in the Thar Desert, most of the positive cases were reported from the hilly and forested terrain of Udaipur in the southern part of Rajasthan. Thus, while in 1986 the Udaipur zone had contributed 44.1% positive cases and 67.7% *P. falciparum* cases as against 24.4% positives and 11.6% P. *falciparum* cases in the desert region, in 1994 only 14.1% positives (*P. falciparum* 26.6%) hailed from Udaipur zone compared to 53.3% positives from the desert districts with a whopping 62.5% *P. falciparum* cases. Amongst the then 11 desert districts, Barmer was the most malarious in 1994 (and even now in 2003!), contributing 31.3% positive cases and 35.8% *P. falciparum* cases (Tyagi, 2002).

Pathways of malaria epidemics in the Thar Desert

Documentable instances of outbreaks of malaria epidemics in the Thar Desert are available since 1975 (Table 2.4), although there is every possibility that mortality might have also occurred in the desert during the nation-wide epidemic in 1908, and perhaps thereafter as well on other such occasions! As a consequence, at least 13 episode-years of malaria epidemics are identifiable since 1975. Save for the massive outbreaks of malaria in 1975, 1976, 1990 and 1994, most other epidemics were focal in nature covering often a part of a district or two. While the epidemics of 1975 and 1976 were predominantly characterized by a resurgence in *P. vivax* cases, the outbreaks of 1983-1984 onwards saw a phenomenal rise in *P. falciparum* (Tyagi, 2002).

Table 2.4 . Frequency of malaria epidemics in the Thar desert districts (1975 – 2002)

District	Years of epidemics	Dominant malaria species at the epicentre of epidemic	Possible reason associated with epidemic	Deaths
Sri Ganganagar (incl. Hanumangarh)	1983, 1984, 1985	*P. falciparum*	Increased vector activity with increase in varied breeding sites	Nil
Bikaner	1975, 1976, 1990, 1992, 1993, 1994	*P. falciparum*	Increased vector activity with increase in varied breeding sites	113
Churu	1975, 1976	*P. falciparum*	Inter-State population migration	5
Jhunjunu	1975, 1976	*P. falciparum*	Inter-State population migration	6
Nagor	1975, 1976, 1990	*P. vivax*	Inter-State population migration	2
Jodhpur	1990, 1992, 1993, 1994, 2001, 2002	*P. vivax and P. falciparum*	Increased vector activity with increase in varied breeding sites; excessive quarry-mining brooding perennial vector breeding	72
Jaisalmer	1975, 1990, 1992 – 1996 1999 – 2001, 2002	*P. vivax and P. falciparum*	Increased vector activity with increase in varied breeding sites; non-immune human population	72
Pali	1975, 1976, 1990	*P. vivax and P. falciparum*	Inundative vectorism	3
Barmer	1975, 1990, 1994	*P. vivax and P. falciparum*	Inundative vectorism	178
Jalor	1975,1976, 1990	*P. falciparum*	Inundative vectorism	20
Sikar	1975, 1976	*P. falciparum*	Flooding	10

(1) Malaria epidemic of 1975-77: Even though no death was brought on record, it was crystal clear that, in view of the steep rise in the passive cases compared to the 1966 level of malaria cases, outbreaks of malaria had also struck the Thar Desert, along with other parts in the country.

(2) *Malaria epidemic of 1983-1984*: Sri Ganganagar district experienced a focal *P. falciparum*-dominated malaria outbreak for the first time in 1983, with 4925 positive cases of which 43.4% were *P. falciparum* cases. In spite of the fact that in 1983 the annual parasite incidence (API) was but second lowest during the last few years, the malaria outbreak in the Nalibed area in the erstwhile Hanumangarh tehsil of the Sri Ganganagar caused high morbidity. Two primary health centers (PHCs), Sangaria and Pilibanga, which were most severely affected, had jointly contributed 72.2% of the total positive cases in the district during the epidemic. It is noteworthy that *P. falciparum* had constantly increased for at least few years since 1987, while the API had kept a low ebb for the same period.

(3) *Malaria epidemic of 1990*: In 1990, a widespread epidemic hit western Rajasthan districts, but particularly the Barmer district claiming 47 lives (Mathur *et al.*, 1992). A total of 19,322 positive cases (*P. falciparum* 66%) were recorded in Barmer district which was 2.5 times higher than what was reported in the preceding year. It was advocated that the epidemic in Barmer district occurred as a sequel to widespread floods in the region. It was demonstrated that the rise in the incidence of malaria was associated with the increase in the rainfall. During 1977, when annual rainfall was 13% above the normal, the annual parasite incidence (API) showed a conspicuous rise at 5.3 and annual *P. falciparum* incidence (AFI) between 0.03 and 0.1. However, 1985 was a year of exception when there was noticeable increase in cases, particularly of *P. falciparum*, despite low rainfall. This increase was, however, attributed to the migratory cases, since the cases did not proliferate locally due to non-conducive transmission conditions. This was substantiated by a significant decline in the incidence in the next 2-3 years.

Both rural and urban areas were equally and simultaneously affected by the epidemic. Though all the eight PHCs presented a rising trend, the PHC Baitu was the worst affected with API touching a peak of 55.3 and AFI 19.6. The highest increase in AFI were in Mandli (4.5 times), Gudamalani (4 times), Ramsar (2.8 times), Ganga (2.3 times), Sindhari (1.75 times) and Chohtan (1.6 times). The Barmer town also recorded a 4-fold rise in API and 3-fold rise in AFI. The district overall API and AFI recorded 2.5 and 3-fold increases, respectively. The mass blood survey carried out during the epidemic showed that all age groups were affected. The slide positivity rate (SPR) was 61%. More than 93% of these had *P. falciparum* infection. A total of 122 infants were found positive of which less than 30% had *P. falciparum* infection. A total of 103 infant positives were from Baitu PHC only. About 88% of the positive infants were detected during September-November, 1990 only.

(4) *Malaria epidemics of 1992-1993:* During late 1992 and early 1993 epidemics of malaria broke out simultaneously in several districts, *i.e.*, Bikaner (Bajju and Kolayat PHCs), Jodhpur (Baap and Phalodi PHCs) and Jaisalmer (Pokaran, Nachna and Sangad PHCs). In Jaisalmer district alone a total of 2128 positive cases (*P. falciparum* 73.6%) were reported during the epidemic as against a combined total of of 1809 cases (*P. falciparum* 27.6%) in the two preceding years, *i.e.*, 1990 and 1991. If the malariometric data from the three PHCs are reviewed it becomes clear that the malaria scenario has drastically changed over the years. A total of 2128 positive cases (*P. falciparum* 73.6%) were reported against a combined total for 1990, and 1991 of 1980 cases (*P. falciparum* 27.6%), clearly indicating a shift towards an ascending order.

During early 1993, the SPR was found to be as high as 85.0% and the *P. falciparum* proportion 88% in two desert villages near the Indira Gandhi canal, Madassar and Awai, (Tyagi, 2002).

(5) Malaria epidemic in 1994: In mid-1994 a major epidemic of malaria broke out in whole of Rajasthan State taking a toll of 452 lives. The worst affected three districts included Barmer, Jaisalmer and Jodhpur, contributing nearly two-third of the total mortality. *Plasmodium falciparum* accounted for over 60% of cases in the desert and was considered as the main cause of death.

In a year-long pilot study during 1993-94, following a rapid mass blood survey immediately after the epidemic in the late autumn in 1992 in the two IGNP command villages (*P. vivax* 11.1%, *P. falciparum* 88.7%) and two unirrigated villages (*P. vivax* 17.1%, *P. falciparum* 82.7%), fever surveys conducted for one full year between April 1993 and March 1994 showed a slightly higher slide positivity rate in the IGNP villages (32.3%) than in the unirrigated villages (25.5%). The most important feature was, however, the higher proportion of *P. falciparum* in the IGNP villages (76.6%) compared to that in the unirrigated villages (16.6%). Influenced by the high percentage of *P. falciparum* in the study villages, the slide positivity rate maintained a high profile throughout. Mixed infections occurred in all villages, although more of them occurred in the unirrigated villages (11.5%) than the other two villages (3.2%). Interestingly, more gametocyte carriers were present in the IGNP non-command villages (16#, 17.8%) compared to the unirrigated villages (2#, 8.3%). Cumulatively, 73.7% of all the positive cases were found in children of less than 14 years of age. The male population (68.4%) suffered more than the females (31.5%).

(6) Malaria epidemic in 1995-1996 : A total of 45 lives (60.7% of the total deaths in Rajasthan) were reported in the Thar Desert region alone, even though there was some debate whether the year of 1995 to be regarded as an year of epidemic or not. Lately 1985 has been rightly characterized, along with 1996, as an epidemic year. The malariometric data of 1995 are comparable with some of the past epidemic years.

(7) Malaria epidemic of 1999: A focal outbreak of malaria occurred in the villages situated close to the main Indira Gandhi canal near Ramgarh in Jaisalmer district, western Rajasthan. Stagnation of water over a month's period in the main canal as well as long standing rain water in the form of large lake-like bodies near these villages formed vast breeding grounds for the vectors like *Anopheles culicifacies*, along with *A. stephensi* already breeding in the 'tanka' and 'beri' in the epidemic-hit villages. A mass blood survey was conducted covering 380 persons with a history of fever in last one month in the epidemic hit villages, namely Seuva, Raghava, Raimala and Sadhana, all within a distance of 2 km from the main Indira Gandhi Cana. Of these 49 (12%) were *P. vivax* and 40 (10.5%) were *P. falciparum*, while 8 (2.1%) had mixed parasitism with both parasites. Results of the mass blood survey point out towards a high slide positivity rate (25.5%) and *P. falciparum* proportion (49.5%) in the villages hit by the epidemic. While Raghwa village has presented the highest SPR (30.4%), Raimala showed the lowest SPR (13.1%). Interestingly, however, while on one hand Raimala singled out with the highest %*P. falciparum* (60%), Seuva offered the lowest *P. falciparum* proportion (41.0%), on the other. Maximum *P. falciparum* cases were nevertheless originated from Sadhna village.

Possible factors responsible for conflagration of malaria, particularly *P. falciparum*-dominated malaria, in the Thar Desert

It is so far an enigma as to how *P. falciparum* ushered in the interior of the Thar Desert and manifested so often in the form of epidemics with high morbidity and mortality (Table 2.5). A few possible interpretations are offered below:

1. With the establishment of extensive canal-based irrigation in the interior parts of the Thar Desert, a far more serious vector of malaria, *An. culicifacies*, ushered in the xeric environments. This species being highly adaptive to new physiographic situations built up its density very fast under given circumstances of a large variety of breeding places in the IGNP area particularly the inevitable and numerous seepage water collections from the canals. What *Anopheles culicifacies* lacks in its susceptibility to infection by *P. falciparum*, it compensates by its relatively much higher density, hence also vigorous and prodigious biting on human host;
2. *Anopheles stephensi*, chronologically the oldeset and most dominant vector of malaria in the typical sand-duned xeric environments of the Thar Desert, which traditionally bred only in the 'tanka', the only major source of reserved potable water invariably present in every desert village and but transmitted 'desert malaria' (Tyagi, 2002) at a low ebb, without epidemics, shifted its breeding habits with the availability of extensive and varied breeding sites due to extensive canalization to compete with that of *An. culicifacies;*
3. The massive Indira Gandhi Nahar Pariyojana, in continuation with the Gang canal system and the Bhakra-Sirhind feeder canal system, together spanning over nearly 70 years of initiation, had attracted large labour forces from distant and often hyperendemic States like Bihar, Uttar Pradesh, Punjab, Haryana, Gujarat and Madhya Pradesh. Such labour forces quite possibly brought along with them

the parasite, possibly a resistant strain of *P. falciparum*, which in the non-immune population had the potential to spread and manifest fast in suitable climatic situations of IGNP-induced environment in a part of the Thar Desert. It had been pointed out that the 1994-utbreak of cerebral malaria in the Thar Desert proved fatal because the affected population had been exposed to the parasite for the first time; and,

Table 2.5. Possible evolutionary correlation amongst the anopheline fauna, malaria incidence and the extent of canal irrigation in the Thar Desert

Period (in years)	No. of anopheline species				Cases reported		Irrigation (in ha x 10^6)
		Vectors		Non-vectors (including occasional or secondary vectors)	*P. vivax*	*P. falciparum* (%)	
Pre-1930	1	*An. stephensi*	-	-	-	-	-
1930-34	1	*An. stephensi*	-	-	-	-	-
1935-39	1	*An. stephensi* *An. culicifacies*[a]	4 -	*An.subpictus, An.annularis, An. vagus, An. splendidus*	-	-	0.24
1940-44	1	*An.stephensi, An. culicifacies*	4	*An.subpictus,An.annularis, An. vagus, An. splendidus*	-	-	0.24
1945-49	1	*An. stephensi, An. culicifacies*	5	*An.subpictus,An.annularis, An. vagus, An. splendidus, An. barbirostris*	-	-	0.24
1950-54	1	*An. stephensi, An. culicifacies*	5	*An.subpictus,An.annularis, An. vagus, An. splendidus, An. barbirostris*	-	-	0.24
1955-59	1	*An. stephensi, An. culicifacies*	5	*An.subpictus,An.annularis, An. vagus, An. splendidus, An. barbirostris*	-	-	0.24
1960-64	1	*An. stephensi, An. culicifacies*	5	*An.subpictus,An.annularis, An. vagus, An. splendidus, An. barbirostris*	-	-	0.32
1965-69	1	*An. stephensi, An. culicifacies*	5	*An.subpictus,An.annularis, An. vagus, An. splendidus, An. barbirostris*	-	-	0.37
1970-74	2	*An. stephensi An. culicfacies*	5	*An.subpictus,An.annularis, An. vagus, An. splendidus, An. barbirostris*	-	-	0.71
1975-79	2	*An. stephensi An. culicifacies*	5	*An.subpictus,An.annularis, An. vagus, An. splendidus, An. barbirostris*	639,754	21,649 (5%)	1.22
1980-84	2	*An. stephensi An. culicifacies*	5	*An.subpictus,An.annularis, An. vagus, An. splendidus, An. barbirostris*	98,242	12,422 (13%)	1.23

1985-89	2	*An. stephensi* *An. culicifacies*	5	*An.subpictus.An.annularis.* *An. vagus. An. splendidus.* *An. barbirostris*	11,746	16,533 (15%)	1.33
1990-94	3	*An. stephensi* *An. culicifacies* *An. fluviatilis*[b]	6	*An.annularis. An. subpictus*[c]*. An. vagus. An. splendidus. An. barbirostris. An. nigerrimus*	386,208	106,963 (29%)	2.17
1995-2000	3	*An. stephensi* *An. culcifacies* *An. fluviatilis*	9	*An.annularis. An. subpictus*[c]*. An. vagus. An. splendidus. An. barbirostris. An. nigerrimus. An.pulchrrimus. An. d'thali*	2,45,960	30,077 (11%)	2.17

[a] Until the late 1980s *An. culicifacies* was recorded only from the desert fringe urban areas.

[b] *An. fluviatilis* is yet to be incriminated with malaria parasites in the Thar Desert, though it is a very efficient vector in southern Rajasthan.

[c] *An. subpictus* is only rarely incriminated in the Thar Desert.

4. The resistant strain of *P. falciparum* inadvertently inveigled into the Thar interiors is uncontrollable by the normal regimen of the prescribed antimalarials.

The above hypothesis can be further aptly understood if the progress of pathways of malaria epidemics in the Thar Desert was traced out carefully (*cf.* Tyagi, 2003).

(2) ***Dengue***: Until massive irrigation programmes under the aegis of Indira Gandhi Nahar Pariyojana (IGNP) were initiated about 3-4 decades ago, the Thar Desert gave a typical desert look with sand dune levees spread all over, dotted with sparce and miniscule hamlets of human habitats (locally called 'dhani'), with invariably one 'tanka' (an underground reservoir of potable water fetched through mostly a camel cart from far-flung areas) per household. Some villages have got a large assemblage of 'tanka' at the periphery for community-use. These 'tanka' exist in thousands all over the Thar Desert. Rarely another type of subterraneous reservoir, called 'beri', which is in fact a collection of several well-like pits in the concavity of a pond, would also be present in the outskirt of the village. Both the 'tanka' and 'beri' are never charged up on their own from the underground water (*cf.* Tyagi, 2002).

In course of application of the IGNP in the Thar Desert, of which one major objective was to provide drinking water to the desert folks, the conduit-supplied drinking water became available in the desert villages during the 1980s. This facility cast one major and, to some extent, disastrous impact on the behaviour of people who changed to consumption of the ducted water supply and leaving behind the now unwanted water-filled 'tanka', in particular those in clusters at the edge of the village, uncared or deserted and much pot-holed. It is highly interesting to note that as long as the water from these 'tanka' was in use, only the malaria vector *Anopheles stephensi* bred there, but on having deserted and dilapidated, with a tinge of turbidity, the same sites became a stronghold for *Aedes aegypti* breeding, replacing altogether the malaria vector. The Thar Desert had the first epidemic of dengue in 1985 in Jalore urban agglomeration, not covered under the IGNP, followed soon with a greater degree of epidemic conflagration in two consecutive years of 2002 and 2003 in Jodhpur's both rural and urban areas, which are largely covered under the IGNP.

(3) ***Filariasis:*** The Thar Desert had remained a *terra incognito* for human lymphatic filariasis till I, in collaboration with Major Dr. R. Srivastava, Officer Commanding, Station Health Organization, Army, Jodhpur, a highly energetic and intelligent medical doctor, detected several cases amongst the army personnels, and their family associates, infected with the microfilariae of *Wuchereria bancrofti*, in 1992-93. The ubiquitous vector, *Culex quinquefasciatus*, prevailed abundantly in Jodhpur and was incriminated, along with a few infected local civilian cases (Tyagi, 2003). It is noteworthy that the mosquito vector species was not inventorized in earlier faunal reports from Jodhpur and it is doubted it existed in any serious proportion till 1971 when the population of Jodhpur city was less than 2 million. The unplanned expansion of the city in the wake of fast-growing population (3.5 million in 1981; 6.5 million in 1991 and 10 million in 2001) led much to a chaotic development, offering in the process a much favourable man-made environment for the prosperity of the mosquito species which specializes in breeding in all kinds of polluted waters, including cess pits, cess pools, ponds and drains etc.

CHAPTER - 3

MOSQUITO VERSUS MAN IN THE EVOLUTION ODYSSEY :

FACE OFF FOR BIOLOGICAL SUPERIORITY

Man, for reasons best known to him alone, is considered to be the latest, but not the ultimate, animal (mammalian) species developed in course of evolution. A human is regarded to be the wisest and strongest creature, belonging to class Mammalia, the most recent vertebrate group, that reign supreme on the planet Earth. This view, prejudiced and rhetorical enough, however, is largely based on the fact that human development supposedly stays atop the evolution tree which comprises simple and unicellular organisms at the root of the metazoan shoot where, too, the invertebrates or non-chordates, who were evolved through vast time-scales and stages from the unicellular organisms, preceded the birth of vertebrates or chordates among whom the cold-blooded (poikilothermic) animals such as the reptiles, amphibians and fishes, in that order, evolved in advance to the birds and mammals, the warm-blooded or homeothermic ones, with the latter embodying man or the present day's human species, *Homo sapiens* (Fig. 3.1). Man considers his species the invincible and capable to command and control other creatures. To a biologist, however, this assumption seems nothing but a case of self-aggrandizement, as all species have been likewise pacing ahead to evolve just as much as humans think they are. The social wisdom of ants and termites, both insect groups and cousins to mosquitoes, is a subject of eternal envy to man; harmonious, disciplined and devoted life by each and every individual, be a soldier, worker or even royal king or queen, without ever an instance of chaos! Had this been not the case, then man alone would be the survivor, and in that case absolutely good for nothing, for there would be none to be ruled! This, however, is not the case and man is pitted in the evolutionary race against as much smarter and dangerous creatures as the mosquitoes, many of which live on his blood!

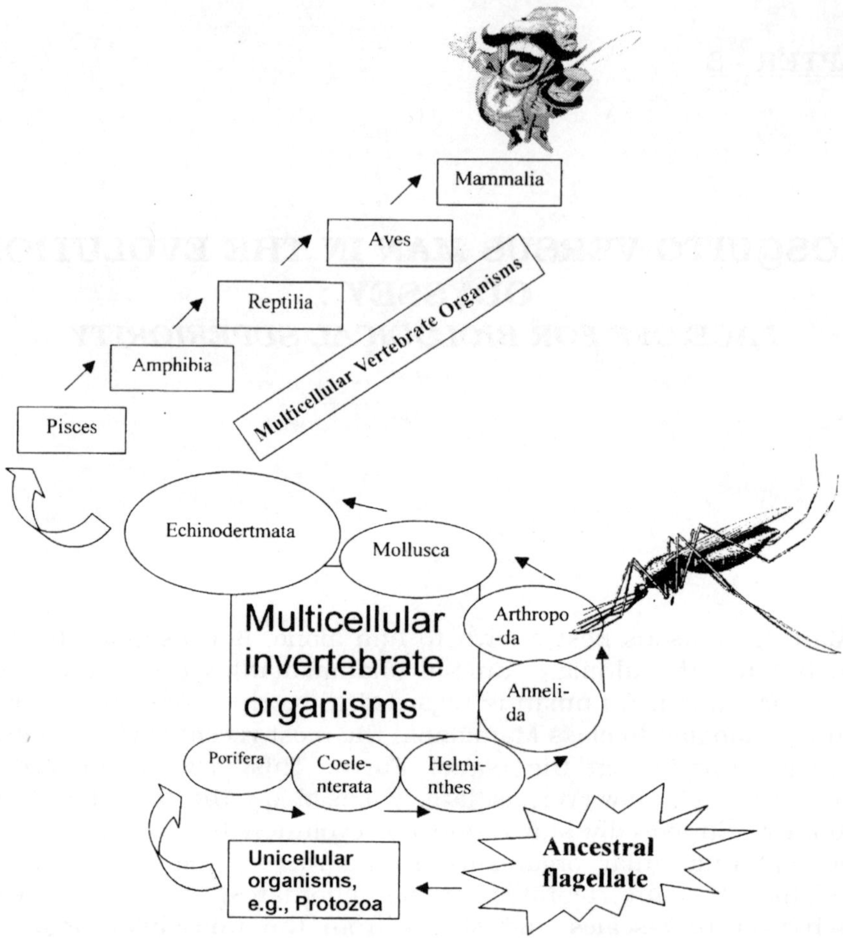

Fig. 3.1. A hypothetical, schematic presentation of evolution of major animal ``groups" (not necessarily in that linear manner) to demonstrate chronological, hierarchical placement of both mosquito (Triassic, 200 million years / Quaternary, 40-60 million years) and man (60 thousand years from now).

There is possibly no argument warranted on the fact that many mosquito-borne diseases, particularly malaria transmitted by certain species of *Anopheles* only, and their propagators are far greatly older and evolutionarily seasoned than humans. It is, therefore, reasonable to suppose that our primate ancestors would have been experiencing malaria attacks before transforming, in Darwinian sense, into the modern human species. Thus, it appears true that both the *Plasmodium* species, responsible for causing malaria fever in humans, and the *Anopheles* mosquitoes, which transfers the pathogen from the infected person to another healthy person, have seen man 'forming' through all his long evolutionary peregrinations.

WHERE DID MOSQUITO COME FROM?

Insects are believed to have been borne in the Devonian period, about 450 million years ago (Martynov, 1932; also *cf.* Kiauta, 1975a). Diptera, the Order of insects, housing mosquitoes, along with several other flies, many of which infamously known in the realms of medico-entomological sciences, are understood to have first appeared no later than the Triassic period, about 200 million years ago, although the first fossil evidence resembling very much to the present day's *Culex* allude towards its evolutionary cradle sometime in the Eocene epoch of the Quaternary period, 40-60 million years ago. *Aedes* mosquitoes however offered the first correctly recognizable fossil belonging to middle Oligocene epoch, about 3 million years ago.

The modern man, or *Homo sapiens*, on the other hand, traces its origin only about 60,000 years ago, out of Africa. With the emergence of the powerful new research tools based on modern molecular genetics, it is now possible to track down human evolution even more accurately.

CYTOPHYLOGENY OF MOSQUITOES IN THE ORDER DIPTERA

With over 100,000 known species in the Order Diptera, the two-winged flies are one of the largest insect orders, second only to Coleoptera, Hymenoptera and Lepidoptera. The order is divided into three suborders, *viz.*, Nematocera, Brachycera and Cyclorrhapha, containing considerably more than hundred families. Its representatives occur in all faunal regions, in all climatic zones and in a great diversity of biotopes. They are amongst the dominant hypsobiont insects on nearly all high mountains, and are abundant even at the highest altitudes at which permanent insect life is still to be found. In the Himalaya the diperans are common still at an elevation of 6300 m (Kiauta, 1975a,b).

E pluralis unum, the order Diptera occupies a unique position in the overall evolution of the phylum Arthropoda. It presents entirely a different type of genetic systems whose relationships to one another present an evolutionary problem of considerable significance. The homopterans are the only other order in which a similar diversity of genetic systems is encountered. Cytogenetic databases are available from about 2000 species, albeit representative only to a few families (Table 3.1). Families of suborder Nematocera are well represented, with a diploid chromosome complement (2n) in the range of 2 to 8; 2n = 6 being the commonest.

In the diploid set of the Order there are 4 to 20 monokinetic elements, usually of considerable size and with varying centromere position. The low number and the morphological diversity of chromosomes generally allow a comparatively easy recognition of homologues. Supernumerary elements were encountered in the families Tipulidae, Syrphidae and Technidae. The modes of chromosomal number spread all over the order are at 2,4, 6, 8 and 12, of which 6 and 8 are due to several nematoceran families, while a huge peak at 12 is due to various acalypterate and calypterate families, including the Drosophilidae. Some families of Nematocera, such as Culicidae with 2n=6 (with rare exceptions), the Simuliidae with 2n=6 (with rare exceptions), the Sciaridae with 2n=8 and the Bibionidae with 2n=10, apparently have a stabilized chromosome number, whereas other families, such as the Chironomidae with 2n = 4,6 and 8, the Psychodidae with 2n=10 and 12, and the Mycetophilidae with 2n = 8, 12 and 14 have developed a variety of numbers. The Brachycera families: Asilidae with 2n = 10, 12 and 14 and Bombyliidae with 2n= 12 and 18 have similar variety as have also the aschizan Syrphidae with 2n=4, 6, 8, 10, 12 and 14, and phoridae withy 2n = 8 and 10. There is a clear shift in chromosome number between the Nematocera (2n = 6 or 8) and the Cyclorrhapha (2n = 12) (Table 3.1).

Table 3.1. Dipterans for which chromosome complements are considerably investigated.

Suborder	Family	Common name by which dipterans are known	General Diploid complement (2n) range in the suborder
Nematocera	Tipulidae	Crane flies	2-8
	Culicidae	Mosquitoes	
	Chironomidae	Nonbiting midges	
	Sciaridae	Gall midges	
	Cecidomyiidae	Gall midges	
Brachycera	Tabanidae	Horse flies	8-18
	Bombyliidae	Bee flies	
	Asilidae	Robber flies	
	Phoridae		
Cyclorrhapha	Syrphidae	Hover flies	4-14
	Sciomyzidae	Marsh flies	
	Drosophilidae	Fruit flies	
	Agromyzidae	Leaf miners	
	Scatophagidae	Dung flies	
	Muscidae	House flies	
	Tachinidae	Parasite flies	
	Calliphoridae	Blow flies	
	Sarcophagidae	Flesh flies	

Save for a few phoride species living in termite nests, all dipterans are bisexual. Generally males are heterogametic, but cases of female heterogamety had also been brought on record in the nematoceran Chironomidae and the acalypterate Tephritidae. Thelytokous species occur in the families Psychodidae, Simulidae, Chironomidae, Lonchopteridae, Chamaemyidae and Agromyzidae. In these families unfertilized females produce exclusively, female progeny.

The order is characterized by a great variety of sex chromosomes mechanisms. The XX/XY is the general mode, met with from the primitive Tipulidae upto advanced Cyclorrhapha, but it can be subdivided in various genetically distinct types. Multiple sex chromosomes were encountered in several families, while the XO/XX mechanism occurs in some representatives of the families Drosophilidae, Asilidae, Tephritidae and Ephydridae. A very peculiar sex determination was recorded in the phorid fly *Megaselia* in which any one of the three chromosomal elements in the haploid karyotype can function as a dominant male-determining chromosome if it bears at one end a so called factor M or "sex realizer". In some species, *e.g.*, in the family Muscidae, no sex elements are recognizable. They have been probably fused with one of the larger autosomal pairs.

Diptera are possibly the only group in animal kingdom with very clear pairing of homologous elements in somatic metaphase. The most primitive dipterans, *viz.* Tipulidae are characterized by the presence of chiasmata in both sexes and by the fact that the sex elements do not form a bivalent in the primary spermatocytes. From this group two main evolutionary lines arose. Of the two one led to mosquito bearing Culicidae, besides Simuliidae and Chironomidae, in which the chiasmata were retained in the male, but the sex chromosomes disappeared as distinct elements.

MAN'S JOURNEY THROUGH TIME

Various theories have been in vogue for the evolution of man (*Homo sapiens*), but a recent synthesis of genetic evidence with current research in geology, palaeontology, archeology, anthropology and linguistics has certainly been able to remove various dark points in our understanding of the human journey through the geological timescale (Wells, 2003).

Genetic samples drawn from human populations in different parts of the world enable scientists to trace their lineages back all the way through an African origin and subsequent colonization of remaining continents. Ironically, however, soon after scientists have gained this remarkable ability to read molecular clocks set within us, the window of opportunity to do so appears to be closing forever. Reason: the unprecedented rapidity with which the hitherto relatively sedentary human populations are inter-mixing across the globe.

In this scientific tour de force, geneticist Wells describes the rise of early hominids in Africa as an evolutionary consequence of global climate change, expansion of grasslands and the consequent adaptive transformation of ape-like *Ardipithecenes* into roaming bands of man-apes that stood erect and hunted cooperatively, thereby developing greater intellectual capacity. Originating from their African cradle nearly two million years ago, early hominids such as the *Homo erectus* soon spread all the way to Indonesia ("the Java man") and China ("the Pekin man"). A later offshoot, the Neanderthala, reached Europe. However, these Homo lineages ultimately lost the evolutionary struggle manifested through environmental changes and competition with other creatures. They left no survivors on earth.

All modern humans trace their direct ancestry to hominids called *Australopithecines* that inhabited Africa over 2 million years ago. Current DNA-based techniques can trace the lineage of *Homo sapiens* back 200,000 years ago to Africa. Although Wells marshals evidence from diverse fields of enquiry to his task of tracing human origins, he primarily writes about a stunning new array of genetic explorations in the last two decades. By tracking a genetic marker called M-168 that is present in the Y-chromosomes of men, Wells demonstrated how the first bands of modern humans, perhaps, not unlike present-day African " bushman", moved out of Africa and colonized the Arabian peninsula about 60,000 years ago. Almost immediately, a sub-group bearing the genetic marker M-130 rapidly moved along the resource-rich coastal fringe of southern Asia, populating Southeast Asia and Australia within a few thousand years. Because of lower sea level that prevailed, the Asian coastline was aligned very differently with far greater connectivity among land-masses, facilitating the rapid migration of M-130 lineage. These people were ancestors to the aboriginal human groups of Australia as well as in south and north-eastern Asia.

Another branch off the M168 colonisers that bore the genetic marker M-89 moved into middle-east. (45,000 years ago) and gave rise to lineage M-9 that moved into southern part of Central Asia (40,000 years ago), following teeming herds of game in those vast grasslands. The M-9 lineage subsequently forked out in three directions: M-45 penetrated deeper into central Asia 35,000 years ago; M-175 moved into the Tibet-southern China region 30,000 years ago and a population bearing the genetic marker M-20 moved into the Indian subcontinent. This M-20 lineage intermixed with the earlier coastal migrant populations of M130. Based on genetic evidence, Wells suggests the encounter between these groups may have been brutal and one-sided, with most men of the M-130 lineage being eliminated and their women being interbred with the more "advanced" (read aggressive) M-20 lineage. Thus, clearly all modern south Asians are a "genetic cocktail" of atleast 30,000-year vintage.

This human odyssey continued across the earth. The bearers of the marker M-242, descendants of M-45, advanced across the frozen north Asian continent, adapting to ice-age conditions to hunt woolly mammoths. They colonized the "new world" 20,000 years before Christopher Columbus "discovered" it. Within a few thousand years, these original Americans had spread all the way to the tip of South America. Around the same time, racing across China was a group of people bearing the marker M-122, descendants of M-175, who eventually populated much of Southeast Asia.

About 10,000 years ago, as ice-sheets retreated across Europe and Northern Asia, modern humans bearing genetic marker M-172 move into Europe. The warm-moist climatic regime that we now live in emerged. The earth was already peopled and ready for the next human revolution:

discovery of agriculture and the consequent ability to modify the environment itself, unlike any other creature before.

It becomes more than plausible from the above account of man's journey through time as to how *Homo sapiens*, as a single species, came to evolutionarily dominate the earth so quickly through a series of steps, albeit a mere 2000 generations! It is nevertheless an unforgiveable blunder to believe that man, proudly credited with the invention of agricultural civilisations about 10,000 years ago, should consider themselves superior to other creatures on earth. Ironically, in the present day world, most of our resources and energies are wasted in trying to prove dominance by one type over the other, both culturally and economically! The prowess of mosquito is not yet fully understood.

WHEN DID MOSQUITO BEGIN TO BLOODFEED?

It is nevertheless interesting to observe that as an Order *per se*, Diptera, to which mosquito belong, they had originated much earlier in the evolution tree than certain other highly significant vector insect groups or orders such as Anoplura (biting lice) and Siphonaptera (fleas) (Martynov, 1938; Kiauta, 1975). It is for this reason perhaps, therefore, scientists have in recent times focused their attention on analyzing the blood from the gum-preserved blood-sucking mosquitoes from the period when the amazing dinosaurs also inhabited the earth, assuming that the blood suckers would have had a feed on the giants, and thus deciphering mysteries and enigmas shrouding dinosaurs. Spielman and D'Antonio (2001) have an interesting observation regarding a particular scene in the blockbuster "The Jurassic Park" - where a mosquito preserved in the soft gum, held in the head of a walking stick of the creator of the park, is referred to having a clue to the enigmatic world of the dinosaurs because the former had sucked blood from the latter and thus kept the "dinosaurous story" encoded in the gene to be isolated and sequenced, but cast doubts if it really was a haematophagous mosquito since it looked more like a *Toxorhynchytes* mosquito which does not feed on blood but on the plant sap! One thing, however, is evident that all mosquitoes are not haematophagous; in fact out of the three subfamilies, namely Toxorhynchytinae, Culicinae and Anophelinae, under the family Culicidae, only culicine and anopheline mosquitoes feed on blood whereas members (toxorhynchytines) of the third subfamily subsist only on plant sap. Among all the three subfamilies, toxorhynchytinae is oldest, followed by culicinae and, lastly, anophelinae. This also offers strength to the hypothesis that haematophagy is a rather recent but highly specialized and effective development amongst the mosquitoes for a successful survival.

Even though there is no direct evidence as to the time when the mosquitoes had developed blood feeding habit, but it is a general belief that mosquitoes had already acquired haematophagous trait much before the dawn of both the birds and mammals in the late Mesozoic era, some 70-230 million years ago, the period best remembered for the giant-sized dinosaurs.

WHY HAVE MOSQUITOES CHOSEN HUMANS FOR BLOOD-FEEDING?

The early species of mosquitoes should have been 'vegetarian', subsisting solely on plant juices obtainable from various tissues of its parts (*e.g.*, members of subfamily Toxorhynchitinae). However, in course of evolution, as they seemed to have searched for a variety of food-source, say those of invertebrates or lower groups in the hierarchy, they began to obtain the needed sugars from such animals as the ants – still quite vegetarian since the mouthparts were so designed as to suck only juices from the formicide's buccal cavity (*e.g.*, *Malaya* spp.). Subsequently, mosquitoes sharing a common ecosystem with several animals, vertebrate and cold-blooded though, for the first time they were engorged on blood of frogs and reptiles, to derive the necessary proteins essential for development of their eggs. Later, mosquitoes found avian and simian blood more easily accessible since both mostly lived at the canopy level of forest trees, and thus entered into a different realm of warm-blooded animals.

More than 2 million years ago in the late Coenozoic era, when the earliest man-like quadruped, or humanoid, was on way to evolutionary transformation, he had learnt to use forest cover for meeting various of its needs including weapons, shelter, faggots (for fire) and food. For this man had had to go deep in the woods, where mosquitoes seemed to have come in contact with him involuntarily. Human body's soft and easily penetrable skin soon became an easy target to secure stomachful blood by the mosquito! A careful scrutiny of the behaviour of all the vector mosquitoes, be it *Aedes*, *Culex* or *Anopheles* species, will prove one point beyond doubt that none of these species were initially anthropophilic or thirsty for human blood. They, however, did become ectoparasites of man when the latter interfered with the natural habitats, mainly through denudation of the forests more often for his greed of securing prosperous development, which resulted in thrashing the mosquitoes out of their innocuous hideouts and look for new shelter points: the human habitations themselves! Soon, mosquito turned synanthropic, almost by force of man and adapted to live and breed in and and around human living and, of course, dining on his blood-meal!

WHO IS MIGHTIER, MAN OR MOSQUITO?

This simple inquiry is bizarre yet highly significant. Man's domination on Earth is unquestionable. But, this feeling of domination or superiority is largely because of a belief that man can 'command' or 'tame' all other creatures for its purpose, but not the *vice versa*. This, however, is a human thought, thought only in a humanly way. Reality takes the mighty-looking man far from the serendipity or sagacity of this self-edificed false myth, when his hierarchy is measured against that of the tiny thing, mosquito, in the time-scale of evolution. Mosquito, like any other insect such as the cockroach or dragonflies, has seen more geological ages and catastrophes in earth's atmosphere than man, and has been able to withstand all these adversaries, including the likes of "the Big Bang" that presumably caused annihilation of the monsterous giants, the unchallenged rulers of the Mesozoic period, called dinosaurs, but continued undisturbed on the path of speciation and meaningful survival. Man, on the other hand, is virtually novice in his experience to witness any major atmospheric vagaries of the type of 'the Big Bang', the fiercely high temperature age or, just the opposite, the ice age. In retrospective comparison, mosquito definitely has an edge over man in its prowess of adapting to varying conditions, and prospectively, too, it occurs, that in the event of a holocaust, broght about by a nuclear war, it is decisively the hexapedan foe who would possibly still continue to live on the Earth, though man might not remain to witness who the more mightier is ?

INGENIOUS MOSQUITO SCIENTISTS

The ruthless strength and prowess of a formidable foe such as the mosquito, who besides being a source of immence nuisance is also a nefarious carrier of several deadly infections, could be conquered, on certain counted occasions, only by the matching character of indefatigability, discipline and knowledge (about the enemy) of a few great entomologists of the likes of Drs. Carlos Juan Finlay, Patrick Manson, Giovanni Battista Grassi, Ronald Ross, Walter Reed, William Crawford Gorgas, Rickard Christophers, J.A. Sinton, G. Covell, D. Viswanathan, T. Ramachandra Rao, P.K. Rajagopalan and V.P. Sharma – all culicidiologists of great virtues who resolved to the brim to defeat the treacherous opponent.

The main basis of giving a brief biographical sketch of some of the prominent culicidiologists here, who are no more with us to narrate the heroic tales rife with physical and mental fatigue and vexation, is to ignite minds of young medical entomologists (or, correctly speaking, medical arthropodologists, since the former term appears a misnomer in present day advancement of this specialized science) with the spirit of unreplaceable characters of perseverance, devotion and application – traits which were common to all the great 'mosquito men' irrespective to vast differences in periods of time, educational background, and professional or otherwise facilities as well as hostilities they endowed. Geniuses, after all, do not do different things, but they certainly do things differently.

(1) DR. PATRICK MANSON

Born at Old Meldrum, Aberdeenshire, NE Scotland, UK, in 1844, Manson, the son of a bank manager, qualified in medicine at Aberdeen in 1865. Graduating from Aberdeen University in 1865, he abandoned a potentially lucrative career in industry for medicine. The following year he responded to the cry 'Go East, young man' and sailed for China. After a few years in Formosa, now known as Taiwan, he moved to Amoy on the mainland. He travelled widely in the vast stretches of the East, becoming medical officer in Formosa, and later moving to Hong Kong. Manson worked in the Far East for 23 years, during which time he addressed himself to the problems of tropical diseases such as tinea, Calabar swelling and blackwater fever. Dr. Manson, who was posted as a British medical officer in Formosa, under China's regime, in the 1870s, became much concerned by the various debilitating forms of gruesomeness and grotesqueness, particularly elephantiasis of legs and hydrosis of the genitals, of the disease called lymphatic filariasis that was caused by a worm and was prevalent widely across the tropical world affecting millions of people! In 1878 he observed that filariae, the worms that cause elephantiasis in man, pass part of their life cycle in the *Culex* mosquito; he thus led the way in the study of the transmission of diseases caused by parasites.

In 1894 he made the deduction that the parasite of malaria passes part of its life cycle in the mosquito, a theory that Ronald Ross was to verify three years later. A founder of two schools devoted to the study of tropical diseases, one at Hong Kong (1886) and the other at London (1898), Manson is often described as the father of tropical medicine. He practised medicine in China (1871) and in Hong Kong (1883), started a school of medicine that became the University of Hong Kong. He became medical adviser to the Colonial Office, and in 1899 helped to found the London School of Tropical Medicine.

Manson retired to Britain in 1889 and, after the failure of his Chinese investments, set up his practice in London. He had felt the want of a school of tropical medicine and accepted an invitation to give a series of lectures on the subject at St George's Hospital. This earned him the appointment of Medical Advisor to the Colonial Office thereby gaining the ear of Joseph Chamberlain, the Colonial Secretary. As a result he persuaded Chamberlain that schools of tropical medicine were a 'Good Thing'. The Treasury was persuaded to give financial support, which was promptly recouped from the indigent colonies, and the London School of Tropical Medicine opened its doors in September 1899 at the Branch Hospital of the Seamen's Hospital, in the far-from-fashionable docklands of the late nineteenth century, and in the teeth of opposition from the War Office, Admiralty and the medical establishment. Manson thus completed his hat-trick of founding medical schools.

To provide students with intellectual stimulus he wrote his *Manual of Tropical Diseases* which promptly became a bestseller in its field. He retired in 1912 to fish in Ireland but returned to London at the beginning of the First World War. Despite crippling attacks of gout he continued to take a lively interest in medical education until his death in 1922. It was his natural affectionate inclination for medical educaion and research that he provided all that help that could have been possible on earth. It goes without mention that it was him whose advice to Ross finally led the latter to discover *Anopheles* mosquito as the carrier for malaria parasite – a work for which Ross was awarded the Nobel Prize for 1902.

(2) MAJOR WALTER REED

On September 13, 1851, one of the world's outstanding physicians and medical research scientists was born in Belroi, Gloucester County, Virginia, in an affluent family of a Methodist minister, Lemuel Sutton

Reed, and his wife, Pharaba White Reed. At an early age, Walter Reed gave evidence of the intellectual brilliancy and earnestness of purpose which distinguished him in later years.

After his basic education at a private school in Charlottesville, Virginia, Walter Reed matriculated at the famed University of Virginia where he completed the two-year medical course in only one year and received his degree in 1869 at the age of eighteen. He was the youngest student ever to graduate from the medical school. Since the University of Virginia had no hospital attached, he took a second degree at Bellevue Medical College in New York in 1870. During the subsequent five years he served his internship at Kings County Hospital in Brooklyn and at the Brooklyn City Hospital, and by this time he earned himself a reputation of a versatile conversant. After further internship at Brooklyn City Hospital with consultant status at Kings County, he was appointed one of the five inspectors on the Brooklyn Board of Health in 1873 at the age of twenty-two. He approached all his duties with enthusiasm and optimism, traits which contributed immeasurably to his success, both social and professional. During all these years he had developed an immense love for music and left no opportunity to attend good lectures on literature and scientific subjects, no matter where they were to be held. It was at this time that the young intern conducted his first research which was the basis of his first scholarly paper, published in 1892, called "The Contagiousness of Erysipelas."

In 1874, having served on the Boards of Health in Brooklyn and New York, he returned to Virginia to visit his family, and soon got married to Emilie Lawrence, the daughter of a North Carolina planter. Because he felt the Army offered a good opportunity for research, he applied for an Army position and was accepted for an appointment of a surgeon in the Medical Department. He passed the required examinations and was appointed Assistant Surgeon with the rank of first lieutenant on June 26, 1875. So began an eighteen-year garrison life. After five years at Fort Lowell and Fort Apache, Arizona, where he served as a beloved family doctor visiting patients in the wild country surrounding his posts, he was promoted to captain and soon thereafter was transferred to Fort McHenry in Baltimore. In this city he became a student of physiology at Johns Hopkins University in his spare time during 1881 and 1882.

After serving at Fort McHenry, the ever perepetating Walter Reed was again assigned to southern and western posts at Fort Omaha, Nebraska, and Mount Vermon Barracks, Alabama. Dr. Reed returned to Baltimore in 1890 as Attending Surgeon and examiner of recruits. This assignment was quite welcome since it provided him with the opportunity to do research. He became a student of bacteriology and pathology under the tutelage of Dr. William Henry Welch, head of the Pathological Laboratory at Johns Hopkins and foremost pathologist and medical bacteriologist in this country. These subjects were not taught as part of the medical curriculum of that day. It is significant that Walter Reed's career coincided with the great flowering of medical science that took place in the 1880's. The germ theory of infectious disease was now accepted as postulated and proved by Pasteur, and Robert Koch had perfected a method for studying bacteria. In the United States George Miller Sternberg, later Army Surgeon General, with whom Walter Reed had a close professional relationship, was one of the founders of bacteriology.

The 1880s were critically important years in Walter Reed's life and in the destiny of the United States. It was during this period, when all areas of the central and northern part of southern Americas were plagued by yellow fever, that the mature scientific investigator was formed. He conducted his own individual research much to the delight and satisfaction of Dr. Welch who had been one of Pasteur's students. His ties with Dr. Welch were strengthened and Dr. Welch's mutually admiring relationship with Surgeon General Sternberg was quite advantageous to Dr. Reed as an Army surgeon protege. Walter Reed, then forty-two years old, was greeted with enthusiasm by an eager group of researchers at the Pathological Laboratory. Captain Reed - promoted on June 26, 1880 - learned the necessary research techniques and was soon working on his own project concerning typhoid fever.

A man of sterling character, religious by nature, prepared for practice and research, a soldier who had learned to endure hardships, a student and pathologist of the highest caliber, Walter Reed was now ready for the great achievements of his lifetime. He would live for only fifty-one years, but between 1892 and 1901, till a year before his death, he was engaged in some of the most important work in the history of medicine. This took the form of research into the etiology and epidemiology of typhoid and yellow fever.

From 1891 to 1893 Walter Reed spent his last western tour in the Dakotas. He was promoted to major in 1893 and then took his post as curator of the Army Medical Museum (now part of the Armed Forces Institute of Pathology) and professor of clinical microscopy in the Army Medical School (now the Walter Reed Army Institute of Research) founded in Washington by Surgeon General Sternberg. At this same time he also held the chair in bacteriology at the Columbian University (now the George Washington University). He worked industriously for five years, teaching and working in his specialty, bacteriology. His work, clinical and academic, was accurate and original. His experience at work in the School gave him an excellent sense of judgement valuable for investigating the causes of epidemic diseases and in making sanitary inspections at military posts. Medicine was making rapid advances and military doctors had to be informed of the new techniques. An anti-toxin for diphtheria had been prepared and there was a race to find the specific agents (bacteria or viruses) responsible for communicable diseases. This was the beginning of medicine as we know it today, an era of great discovery. At the time of the Spanish American War, however, he was casting about for a position in the field which would allow him to put his knowledge of Army routines and sanitation to use for the benefit of soldiers in Cuba.

It was not until mid-August in 1898 when the war was already over that he had an opportunity to perform this valuable service. General Sternberg was then able to secure a board of officers headed by Major Reed to investigate the spread of typhoid fever. The Typhoid Board performed its greatest service through the discovery that this dread disease, prevalent at almost all U.S. Army encampments, was spread most commonly and disastrously by contact between persons and flies soiled with human excrement containing typhoid bacilli, by human carriers who shed bacilli by the billions, and by impure drinking water. This triumph for Army medicine demonstrated for the first time the effects of intestinal disease-producing agents or the pathogens. This victory also served to dispel old notions that such diseases were caused by miasmas or foul emanations from swamps and rivers.

In May 1900 Major Reed was appointed president of a board whose purpose it was to study infectious diseases in Cuba paying particular attention to yellow fever. The other members of the board were Acting Assistant Surgeons Major James Carroll, Major Jesse W. Lazear and Major Aristides Agramonte of Havana. As a result of this Yellow fever Board, very few people living today have any knowledge of this dread disease. The onset of yellow fever came with chills and a headache. Then followed severe pains in the back, arms and legs accompanied by high fever and vomiting. The feverish stage might last hours, days, or weeks. Jaundice, from which the fever derives its name, might then appear. Then came the so-called "stage of calm" when the severity of the symptoms subsided and the fever dropped. In less serious cases this stage indicated recovery. But in the main, this stage was followed by a return of the fever accompanied by internal bleeding which caused the dreaded "black vomit" when blood released into the stomach was ejected.

Yellow fever killed more men in the Spanish-American War than did the enemy. It appeared in Central America in 1596, probably imported from Africa by slave ships. It may have been the disease from which members of Columbus' second expedition suffered in 1495. Ninety epidemics struck the United States between 1596 and 1900. In 1793 an epidemic first hit Philadelphia, then the U.S. capital, causing the Government to flee as ten per cent of the population perished. Washington went to Mount Vernon while Jefferson fled the disorder caused by the onslaught of the disease. Because of frequent epidemics which destroyed ninety per cent of his expeditionary forces in 1802, Napoleon was influenced to sell the Louisianna Territory. It was chiefly because of "yellow jack," as the disease was nicknamed from the penant which was flown during quarantine, that the French were unable to complete the Panama Canal. The danger of contaminating the southern states was considered to be a major factor in the annexation of Cuba. Reed and Carroll had estimated that there were 300,000 cases in the

United States between 1793 and 1900, which cost the nation almost $500,000,000 with a mortality rate usually at forty per cent but sometimes as high as eighty five per cent. The scourge of yellow fever had plagued the southeastern United States for two hundred years, but nowhere was it more prevalent than in Havana. Surgeon General George Miller Sternberg was this country's leading expert on yellow fever. Because neither he nor other researchers had been able to pinpoint the specific cause, he was astounded to hear the claim of Dr. Giuseppe Sanarelli, who in 1897 stated quite conclusively that the fever was caused by *Bacillus icteroides*.

On June 25, 1900 Walter Reed arrived at Columbia Barracks in Quemados about six miles from Havana. Major Reed attended Major Jefferson R. Kean, Chief Surgeon of the Department of Western Cuba, who had himself contracted the disease and fortunately recovered. Then Doctor James Carroll and Major Reed set out enthusiastically to prove Sanarelli's theory. By August, 1901 however they had found no causal relationship between *Bacillus icteroides*, a member of the hog-cholera group, and yellow fever. The Board then turned its attention to the theory of Dr. Carlos Juan Finlay and examined it more carefully. For nineteen years this resident of Havana had contended that yellow fever was carried in the body of a common house mosquito which at that time was called *Culex fasciatus*, later *Stegomyia fasciata*, and is now known as *Aedes aegypti*. This theory had been expounded even earlier, but it was Finlay who was its staunchest exponent. However, after some 100 experimental inoculations had failed to produce any cases of the disease under strict laboratory control, Finlay was scoffed at and even ridiculed as the "mosquito man". There was evidence, however, that tended to lend credence to this theory which even the Yellow Fever Board, optimistic though it was, had doubted. First of all, the disease skipped erratically from house to house, jumping around corners. One member of a household might contract the disease while others in close contact never became ill or did so after a period of about two weeks had elapsed. This was quite unlike any other infectious disease except malaria which had already been shown to be spread by the *Anopheles* mosquito, by Ronald Ross in India (*vide infra*). The Yellow Fever Board then decided that the best way to study yellow fever was not by searching for a specific agent but rather by identifying the means by which the fever was transmitted. For this purpose Major Reed organized the Board in the following manner: Major Reed himself was in charge of the entire project, Dr. James Carroll was in charge of bacteriology; Dr. Jesse W. Lazear was in charge of the experimental mosquitoes, and Dr. Aristedes Agramonte was in charge of pathology.

Dr. Lazear had recently been working with malarial mosquitoes and attended his duties with great enthusiasm in view of the information he read concerning the observations of Dr. Henry R. Carter of the Marine Hospital Service (now the Public Health Service). Dr. Carter had observed that it took two or three weeks for the first case of yellow fever to produce the next case in a community. On the basis of this observation, he suspected that an insect might be the intermediary since this would account for the delay in transmission as the disease ripened in the mosquito. Dr. Finlay had given some of the black, cigar-shaped eggs to the Board, and Lazear allowed them to hatch. It wasn't difficult to maintain a supply since the mosquitoes bred in any clean, still water. Dr. Carroll then volunteered to be bitten and promptly developed a successful case of yellow fever which was experimentally defective since there may have been other sources of contamination. Carroll luckily recovered and went on with his work in bacteriology. Next, Lazear got Private William Dean of Ohio bitten by the female *Aedes aegypti.* He developed the first successful case and recovered. Finally, Dr. Lazear allowed himself to be bitten and, after several days of delirium and black vomit, died - a true martyr to science.

Major Reed, although grieved at Lazear's death, did not lose heart and picked up meticulously the threads of Dr. Lazear's highly exciting experiments (minutest details of which he scrupulously noted in a diary), tracking down the secret of the fever. Walter Reed found that in order for a mosquito to become infected, it had to bite a yellow fever patient during the first three days of his illness as only during that time was the agent present in the bloodstream. Further, it required at least twelve days for the agent to ripen in the female mosquito (only the female *Ae. aegypti* draws blood) and migrate to her salivary glands before the fever could be passed to another person. He solved the puzzle on 13th October 1900, announcing on his results to the annual meeting of the American Public Health Association that "the mosquito serves as the intermediate host for the parasite of yellow fever." Thus, by an independent move, Dr. Reed also disproved the fallacious belief that yellow fever could be transmitted and induced from clothing and bedding soiled by the excrement of yellow fever sufferers. These articles were known as fomites and were commonly thought to carry the disease. Just as "everybody knew" that the mosquito theory was foolish, so "everybody knew" that fomites were dangerous.

These two cases, although sufficient to convince the Yellow Fever Board that they were at last experiencing some success, were not enough for the thorough scientific mind of Walter Reed, nor would they be for a public which the press had instructed in the "foolishness" of the mosquito theory. With the express permission of General Leonard Wood, Governor General of Cuba, Camp Lazear, was established on November 20, 1900. Moreover, General Wood authorized the Board to use and pay American and Spanish volunteers for the experiments since at this time yellow fever was thought to be a disease afflicting only humans. Dr.

Carroll had exhausted the list of experimental animals, rats and the like normally used for scientific research, failing to produce any cases of the fever in them. In addition to the mosquito theory, "bacillary theory" was also contemplated.

Had it not been for Major Reed's fair and thoroughly scientific approach to the problem and misconceptions concerning the disease, especially the whole contagion theory, yellow fever might have continued for years. His efforts were providentially assisted by the inestimable courage of a large number of volunteers, just like Hussein proved for Ross's malaria research.

In February, 1901 Walter Reed returned to the United States where he was an instant success in medical circles. All through the acclaim he remained modest and reserved. His constant hope of doing something to relieve the suffering of mankind had been fulfilled; his dedication to duty, sound judgment, and thorough scientific methods was an inspiration to the deans of medical research. Soon, he took up an assignment of professor of pathology and bacteriology at the Columbian (George Washington) University Medical School and as professor of bacteriology in the Army Medical School (the present Walter Reed Army Institute of Research). In the summer of 1901 he was awarded two honorary degrees: a Master of Arts from Harvard University and a Doctor of Laws degree from the University of Michigan. He was appointed librarian of the Surgeon General's library on November 1, 1902.

Dr. Reed died of peritonitis on November 23, 1902, following an appendectomy. He was buried in Arlington National Cemetery. On his simple monument is inscribed the following epitaph, taken from the remarks of President Eliot when Harvard University conferred the Master of Arts degree: "He gave to man control over that dreadful scourge, yellow fever." Today a great hospital and medical center stand in constant tribute to Walter Reed. In 1945 he was elected to the Hall of Fame at New York University, the first physician to be so honored. On November 21, 1966, a bronze bust of Major Reed was unveiled at Walter Reed Army Medical Center by the Walter Reed Memorial Association, an organization which, since its inception in 1903, had resolved to erect a memorial in Washington to perpetuate his fame and memory.

During his career in army Dr. Reed travelled extensively mostly due to frequent transfers. Truly, one of the several marvels of his life is that his relegation to frontier garrisons, unfavourable for intellectual contacts, did not ruin him. On the contrary, they provided him infinite opportunities to accomplish the rarest tasks of great human values.

(3) DR. WILLIAM CRAWFORD GORGAS

William Crawford Gorgas (October 3, 1854 - July 3, 1920), Surgeon General, January 16, 1914-October 3, 1918, was the son of General Josiah Gorgas, a native of Pennsylvania who graduated from the United States Military Academy in the class of 1841. Assigned to the ordnance service, he was in command of Mount Vernon Arsenal, near Mobile, Alabama, in 1853. The Junior Gorgas was married in 1885 to Marie Cook Doughty in Cincinnati. They had one child, Aileen Tyster. His varied duties took him to all parts of the world and he received honors too numerous to list. A person of initiative, vision, purpose, tenacity and energy, he was probably best known for translating the discoveries of Dr. Walter Reed concerning yellow fever and malaria to practical application. Gorgas' work conquered these diseases by destroying the disease-bearing mosquitoes of the tropics.

Gorgas' early education had been quite irregular. He had had the advantages of a private school in Richmond but the times were too distracting for satisfactory progress. He spent six years at the Sewanee school and in 1875 graduated with the degree of bachelor of arts at the University of the South, and his MD from Bellevue Medical College in New York.

From the experiences of his childhood Gorgas had acquired a strong desire for a military career. Every effort was made to obtain for him an appointment to the West Point academy but without avail. Much against his father's wishes he determined to enter the service by way of a medical degree. He entered Bellevue Hospital Medical College in New York in 1876 and was graduated in 1879 after three years of financial difficulties. Following an internship in Bellevue Hospital he was appointed, on June 16, 1880, an assistant surgeon in the medical corps of the army. For nearly two decades thereafter, Gorgas' life was that of the average medical officer of the period. Following several years in Texas posts and a tour of duty in North Dakota he spent practically the entire decade preceding the Spanish-American War at Fort Barrancas in Pensacola Bay, Florida.

Shortly after the beginning of his army career he went through an epidemic of yellow fever at Fort Brown, Texas, and was himself stricken with the disease. Thereafter, as an immune, he was frequently drafted for service where yellow fever existed. This accounts for his long service at Fort Barrancas, a post in a section notorious for its epidemics and itself frequently subject to visitations of the disease. He was promoted to the rank of captain on June 16, 1885, and then to major on July 6, 1898. To Gorgas, as to others, yellow fever was an enigma. Its suddenness of appearance, its puzzling choice of victims, and the inutility of ordinary means of disease prevention were quite beyond understanding. With the Spanish-American War in progress, Gorgas arrived at Siboney, in Cuba, on July 7, 1898. While negotiations for the capitulation of Santiago were in progress yellow fever broke out in the American forces. Gorgas was assigned to the yellow fever section of the Siboney hospital and shortly thereafter took over the command of the hospital from Major Louis A. La Garde. It is significant of the view of the disease then currently held that it was recommended that the village of Siboney be destroyed by fire. Not only was this recommendation carried out, but in September the Siboney group of hospitals with much of their equipment was similarly destroyed.

Major Gorgas was returned to the United States in September, convalescent from typhoid fever. Later in 1898 he returned to Cuba and early in 1899 when he became chief surgeon of the Department of Havana. Following the appointment of General Leonard Wood as military governor of Cuba in December 1899, Gorgas was made chief sanitary officer for the city of Havana. Though yellow fever at this time showed only as sporadic cases it was, as always, the chief concern. Gorgas applied to the city the generally accepted methods of disease control. He cleaned up the city, segregated the sick, and quarantined infected localities. Though a friend of Dr. Carlos J. Finlay and familiar with his theory of the mosquito transmission of yellow fever Gorgas placed no confidence in the idea. Despite the greatly improved sanitary condition of Havana the yellow fever situation, instead of improving, became much more serious. It was not until the board, of which Major Walter Reed was the head, furnished proof that the *Stegomyia* mosquito (which we now know as *Aedes aegypti*) was the carrier of the disease that truly effective methods could be instituted. *Aedes Aegypti* was the common mosquito pest of the city. It is a highly domesticated insect, breeding in all kinds of water containers in and around habitations. The surest control of the insect was deemed to be the elimination of its breeding places. This plan was adopted and, though the task had many difficulties, Havana was not only freed of its mosquitoes but was permanently rid of yellow fever. The results obtained by his work in Havana brought him an international reputation as a sanitarian.

The years from 1900 to 1904 brought the gradual development of plans for digging the Panama Canal. There was early recognition of the necessity for expert sanitary advice upon the project and in 1902 Gorgas was transferred from Havana to Washington and assigned to this work.

In March 1903 Congress raised him to the grade of colonel in recognition of his services in Havana. For two years he studied the canal problem, reviewing the experience of the French on the isthmus and making visits to the Suez Canal and to Panama. Actual work upon the canal commenced in 1904, and Gorgas with his staff of assistants arrived in June of that year. He early encountered administration difficulties. Despite the positive knowledge that the French failure had been due to disease the American administration was disinclined to support adequate measures for preventing a repetition of that experience. The first Canal Commission, headed by Admiral John C. Walker, had strongly in mind the prevention of graft and extravagance that, with of course the scourges of malaria and yellow fever, had caused much of the disaster of the French effort, particularly of De Lesseps (*vide infra*). Expenditures for sanitary improvements were regarded as falling under the latter head. It required a visitation of yellow fever, starting in November 1904, to obtain for Gorgas any substantial support for his work. He began in the Canal Zone the measures which had been highly successful in Havana. Again the mosquito was to be deprived of breeding places and cases of yellow fever segregated and protected from mosquitoes. The situation in Panama presented more difficulties than that in Havana and results were far less prompt. It was well into 1905 before yellow fever had been eradicated, and in the meantime determined effort were being made to discredit Gorgas' work and to supplant him. It is probable that these could have been successful but for the interest aroused by a report made by Dr. Charles A. L. Reed of Cincinnati to the American Medical Association in March 1905, in which the obstructive hand of Commissioner Carl E. Grunsky was so largely featured. The discharge of the Walker Commission at about this time and the appointment of another headed by Theodore P. Shonts did little to mitigate Gorgas' troubles. Yellow fever was still prevalent and the new commissioners were dissatisfied that the first interest of the sanitary service was the elimination of mosquitoes rather than the general improvement of the cities of Panama and Colon. They recommended the removal of Gorgas which not only drew the disapproval of President Roosevelt but caused an order for active support of his work. In November 1906, the President paid a visit to Panama and shortly thereafter Gorgas was made a member of the canal commission.

For a time he had practically a free hand, but after the reorganization of the commission in 1908, with Colonel George A. Goethals as chairman and chief engineer, his troubles began anew. Goethals, given unusual powers by executive order, ruled the Canal Zone with a despotic control. He was free in criticism and centered his attacks upon the expense of the sanitary service. Despite the difficulties thrown around his work, due to lack of cooperation from the chief commissioner, Gorgas not only freed the Canal Zone from yellow fever but he made the cities of Panama and Colon models of sanitation comparable with any city of the United States. In the meantime his reputation had extended until he was generally regarded as the world's foremost sanitary expert. In 1913 he was asked by the Transvaal Chamber of Mines to visit South Africa and make recommendation for the control of pneumonia among the negro mine workers. It was while engaged in this work that he received the notification of his appointment as Surgeon General of the Army, with the rank of brigadier general, on January 16, 1914. He returned to the United States in April to take up his new duties, and on March 4, 1915, he was advanced to the grade of major general. The recently organized International Health Board enlisted him as an advisor, and in 1916 sent him with a staff of assistants for a tour of South and Central America with a view to continuing the fight on yellow fever in these sections. Following this trip, a plan for the elimination of yellow fever was adopted and Gorgas was made director of the work.

General Gorgas had always been more interested in disease prevention than in office administration. In January 1917 he informed the Secretary of War that he wished to retire. During war days, General Gorgas' time was largely taken up by consultations with representatives of the Secretary of War, the Council of National Defense, the General Medical Board, and with Congressional committees. General Gorgas was retired on account of age on October 3, 1918, shortly before the Armistice and again became available for work with the International Health Board.

General Gorgas was commissioned to investigate the yellow fever situation on the west coast of Africa and in May, 1920 he sailed with his staff for London. After attending the meeting of the International Hygiene Congress in Brussels he returned to London where he experienced a stroke of apoplexy, and died a month later, on July 3, 1920, in the Queen Alexandria Military Hospital at Millbank. The funeral was held in St. Paul's Cathedral and the body returned to the United States to rest in the Arlington National Cemetery. He was survived by his wife, Marie Cook, and a daughter.

Many books and articles have been written about Dr. Gorgas recognizing him as a benefactor of all mankind. General Gorgas has been memorialized at the University of Alabama, in Mobile, and in many other places around the world. His statue is in the New York University Hall of Fame. General Gorgas had been the recipient of many honours. Honorary degrees had been bestowed upon him by the University of the South, by Harvard, Brown, Pennsylvania, Johns Hopkins, and by Oxford Universities. He had been decorated by a number of foreign countries. The University of the South and the University of Alabama, both of which his father had served, offered him their presidencies. In 1908, the American Medical Association elected him its president. During his last illness he was visited by King George and knighted.

A lesser man than Gorgas might have failed before the opposition he encountered even after the success at Havana gained world-wide publicity and acclaim. United States officials in the Canal Zone refused to believe in Gorgas. They pigeon-holed his requests for equipment, refused him authority to deal with the people, discredited him to higher authorities and frequently clamored for his dismissal. His greatest gift to the world was a dengue/mosquito-free Panama. Someone thus truly said of him as, "*His application of sanitary measures eliminated yellow fever as a blight on the health of the people of the world and made possible the construction of the Panama Canal".*

(4) PROFESSOR GIOVANNI BATTISTA GRASSI

Giovanni Battista Grassi was born on 4th March, 1854 in Como, Italy. He obtained his doctorate in 1878 and became Professor of Zoology at the University of Catania till 1896. He demonstrated (1898) that the *Anopheles* mosquito carries the *Plasmodium* parasite of malaria in its digestive tract. He is known also for his researches on the life history and societies of the Termitidae, and on the developmental relationship between *Leptocephalus* and the common eel and other muraenidae. He died in 1925.

Even though the balance of pioneering research work on *Anopheles* mosquito as the carrier for malaria weighed heavier on Ross's side, nevertheless no one could ever deny the genius of Professor Grassi, whom many considered as the most authoritative zoological scientist of his time. Dr. Patrick Manson, who mentored Ronald Ross and played a key role in his achieving the most significant discovery of the 19th century, was even in favour of Grassi when a decision about the 1902 Nobel Prize was to be taken! The Italian Government had duly recognized the genius, however, and issued a stamp (25 L) in his honour on November 19, 1955, on the occasion of his 30th death anniversary, bearing portrait of Battista Grassi, mosquito and microscope. The people of Italy also dedicated a boulevard in Milan in his memory and named it as Giovanni Battista Grassi Street 74, harbouring the famous Luigi Sacco Hospital. Professor Grassi received many honours and distinctions among which the Darwin Medal presented to him in 1896 was very dear to him.

(5) DR. RONALD ROSS

Ronald Ross (1857-1932), a son of an army officer, was born in Almora, India, on May 13, 1857. He did his schooling, from the age of eight, in England. His record was nothing much remarkable, but he showed a distinct talent for subjects such as art and literature. Young Ross thought of becoming anything but a doctor which his father was determined to make him and join the Indian Medical Service, largely for his awareness of the comfort and ease with which a British officer lived in India. Ronald Ross, almost reluctantly, did just enough to pass his examinations at St. Bartholomew's Hospital Medical School in London. However, having failed the examination of the Society of Apthecaries, he was ineligible to join the Indian Medical Service, so he spent a year as a ship's surgeon, during which time he started penning his first novel. Having passed the Society of Apothecaries examination at the second attempt, his performance was such that he received, in 1881, a commission only for the Madras service, considered by many the least prestigious of the three Indian Presidencies; Bengal and Bombay were the far more desirable appointments by the incumbents.

Ross's first tour to India lasted eight years, from 1881 to 1888. As for most of the young British officers, with only a couple of hours duty every day, Dr. Ross too had plenty of leisure time, golfing, fishing and hunting, and only disinterestedly performing his medical duties. He had finished his first novel, as well as writing some poems and short plays, and was developing a long-standing interest in mathematics. *Inter alia*, somehow he continued to feel a big void in his strong desire to apply life to fruition. With a pinch of salt in his day-to-day medical duties at the hands of his superiors on one or another reason, he was exacerbated to the point of being a paranoid. He began to think to do something extra-ordinary, something which others could not do and what could carry him higher than others.

Amidst the pressure of his everyday medical duties, Ross spent three years studying mosquitoes and their parasitres. In doing so, he naturally faced enormous problems since he was unable to identify mosquitoes and had never seen a malaria parasite earlier. Ronad Ross became a medical scientist almost by accident.

During his first furlough (1888), he obtained further training in public health, bacteriology and microscopy. He searched for the parasite which Laveran, a French army surgeon, had found in 1880 in the blood of victims of malaria, but he was unsuccesful. He returned to India, with a bride and a microscope. In the second furlough (1894), he met Patrick Manson in London, who had by then already demonstrated mosquito's role in transmitting lymphatic filariasis and, rightly so, earned the reputation of the Father of Tropical Medicine. Manson showed to Ross malaria parasites for the first time, and demonstrated to him how to find the parasites in blood smears. Manson had inferred, based on his earlier experiences in China in respect of filaria, that mosquitoes could be responsible for transmitting malaria, too, although his hypothesis remained merely a conjecture. Both became friendly and, in fact, Ross had even taken Manson, a true giant of medical science and a father figure, for his mentor! Ross took note of his advice quite seriously and was determined to prove Manson's "Grand Induction". The 173 letters that the two exchanged between 1895 and 1899 are rich in both scientific and human terms, as well as remain one of the celebrated scientific correspondences. These letters record, in great detail, Ross's research in India elucidating the role of mosquitoes in the transmission of malaria and worked to firmly cement their intimacy. Ross's regular letters, taking as much as three weeks in transit at that time, to his mentor document the frustrations and false trails as well as the excitement of discovery. Manson, in turn, acted as a kind of agent in London, publicising his findings, offering advice and seeking to use his influence to secure for Ross the working conditions he so desired.

> *"This massive correspondence is also a testimony to the rise and fall of a relationship, as Ross's preoccupation with his place in the history of malariology led to a breach between the two men. That Ross was completely obsessed with garnering every possible laurel in the wake of his "mosquito malaria theory", it ultimately spoiled his relationship with his mentor. Themes of priority, nationalism, and personal vanity punctuate this latter correspondence, which also reveals new insights about the golden years of tropical medicine. This happened primarily because Manson thought to credit Grassi, who was a Professor of Zoology with extensive mosquito/malaria-related research experience and was considered, rightly so, by many scientists to be the preeminent medical zoologist of his time, for his commendable contributions, for which Ross turned cool toward his former teacher. Ross continued to critically despise Manson, ignoring latter's immense assistance and guidance in his pursuing malaria research as evidenced from scores of affectionate letter that passed between them. This bitterness culminated to its zenith when Ross, recuperating from a heart stroke by the end of his life, wrote a highly unpleasant book charging that Manson even hindered rather than helped his malaria work".*

Ross had a tumultuous career since many of his medical superiors and colleagues did not see eye-to-eye. When his service bosses were unsympathetic to his requests for leave, at times he threatened to take early retirement from the I.M.S. and return to his first love, namely, literature. It was the rock-like support of Manson, his mentor till that time, who, on one hand, encouraged him to continue his research during these periods of discouragement and despair, but, on the other hand, also endeavoured hard to convince his I.M.S. bosses to think favourably about his efforts.

The most significant step forward occurred on 20th August, 1897, when he got the first sight of 'pigmented cysts' in the stomach of a mosquito. Ross celebrated this historical event as "Mosquito Day", and the jubilation showed up in his memorable poem that he wrote soon after the discovery was made.

This day relenting God
Hath placed within my hand
A wonderous thing; and God
Be praised. At his command,
Seeking his secret deeds
With tears and toiling breath,
I find thy cunning seeds,
O million-murdering death.

I find this little thing
A myriad man will save,
O Death, where is thy string?
The victory, O Grave?

Later, he succeeded to demonstrate completion of cycle of avian malaria parasites in the *Anopheles* mosquito. For his this pioneer work, Ross was awarded the 1902 Nobel Prize in Physiology or Medicine. He was knighted in 1911.

Ross served in the Indian Army from 1881 to 1899, before he returned to England. During the later part of this period of his service in India he studied malaria. He subsequently held various positions notable amongst which were (i) Professor of Tropical Medicine at University College, Liverpool (from 1902), (ii) Physician at King's College, Hospital, London, (iii) Consultant on malaria to the War Office, and (iv) Director of the Ross Institute and Hospital for Tropical Diseases, London (from 1926).

Ronald Ross is often said to have become a great medical scientist almost by accident, but the reality is far from truth; he, for several reasons pressing constantly on him irresistably during the course of his service in India, yearned to do something extra-ordinary and remarkable which could carry him on the zenith of pyramidal success, and malaria, a subject then was pursued all over the world with frenzied vigour and enthusiasm, gave him the right launching pad.

6. DR. AMICO BIGNAMI

Amico Bignami graduated in Rome in 1882. He first worked in Rome as professor incaric, and for some years was assistant to Tommasi Crudelli in pathology. In 1883 he was appointed professor extraordinary of pathology, and in 1906 became full professor at the Royal University of Rome. In 1917 he was appointed professor of medicine, holding that position until his retirement in 1921. From 1896 he was primary physician at the Ospedale riuniti di Roma.

Bignami's scientific works concern the pathological anatomy of the brain, acromegaly, the leukaemic diseases, but his particular interest was in malaria research. In 1896 he put forth the hypothesis of infection by mosquitoes and attempted to prove the mosquito theory in man. Bignami had captured mosquitoes from regions with a high incidence of malaria and allowed them to bite healthy human beings. But like Ronald Ross in his early work, Bignami failed to appreciate that only one type of mosquito could transmit human malaria. The next year, however, Sir Ronald Ross (1857-1932), a British Major in the Indian Medical Service, demonstrated the transmission of malaria by mosquitoes in an avian model. His experiments confirmed the 'mosquito theory' of malaria. In 1898, Bignami, Giovanni Battista Grassi and Giuseppe Bastianelli first infected humans with malaria by mosquitoes. *In this experiment Bignami produced malaria in himself, allowing an infected mosquito to bite him.* On 28th November, 1898 these observers were able to report to the Accademia dei Lincei their demonstration of the development of human malaria parasites on the gut wall of *Anopheles claviger*. This Italian group of investigators thus proved that the mosquitoes of the genus *Anopheles* are the vector for human malaria.

7. DR. T. RAMACHANDRA RAO

One of the great Indian stalwarts in the science of medical entomology in post-independence period, Dr. Ramachandra Rao, son of

Shri Thammajee Rao, a member of the Mysore State Civil Service, was born on 9th October, 1907. Having completed his early education and graduation in science from the famous Central College, Bangalore, young Ramachandra Rao moved to Calcutta University where he earned his M.Sc. in Zoology, with distinction. Subsequently, he took up lecturership at the Mzysore Medical College for a couple of years, however all the while concentrating on further higher education. As a result, in 1951, he was bestowed with the degree of Doctorate in Science from his *alma mater*, the Calcutta University.

After completing education in Bangalore, Karnataka he earned a Rockefeller Foundation Fellowship in 1936 and worked under the guidance of Prof. P.A. Buxton at the famous London School of Hygiene and Tropical Medicine. He started his research career with interests in the insect cytogenetics, mostly working on grasshoppers. Soon, however, with the ever mounting importance of the medical parasitology or entomology, he before long joined Dr. Paul Russell, a world renouned malariologist, in 1932, and undertook investigations on mosquitoes' bioecology and control. The team of Dr. Rao and Dr. Russell conducted a very large number of innovative and pioneering investigations, particularly on the main malaria vector, *Anopheles culicifacies*, in various different States in the country endemic or hyperendemic for malaria. His love for malaria control brought him as an entomologist to the the newly established Malaria Organization of Bombay State (1942), where he amply shone his remarkable mosquito/malaria-based knowledge and got promoted to the rank of Deputy Director of Public Health of the Maharashtra State, and in-Charge of Malaria and Filariasis. Here, he had another marvelous opportunity to prove his grip on mosquitoes, and, with Dr. D.K. Viswanathan, carried out numerous studies on the bionomics and control of *An. fluviatilis*, another important vector of malaria, in the hilly and sylvan environs. The early breakthrough came in his life when he demonstrated effectively the usefulness of DDT in controlling malaria through indoor residual spraying in rural India. These investigations on DDT application against malaria mosquitoes became the hallmark for both the National Malaria Control Programme (1953-1958) and the National Malaria Eradication Programme (1958-2003), besides fetching him international acclaim.

A man of great natural instincts, he never seized to learn new things in biomedical research. Therefore, when Virus Research Centre (now the National Institute of Virology) was established at Pune, under joint collaboration of the Indian Council of Medical Research and the Rockefeller Foundation, he was naturally the first choice to head this premier institution. There he was the Director between 1961 and 1970, a position that he held till his retirement. An able medical biomedist and an astute administrator, Dr. Ramachandra Rao developed the VRC into an international reference institute on virological researches.

A vastly travelled scientist, he was a member to several committees of the World Health Organization, particularly on matters relating communicable diseases such as malaria, insecticides, vector biology and control, and virus diseases. While he served as a consultant and an advisor to the Indian Council of Medical Research for 35 long years, the WHO also availed his vast knowledge and expertige as the Team Leader in Afghanistan and Tanzania. He also served as the Officer on Special Duty with the ICMR. He was a Fellow of some highly reputed international societies, besides having chaired as President of Zoology and Entomology Section of the Indian Science Congress Association in 1972. His *magnum opus*, The Anophelines of India, is a major reference work on the subject for the Indian researchers.

Dr. Ramachandra Rao was an epitome of hard work and dedication throughout his life, albeit failing eye-sight in the wee years. He was ever vigilant and keenly interested in all malariological researches in the country till his last breath, on 8th November, 1984. I have a vivid memory of Dr. Ramachandra Rao, whom I met the only time in early 1981 in Delhi, in the ICMR's Malaria Research Centre, when he took the opportunity to give me my first malariological lesson. "Think, Dr. Tyagi, how you can eradicate this menace, malaria?", argued Dr. Ramachandra Rao, which of course became my guiding force all these years (*cf.* Tyagi, 2002).

CHAPTER - 4

THE BIOLOGY OF MOSQUITOES : *THERE IS ALWAYS SOME LITTLE THING THAT IS TOO BIG FOR US*

Mosquitoes, classified under the family Culicidae of the order Diptera, are by far the most thoroughly researched haematophagous insects. They are able to cast a much greater influence on human health and well-being throughout the world than any other arthropod, mainly because of their involvement in both transmitting a number of dreadful diseases (malaria, filariasis, dengue, Japanese encephalitis, Rift valley fever, Chikungunya, West Nile virus and O'Nyong-nyong) and creating biting nuisance of great public health significance. Consequently, of the many species of blood-sucking insects, mosquitoes particularly belonging to genera *Anopheles, Culex, Mansonia* and *Aedes* are amongst the most important insects from medical standpoint as they are responsible for nearly 10% of all the sickness of humans. Although, there are about 3500 species all over the world, however only fewer than 100 species under eight genera only are vectors of diseases (Service, 1996). These vectors are organized in two of the three subfamilies, Anophelinae and Culicinae, while the third subfamily Toxorhynchitinae has members with mouthparts suited to feed on plant sap only. Mosquitoes are cosmopolitan in distribution and are found in all climatic zones and zoogeographical regions.

LIFE OF A MOSQUITO

Mosquitoes are very slender flies, generally with an elongated piercing proboscis and no ocelli. The palpi is stiff and not pendulous. Legs are long, antennae densely plumose in the males, pilose in females. Wings are fringed with scales along the posterior margin and the veins. The larvae and pupae are aquatic and very active, though only the former feed. The larvae are metapneustic, with an elongated thoracic mass (Fig. 4.1).

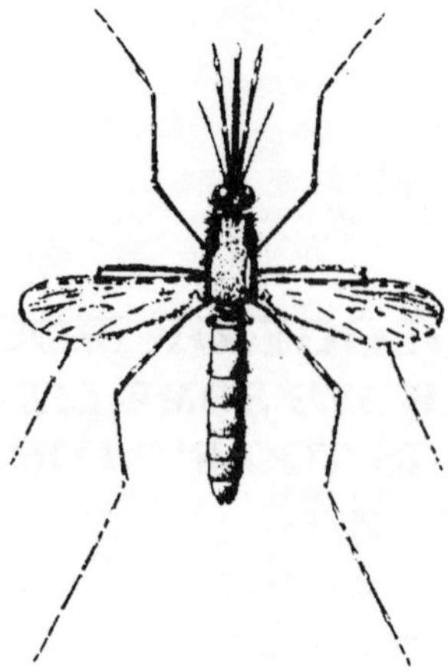

Fig. 4.1. General morphology of a mosquito.

Mosquitoes have the capability to bite by day or by night, depending on species. They breed in all kinds of water fresh and salt, flowing or stagnant, foul or potable, water in tin cans, tree holes (phytotelmata) and salt marshes. Large coastal areas are occasionally uninhabitable because of salt marsh mosquitoes. Mosquitoes may render recreation centers unusable and make normal living in and around homes extremely unbearable and vexing. Ironically, real estate values may suffer where mosquitoes are present, and losses caused by reduced industrial efficiency can be considerable because of the mosquito menace (Spielman & D'Antonio, 2001).

As a principle, this is only the female of the vector mosquito which feeds on the blood of human or animal host, although females of some species are known to supplement their meals with plant sugars, too. Female mosquitoes may feed on the blood of a wide variety of mammals, birds, reptiles and amphibians. Mosquitoes are generally host-specific for their blood-feeding requirements but they may quickly diverse in case of any exigency.

Mosquitoes have their body clearly divided in three parts: head, thorax and abdomen. The head bears a pair of large compound eyes, a pair of antennae and mouthparts which are piercing and sucking type in females suited primarily for blood feeding, while in males they are developed to suck sap from the plants. Antennae too differ being shorter and sparsely haired in females while in males they are thickly haired and bushy. There is a pair of fully developed forewings, while the hind wings are reduced to rudimentary halteres or balancers. The males are equipped with external genitalia, called hypopygium. Following mating, males pass spermatozoa to the females where they are stored in spermatheca. Eggs are fertilized as they pass out of the ovary of the female. Females generally lay eggs in mildly moving or stagnant fresh (e.g., *Anopheles*) or polluted water bodies (e.g., *Culex*). Eggs are laid singly directly on water surface (*Anopheles*), in batches/groups cohesced in a boatraft-like shape (*Culex, Culiseta, Mansonia and Uranotaenia*) or along the moistened surfaces, like soil etc., close to water or on damp soil that will be flooded by water (*Aedes* and *Ochlerotatus*). Eggs of the genus *Toxorhynchites* are also laid singly on the water surface. Most eggs hatch into larvae within 48 hours; others might withstand subzero winters before hatching. Some mosquitoes (e.g. *Aedes, Toxorhynchites*) are specialized to lay eggs in plant materials or tree-holes (phytotelmata). Some species may survive in the egg stage for 1-2 years before flooding occurs. *Aedes, Orthopodomyia* and *Psorophora* eggs are commonly laid in this manner. Water is a necessary part of their habitat. In general, development of mosquitoes is essentially the same.

There are four stages in the life cycle of mosquitoes; egg, larva, pupa and adult. The immature stages are aquatic in nature. The eggs are usually white when first deposited, becoming dark within an hour or two. Anopheline eggs are oval-shaped and float horizontally. They are usually pointed on one end and have a pair of lateral floats. The larvae of all mosquitoes get their food from the water where they live, but to respire they have to come to the surface. Most larvae have siphon tubes for breathing and hang upside down from the water surface. *Anopheles* larvae do not have a siphon and lie parallel to the water surface to get a supply of oxygen through a breathing opening or spiracle. Larvae of the genus *Mansonia* and *Coquillettidia* are adapted to take oxygen under water from the air-conducting parts of plants. The larval developmental period includes four instars (*i.e.*, during growth the larva moults, or sheds its skin, four times and the stages between moults are called instars) which usually take 5-15 days to complete. At the end of the larval development pupa is formed by discarding the skin of the fourth instar larva. Only larvae feed while pupae do not feed. The larvae feed on microorganisms and organic matter in the water. One mosquito species larva feeds on larvae of other mosquitoes: *Toxorhynchites*, the largest known, are predators of other mosquito larvae sharing their habitat. *Toxorhynchites* larvae are much larger than other mosquito larvae. The larval body is characteristically divisible in three parts, but the pupa has only two parts; the fused cephalothorax and the curled and segmented abdomen. Pupae, commonly called "tumblers", too occur in water and are actively motile. When disturbed, it dives in a jerking, tumbling motion toward protection and then floats back to the surface. For respiring all pupae, except those of *Mansonia* and *Coquillettidia*, move (tumble) with a flip of their tails (caudal fins) to the surface where with the help of a pair of trumpets atop head region the oxygen is taken in. The pupal stage lasts for one day to several days. The metamorphosis of the mosquito into an adult is completed within the pupal case. When the pupal growth is complete, the pupal skin or exuvium breaks, the adult works its way out, crawls on to the surface of the water for a short time to allow itself to dry and all its body parts to harden. The wings have to spread out and dry properly before the newly emerged adult can fly. Though not a general rule, blood feeding and mating does not occur for a couple of days after the adult emerges. The adult mosquito, in most cases of species, is a small fragile insect with a slender abdomen, one pair of narrow and delicate wings, covered or thinly so with scales, and three pairs of long slender legs. As the case may be, species may greatly vary in their size varying from slightly over 4 mm to about 12 mm. The three body parts (i.e., head, thorax and abdomen) are clearly visible (Fig. 4.2).

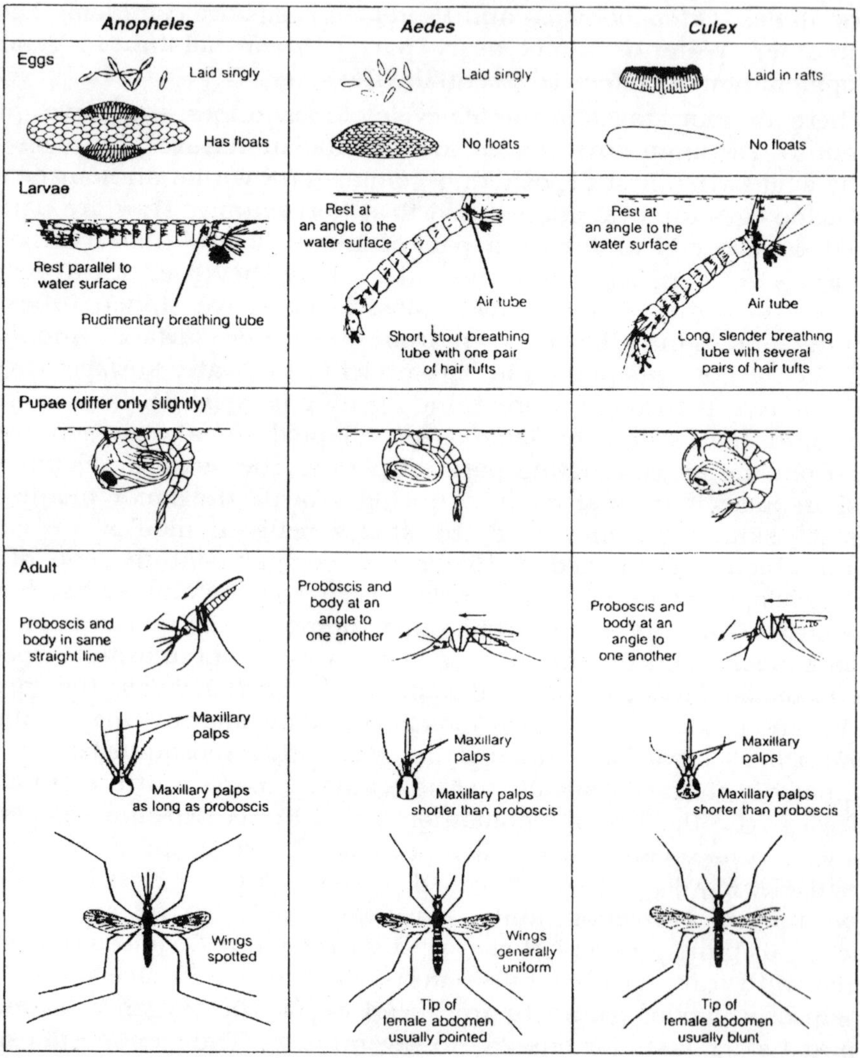

Fig. 4.2. Some of the main characteristics for differentiating *Anopheles*, *Aedes* and *Culex* mosquitoes.

The egg, larva and pupa stages depend on temperature and species characteristics to determine how long they take for development. For instance, *Culex tarsalis*, a common mosquito in USA might go through its life cycle in 14 days at 70°F and take 10 days at 80°F. On the other hand, some species have naturally adapted to go through their entire life cycle in as little as four days or as long as one month (Fig. 4.3).

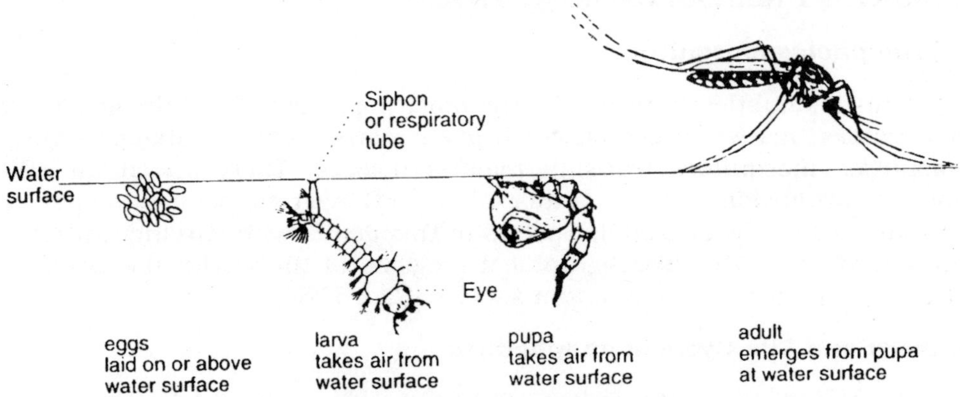

Fig. 4.3. The life cycle of the mosquito.

Only female mosquitoes require a blood meal (protein) and bite animals warm or cold blooded. Stimuli that influence biting (blood feeding) include a combination of carbon dioxide, temperature, moisture, smell, colour and movement. Most mosquitoes are nocturnal in their feeding behaviour (e.g., *Anopheles*), but some are diurnal or crepuscular (e.g., *Aedes*). When the female finds a suitable host, such as the man, it introduces the piercing mouthparts through skin deep into body and penetrates a capillary, to suck the blood directly from the capillary or from the pool of the blood oozed outside the punctured capillary. Females of certain mosquitoes are single-time feeder while many species tend to feed in shift or multiple occasions on a given night or day-time. The number of the egg formation and development in ovary of the female is directly dependent on the quantum and nature of supply of blood meal received by the female mosquito. Depending on a particular species, an engorged female may like to rest inside house or prefer outside environs. This characteristic of the vector species has for long been exploited for their control through indoor spraying of residual insecticides.

Mosquitoes are the sole vectors of malaria, yellow fever and dengue, and they participate significantly in transmitting lymphatic filariasis and the encephalitides of various different kinds. Besides, their bites may itch for days, causing much loss of sleep and tranquility, and serious nervous irritation and restlessness. Their saliva contains proteins alien to the human body and repeated bites result in sensitivity to it. Mosquitoes often cause serious economic losses limiting outdoor movements.

Male mosquitoes do not bite, but feed on the nectar of flowers or other suitable sugar source. Acquiring a blood meal, *i.e.*, protein, is essential for egg production in females, but mostly both male and female mosquitoes are nectar feeders. Of those female mosquitoes capable of blood feeding, human blood meals are first or second choices. Horses, cattle, smaller mammals and/or birds are preferred for blood feeding.

Female *Toxorhynchites* mosquitoes actually can not obtain a blood meal and are restricted to a nectar diet.

IMPORTANT GENERA OF MOSQUITOES

(1) *Anopheles* mosquitoes

This is doubtlessly medically the most important of all the genera of mosquitoes, many species of which are serious vectors of malaria. Only *Anopheles* mosquitoes transmit malaria disease. There are about 400 species worldwide, and in India about 60 species occur (Nagpal & Sharma, 1995; Tyagi, 2002). Species of this genus with varying numbers are found in all the zoogeographical regions of the world; the Oriental Region harbouring the maximum species (Rao, 1984).

Biology and Life Cycle of anophelines

As stated above, there are four stages through which anopheline mosquitoes complete their life cycle. Important taxonomic and biological features in each of these are described below (Fig. 4.4):—

(i) *The Egg*: Eggs of *Anopheles* mosquitoes are boat-shaped with an upper deck having a micropile at one end and a fringe all around, and usually with lateral floats with transverse striations called ridges. Eggs are laid singly. In optimal temperature it takes about 36-48 hours for eggs to hatch.

(ii) *The Larva*: This is very agile from the time it hatches out. There are four larval instars before it converts into pupal stage. It rests characteristically parallel to the surface of water. The head of a larva is capable to rotate at 180° to enable it feed more intensely and vigorously. Characteristic palmate hairs, present in many species, occur on both thorax and abdomen. Two (one on each side) reversible organs, called Nuttal's organs, occur on the first segment of the thorax. These are best seen in the live specimens as they are fleshy and protrusible. Larvae have a dorsal spiracular plate supporting two spiracles meant for bringing about respiration. The total duration taken by larvae under normal conditions of temperature ranges from 5-7 days.

(iii) *The Pupa:* This has a peculiar 'comma' form with highly active up and down vertical movements. The paddle is oval in outline. It may be either undivided or only slightly invaginated, with the tip having a pair of setae one of which is ventral and sub-apical. On the abdominal segments 3rd to 7th there are located spine-like seta No. 9 (Belkin's system) at the postero-lateral corners usually with branches on 8th abdominal segment. The trumpet, located on the top of the cephalothorax, often bears at least one cleft. Pupal development completes in about 48 hours.

(iv) *The Adult:* Soon after the adult emerges from pupae it flies away to take shelter in some suitable place, where it rests for sometime before taking its first blood meal. Males generally emerge earlier than females, and have shorter lifespan than their counterparts.

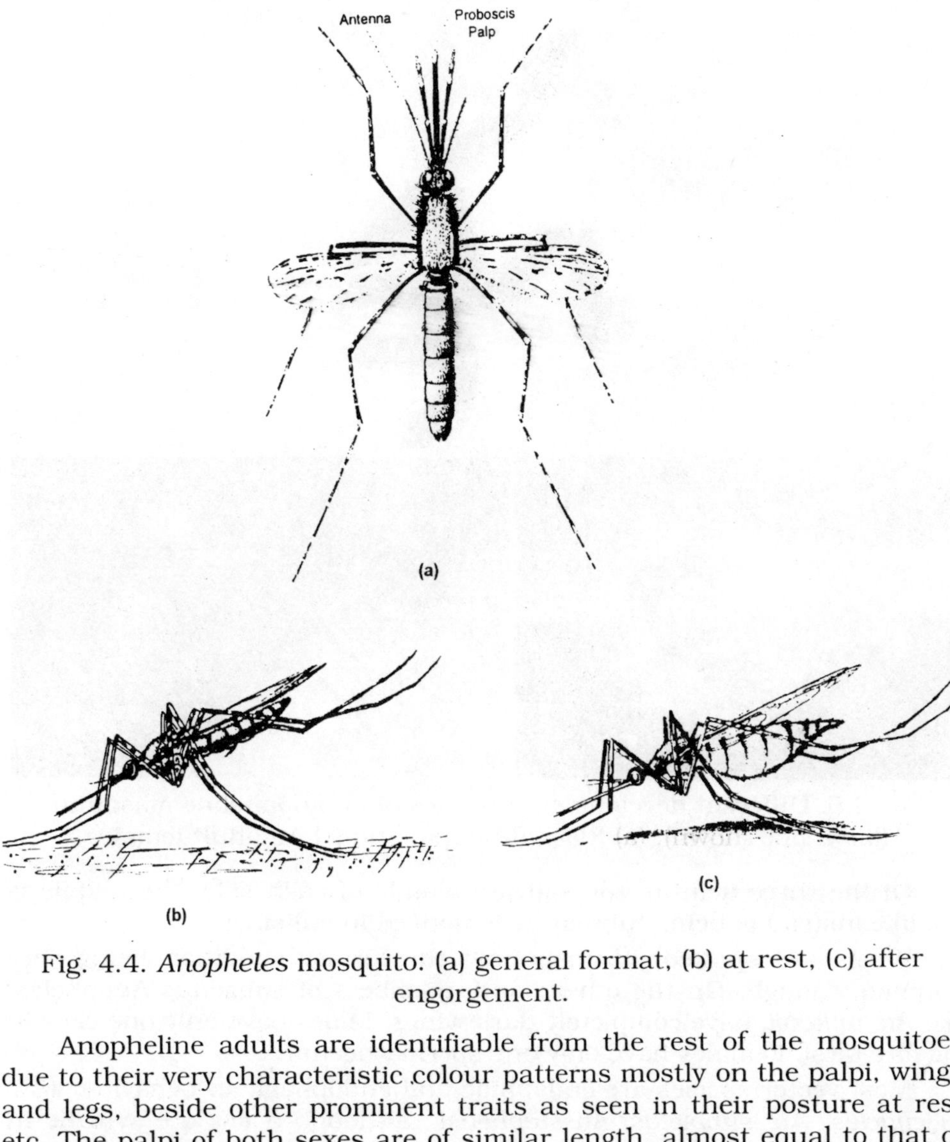

Fig. 4.4. *Anopheles* mosquito: (a) general format, (b) at rest, (c) after engorgement.

Anopheline adults are identifiable from the rest of the mosquitoes due to their very characteristic colour patterns mostly on the palpi, wings and legs, beside other prominent traits as seen in their posture at rest etc. The palpi of both sexes are of similar length, almost equal to that of the proboscis. In males, however, the two distal segments in palpi are enlarged giving them a club-like appearance. The posterior edge of scutellum is only slightly curved and is continuous without any marked division into lobes. Hairs on these lobes are in continuous series.

The axes of head, thorax and abdomen are nearly in a straight line. Anopheline mosquitoes rest on an object in a characteristic posture. The angle of resting, varying from species to species, is close to 60° with the tip of the abdomen directed away from the surface (e.g., *An. hyrcanus*, *An. barbirostris*). In a few species like *An. culicifacies*, *An. culiciformis*, *An. sintoni* etc. the body is almost parallel to the surface, giving adult a semblance of a *Culex*-like posture.

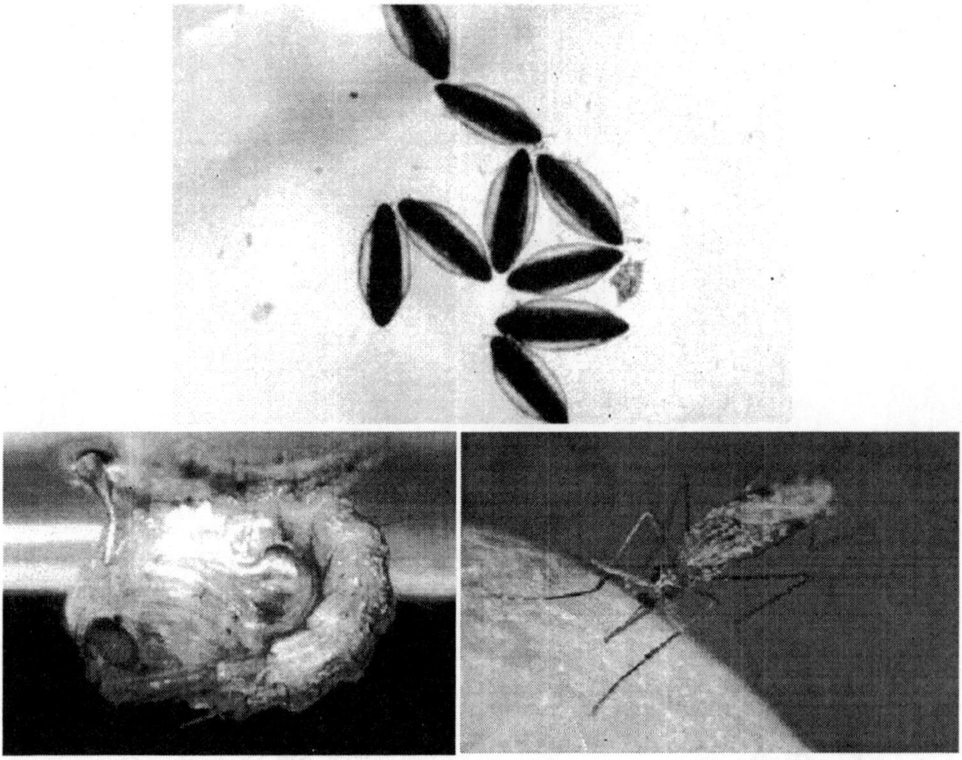

Fig. 4.5. Different developmental stages of an anopheline mosquito (larva not shown): (a) Eggs, (b) Pupa, and (c) An adult female.

Of the three acini of the salivary glands of each side, the middle is sac-like instead of being tubular as is noticed in culicines.

Wings are spotted or unspotted; the former condition being very common, though. On the other hand, members of subgenus Anopheles, like *An. aitkenii*, have completely dark wings. Males have only one claw in the fore tarsi. Females have only one spermathecum.

Most vector species are endophilic and endophagic in behaviour (*An. culicifacies*, *An. subpictus*, *An stephensi*), although a few are sylvatic in nature, too (*An. dirus*, *An. pulcherrimus*). Although, there is no direct proof of hibernation exhibited by any Indian species, there are species in colder temperate regions elsewhere in the world which manifest overwintering. In India, yet, *An. subpictus* is suspected to possess the potential to overwinter.

Some species exhibit "sibling species complexes", a phenomenon which carry tremendous medical significance in the sense that while some of them may be vectors, others are either less significantly important or are not able to transmit the pathogen at all. The most widely examined species-complexes belong to *An. gambiae* in Africa and *An. culicifacies* in India (Subbarao *et al.*, 1999).

(2) *Aedes* mosquitoes

Several species of the genus *Aedes* are of tremendous medical and public health importance, for instance *Ae. aegypti*, *Ae. albopictus*, *Ae. sollicitans*, *Ae. nigramaculis*, *Ae. vexans* etc. In as far as India is concerned, the most important vector species are *Ae. aegypti* and *Ae. albopictus*, due to their involvement in transmission of dengue fever. *Aedes aegypti* is also the vector for the deadly Yellow Fever prevalent in Africa and South America.

Biology and life cycle of *Aedes aegypti*

Main biological and taxonomical features are described for the two important vector species (i.e., *Ae. aegypti* and *Ae. albopictus*). In general, the biology of *Ae. aegypti* follows the pattern of other culicid mosquitoes. The life cycle is completed through four distinct stages; egg, larva, pupa and adult. The larval development passes through four instars, each undergoing progressive increase in size till the pupal stage. Pupae and adults do not show any growth. Only female adults of *Ae. aegypti* feed on blood, sugar and water for maturation of eggs, energy requirements and maintenance of the osmotic balance of the body (Fig. 4.6 and 4.7).

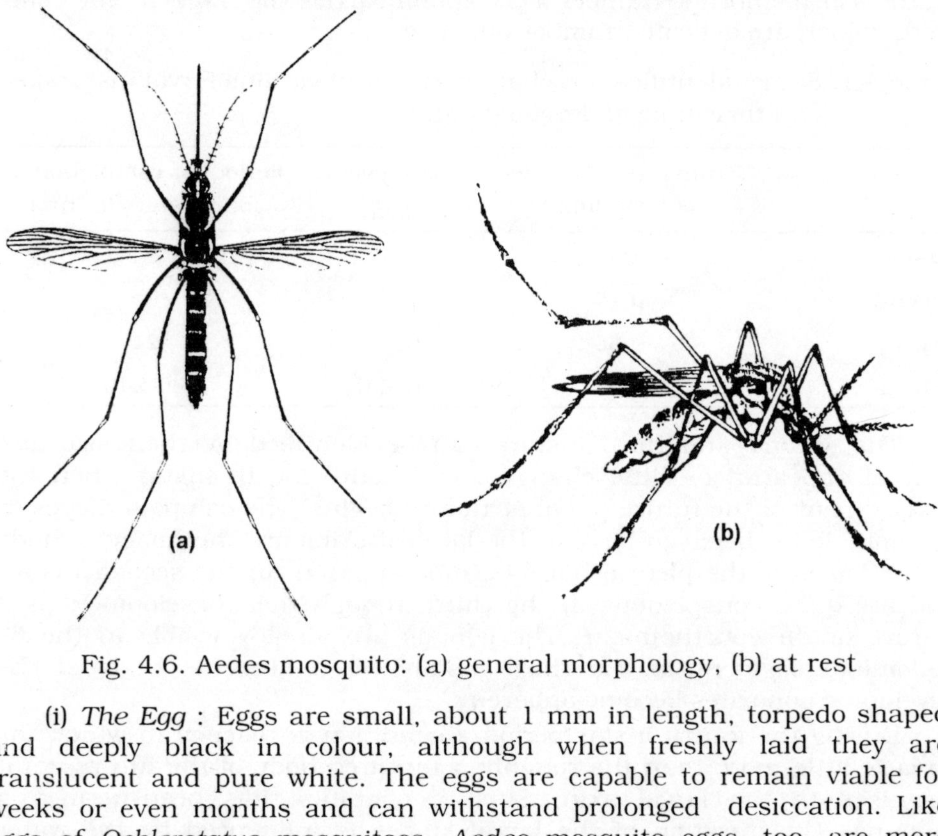

Fig. 4.6. Aedes mosquito: (a) general morphology, (b) at rest

(i) *The Egg* : Eggs are small, about 1 mm in length, torpedo shaped and deeply black in colour, although when freshly laid they are translucent and pure white. The eggs are capable to remain viable for weeks or even months and can withstand prolonged desiccation. Like eggs of *Ochlerotatus* mosquitoes, *Aedes* mosquito eggs, too, are more resistant to drying out (in fact, some complete drying out before the eggs will hatch) and hatch only when flooded with water (salt water high tides, irrigated pastures, treeholes flooded by rains, flooded stream bottoms). Eggs house fully developed larvae ready to hatch. The head of the larva occupies nearly one-third of the egg-volume, while the siphon and other terminal parts fill most of the remaining posterior pole. The egg has very characteristic polygonal markings, lined with a network of milky white membranous channels filled with air at the base of the exochorion. Beneath exochorion is a dense, tough and hard endochorion membrane which gives the egg its shape and rigidity.

(ii) *The Larva*: As in all other mosquitoes, there are four larval instars, each instar ending in a moult or ecdysis. The first instar larva hatches out from the egg by first widening the transverse crack at the dorsal side anteriorly. Finally the larva breaks the eggs by means of a thorne-shaped egg breaker. Initially the freshly hatched larva is unable to move but shortly the young larva wriggles its way to the water surface to make the contact with the free air. It takes about 2 –3 days at 27-29°C for the larva to hatch out from the egg. However, the larva may take few minutes to several hours to emerge at the surface. The first instar larva is identifiable on the basis of the presence of the egg breaker on the head. The larva carries a characteristic siphon. The pectin spine arising on the soft membrane number 4-5. Siphon carries the trachea. The comb teeth, which are delicate, number only five.

Table 4.1. Some identification characteristics of various larval instars as well as their time of development.

Larval instar	Transverse diameter of head (mm)	No. of pectin teeth	Period for completion of four instars (in hrs)
First	0.3	6	24
Second	0.45	7-10	24
Third	0.65	12-15	24
Fourth	0.95	12-20	96-240

The second and third instars can be identified by their size and general appearance. Little change occurs until fourth instar when the development of the future pupal structure begins. The compound eyes or the optic lobes begin to form in the late third instar. The imaginal buds at the bases of the pleural hairs become apparent in the second instar, and are quite conspicuous in the third, though their development is a feature of the fourth instar. The gonads are already visible in the 6th abdominal segment in the third instar, when also the buds of the developing gonoxites become apparent.

During the fourth instar period a condition is reached in which the larva is little more than the roughly sculptured body of the future adult disguised in the larval skin. Such a condition has been termed a prepupa. The duration of the larval stages is influenced by the water temperature. The most common period lies between 6 and 8 days.

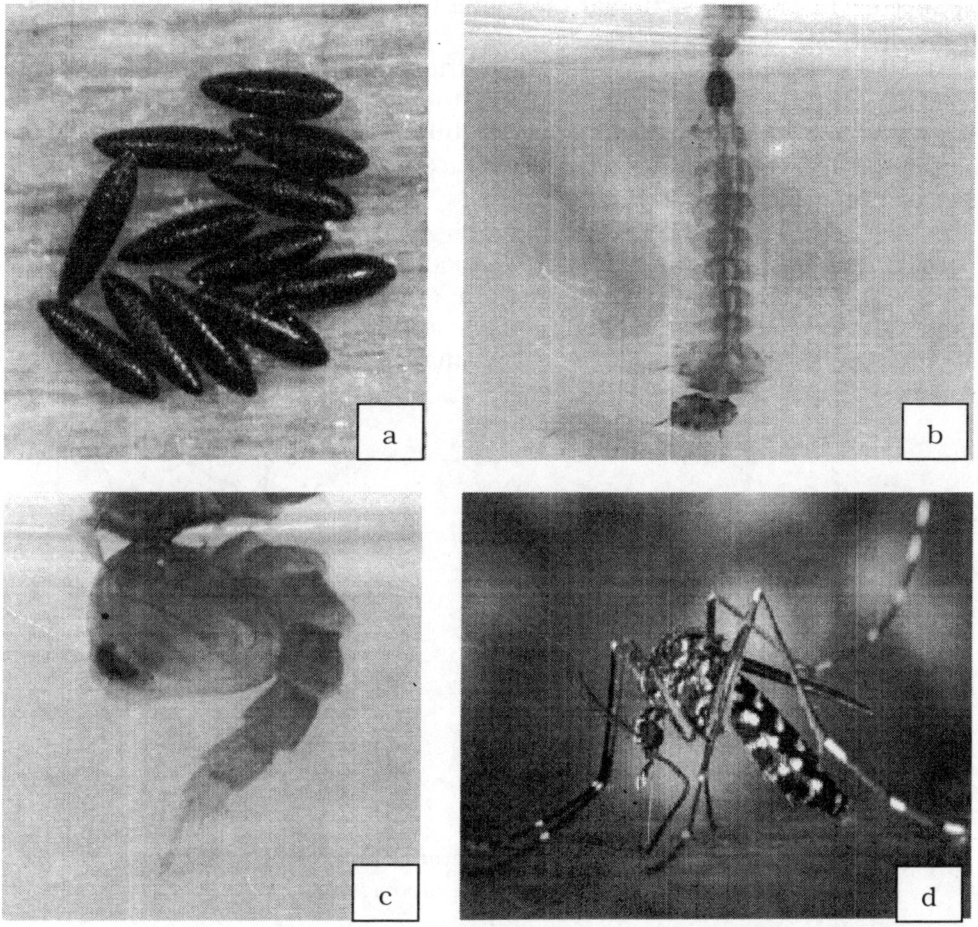

Fig. 4.7. Different developmental stages of an aedine mosquito, *Aedes albopictus*: (a) Eggs, (b) Larva, (c) Pupa, and (d) Adult female.

(iii) *The Pupa*: The pupa of *Ae. Aegypti* resembles most of other culicines. However, it differs from the anopeline pupa in having more rounded body and cylindrical breathing trumpets. The main characters used to identify the pupa of *Ae. Aegypti* from other culicine pupae are: (i) form of the respiratory trumpets, (ii) arrangement and characters of the hairs, particularly those in the abdominal region, and (iii) characters of the tail fins or paddles. The pupal body consists of a large globular anterior region, the cephalothorax, and a narrower abdomen which is normally kept flexed under the former to propel the insect when swimming.

(iv) *The Adult:* It is a small dark structure easily identifiable by the lyre-shaped, silvery white lines on the thorax and the white bands on the tarsal segments. The adult is both vector as well as pest when present in large numbers and therefore deserve special attention in medical entomology. This is a thoroughly domesticated mosquito and breeds almost exclusively in artificial containers in and around human habitations. The females lay their eggs singly on the water just at the margin, or on the sides of the container above the water line. Given the opportunity, the adult prefers human blood to that of other animals. Adults readily enter houses to find suitable hosts. The adult quietly hails to settle for feeding, preferring to bite around the ankles, under shirt-sleeves, or in the back of the neck.

Like *Ochlerotatus*, the *Aedes* mosquitoes are also painful and persistent biters. They search for a blood meal early in the morning, at dusk (crepuscular feeders) and into the evening. Some are diurnal (day-time biters) especially on cloudy days and in shaded areas. They are strong flyers and are known to fly many miles from their breeding sources.

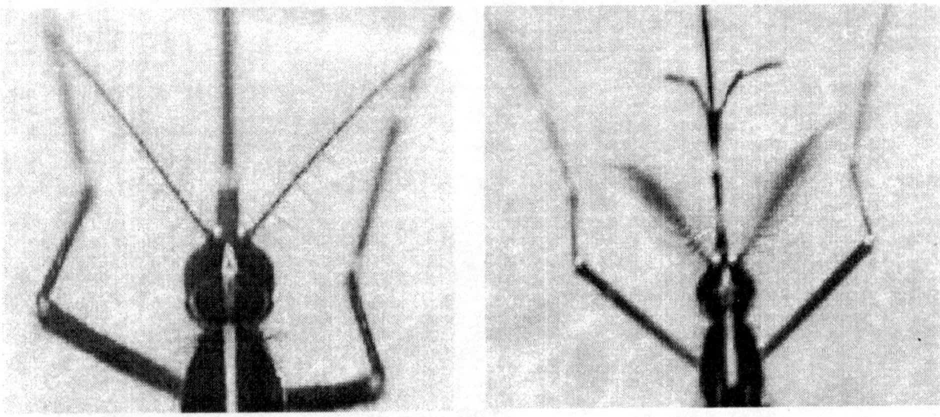

Fig 4.8. Mouthparts and antennal differences between female and male mosquitoes of *Ae. albopictus*.

The adult of *Ae. aegypti* resembles markedly to that of *Ae. albopictus*, commonly called as the Asian tiger mosquito, another mosquito of great medical significance, except that the lyre-shaped markings on the thorax are replaced by a stripe of silvery-white scales on the dorsum of thorax. *Aedes albopictus* is a forest dwelling species and has adequately adapted itself to urban environments as well. A native of Asia, it has spread to most tropical parts of the globe. In Kerala State, southern India, it breeds abundantly in the latex collecting cups associated with rubber plantations, besides cocoa and coconut shells. It has the nasty habit of displacing other species, including *Ae. aegypti*, from their breeding strongholds and has fast adapted to breed in all types of discarded containers, including tyres and littered plastic tea-cups (Hiriyan *et al.*, 2003).

(3) *Culex* mosquitoes

The genus *Culex comprises* several medically important species, such as *Cx. pipiens, Cx. quinquefasciatus, Cx. tarsalis, Cx. nigripalpus, Cx. tritaeniorhynchus, Cx. bitaeniorhynchus, Cx. pseudovishnui, Cx. restans* and *Cx. salinarius*. In Indian context, of greatest importance are

Cx. quinquefasciatus and *Cx. tritaeniorhynchus* which, besides being serious pests, are in the first place responsible for transmitting bancroftian filariasis and Japanese encephalitis, respectively. These are brown coloured mosquitoes of medium size (Fig. 4.9 and 4.10).

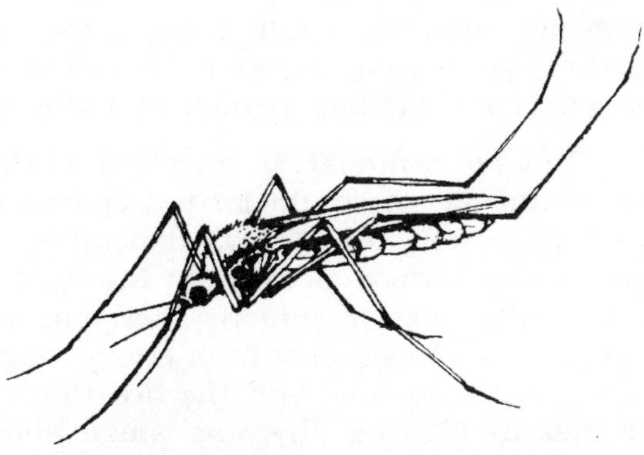

Fig. 4.9. A culex mosquito, at rest.

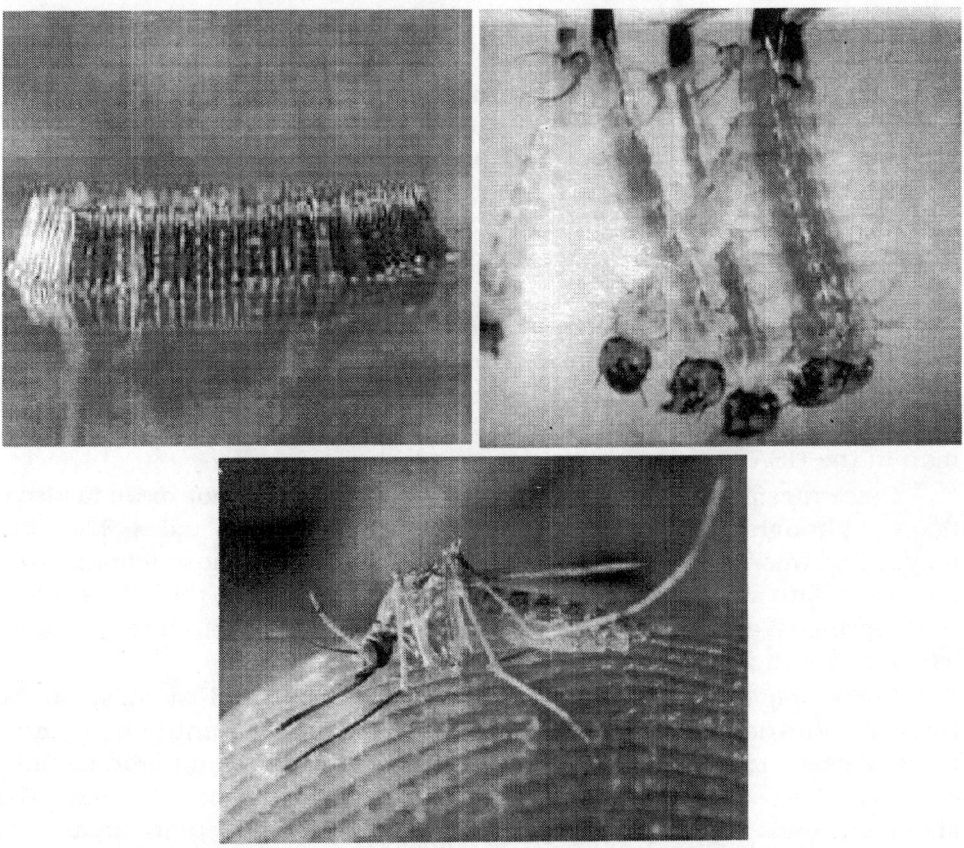

Fig. 4.10. Different developmental stages of a culicine mosquito (pupa not shown): (a) Eggs, (b) Larvae and (c) Adult female.

(i) *The Egg*: Many mosquitoes, such as *Culex quinquefasciatus*, lay their eggs on the surface of fresh or stagnant water. *Culex* mosquitoes usually lay their eggs at night over a period of time, sticking them together to form a raft of 100 to 300 eggs. A raft of eggs look like a speck of soot floating on the water and is about 1/4 inch long and 1/8 inch wide. A female mosquito may lay a raft of eggs every third night during its life span. Like the eggs of *Anopheles* and *Mansonia* mosquitoes, *Culex* eggs too are more susceptible to long periods of drying out.

(ii) *The Larva*: Tiny mosquito larvae emerge from the eggs within 24-48 hours almost in unison. Mosquito larvae, commonly called "wriggglers", live in water from four to 14 days depending on water temperature. They come to the surface at frequent intervals to obtain oxygen through a breathing tube called a siphon. They tend to feed constantly since maturation requires a huge amount of energy and food. They hang upside down at the water surface with the breathing tube up and the brushes by their mouth filtering anything small enough to be eaten toward their mouths to nourish the growing larva. They feed on algae, plankton, fungi and bacteria and other microorganisms.

(iii) *The Pupa*: A typical culicine pupa (cf. *Aedes* mosquito pupa).

(iv) *The Adult*: *Culex quinquefasciatus* is fondly referred to as house mosquito due to its abundant occurrence in houses. It breeds in cesspools, storm-sewer catch basins, street gutters, open septic tanks and effluent from sewage disposal plants. Adults are active only at night. When disturbed, they can be easily spotted flying out of their hideouts in and around houses, outbuildings and other shelters. *Culex tritaeniorhynhus* breeds mainly in rice fields and it is both an exophilic and exophagic mosquito.

Culex mosquitoes are painful and persistent biters also, but prefer to attack at dusk and after dark. They readily enter dwellings for blood meals. Domestic and wild birds usually are preferred over man, cows and horses. *Culex nigripalpus* is known to transmit St. Louis encephalitis to man in the USA.

Culex mosquitoes are generally weak flyers and do not move far from home, although they have been known to fly up to two miles. They live only a few weeks during the warm summer months. Those females that emerge in late summer search for sheltered areas where they "hibernate" until spring. Warm weather brings them out again in search of water, on which to lay their eggs.

Culex pipiens and *Cx. quinquefasciatus* are found in most of the tropical countries. All culicines generally rest with their entire body more or less close to the surface. The scales are narrow and not metallic coloured. There are no distinct patterns or markings on the body. The abdomen ends bluntly. Prespiracular and postspiracular areas are without bristles. Forelegs in males bear toothed claws. In females, the claws are there but simple. The tarsi in females have large pulvilli.

(4) *Mansonia* and *Coquillettidia* mosquitoes

Large sized, yellowish brown to dark brown mosquitoes with very painful bites. *Mansonia* occurs in all zoogeographical regions, while *Coquillettidia* occurs mainly in Afro-tropical region but also in Neartic, Neotropical, Oriental and Australasian regions.

Mansonia mosquitoes are characterized by possessing postspiracular setae, but no prespiracular setae are found. They are in the habit of biting and resting both inside and outside the houses. They breed in the heavily vegetated pond water growing aquatic plants such as *Pistia* spp. etc. Larvae respire on air-conducting parts of the aquatic plants. *Coquillettidia* mosquitoes have neither pre- or postspiracular setae, but have an overall yellow colour. The adult body has a covering of broad scales, which may form fine speckled patterns of colour on the wings. The abdomen is blunt ended. The scales on the wings are very broad. The claws are simple and pulvilli are small or even absent. Some species like *Coquillettidia perturbans* breeds in marshes, ponds and lakes having a thick growth of aquatic vegetation. Larval development is generally slow, lasting several months. Just like that in *Mansonia* species, larvae in this species too breathe by penetrating air conducting plant tissues with their breathing tubes.

They readily enter the house and bite viciously. Some of them, such as *Ma. annulifera* and *Ma. uniformis* are vectors of human lymphatic filariasis caused by *Brugia malayi.* In India, this type of lymphatic filariasis is mostly restricted in Kerala State.

(5) *Haemagogus* mosquitoes

These mosquitoes occur in Neotropical region only where they are of much importance in public heath. Two Indian mosquitoes, *H. discrepans* and *H. tripunctatus*, have been provisionally referred to this tropical American genus, but as no males of Indian species have ever been discovered, their true generic position is still pending settlement. Similarly, a third oriental species, *H. achaetae* is also likewise known from the female (Barraud, 1934). Both Indian species appear to be confined to heavily forested localities, and are rare. The females are day-flying and most probably suck human blood.

Haemagogus mosquitoes are identifiable due to certain specific characteristics. Adults are densely scaly mosquitoes, whose general colouration is black with brilliant silvery markings. Scales on their wings are narrow, but they are fairly broad on the body. The anterior pronotal lobes are large, collar like, closely approximated behind the head and nearly meet at the top. The abdomen of these mosquitoes is blunt ended. Large part of dorsum of thorax without bristles; none present even immediately in front of scutellum. Forelegs of the sexes possess toothed claws. There are no prespiracular or postspiracular areas. Proboscis is slender, distinctly longer than fore femur. The palpi not more than $1/6^{th}$ of length of proboscis; in males these likewise are shorter than the proboscis.

(6) *Sabethes* mosquitoes

Members of this genus occur only in Neotropical region. They are identifiable due to scutum and abdomen being covered with scales having a metallic colour. The scales of wings are narrow, but on the body they are fairly broad. The abdomen is blunt ended. The anterior pronotal lobes are very large. The mid or hindlegs have a conspicuous area of long scales in the shape of a paddle. Prespiracular bristles are present but there are no postspiracular bristles. There is a group of bristles on the postnotum. The palps of males are shorter than the proboscis.

(7) *Psorophora* mosquitoes

This genus has several species which are of immense public health importance largely due to their nuisance value particularly when in large numbers. Many species like *P. ciliata* and *P. columbaie* are large sized with shaggy legs. The scales on body and wings are narrow and without metallic colour. Prespiracular and postspiracular bristles are present. The abdomen is pointed at the tip. Both sexes have forelegs with toothed claws. The male palps are longer than the proboscis. They are vicious biter and because of their gigantic appearance they often offer a frightening look. Some species do not lay eggs directly in water but on the ground that is subject to flooding from rainfall, overflow or irrigation. They breed in temporary pools, irrigation ditches and seepage pools, and larvae survive predating on other aquatic insects including mosquito larvae. Members of this genus occur only in the Neartic and Neotropical regions. Like *Coquillittidia* and *Mansonia* mosquitoes, the *Psorophora* mosquitoes too are becoming more pestiferous as an ever-expanding human population invades their natural habitats.

(8) *Culiseta* mosquitoes

Found in all zoogeographical regions in the world, they are identifiable due to presence of narrow scales on the body and wings which are non-metallic in colour; there are, however, coloured patterns on the body. The abdomen is blunt ended. Prespiracular bristles alone are present. Forelegs of males have toothed claws; in female's forelegs they are simple. The stem vein (base of subcostal vein) of the wing usually has bristles on the ventral surface.

Most species are moderate to highly aggressive biters, attacking in the evening hours or in the shade during the day.

(9) *Armigeres* mosquitoes

This is a very painful biter, both in day and at dusk, and occurs mainly in the Oriental region, but a few species are found in Australian region and one as far as Japan. A medium-sized to rather large-sized mosquito it tends to seek blood by attacking and obstinate feeding. Mosquitoes of this genus are not overtly ornamental and resemble in general appearance and structure those of *Aedes* (Subgenus *Stegomyia*).

Scales of head broad and flat, a few upright scales on nape, at least in some species, scutellar scales are all broad and flat. Proboscis rather stout, slightly curved downwards towards tip, the curved part somewhat laterally flattened. Palpi of male as long as, or longer than, proboscis, of 3-segments, the last two of which are long, slender, turned upwards, and without hair-tufts. Palpi of male of variable length, form about ½ to ¾ of length of proboscis. Antennae of male plumose, the plume-hairs projecting fairly evenly round shaft. Abdomen of female somewhat tapering, segment 8 partially retractile, cerci short. Coxite of male hypopygium with lobe on inner surface bearing spines. Style of variable form, but usually with spines along apical part or at tip. Phallosome divided into small lateral plates with teeth. Paraprocts without teeth or hairs at crown. The only character distinguishing the genus, as a whole, from *Aedes* appears to be the form of the proboscis.

Of the nearly 30 species known under the genus, 13 have so far been recorded in India, of which *Ar. subalbatus* is very common in the Western Ghat ranges in Kerala State. All species chiefly occur in forested areas with high rainfall, and the larvae live mainly in bamboos, coconut shells etc. In Kerala's Western Ghat region where latex producing trees are afforested, *Ar. subalbatus* has often been collected in the latex collecting cup fastened along with the trunk of the tree. Two species, *Ar. subalbatus* and *Ar. flavus*, have been occasionally incriminated with the Japanese encephalitis virus.

CHAPTER - 5

PESTILENT AND PHLEBOTOMIST MOSQUITOES: *THE ULTIMATE VAMPIRE*

Of all the predators that stalk the planet, mosquito has taken more lives than any other organism. Consequently, other than mosquito, no animal on earth has touched so directly and profoundly the lives of so many human beings with respect to their health, economy, social activities and tranquility. Each year some 500 million people are affected by mosquito-borne diseases and nearly 3 million die from these. With a tiny little bite this pest, with the skill of a phlebotomist, transmits more than a hundred diseases, from protozoan (*e.g.*, malaria) and helminthic (*e.g.*, lymphatic filariasis) to viral infections (dengue, yellow fever, several different kinds of encephalitides). There are over 3500 species of mosquitoes but only a small percentage of these phlebotomize and/or carry disease pathogens to humans.

In fact there are more pestilent or nuisance mosquitoes than vectors. Although, in public health arena, a pest mosquito is generally accorded much lesser importance, yet in view of multifarious repercussions associated with the pestilence, its importance can in no way be undermined. A pest mosquito causes harm to human being in several ways, for instance, by attacking for feeding blood, disturbing sleep through attacking for feeding and/or making unpleasant whining sound near ears, causing painful welts or wheals at the site of bite, and, in certain situations like recreation parks near marshy beach areas, by causing disturbance and much loss of tranquility through their efforts to enter into ears, eyes, nose and even mouth fearing induction of some kind of allergy due to their scales and hairs! Considering bare facts that a mosquito weighing 2.5 mg can take blood to the tune of 250% (*i.e.*, approximately 6-7 mg) of its weight and there are over 100 trillion mosquitoes worldwide, then on a given day if all these are allowed to feed on human blood then the mankind loses about 625 x 10^{12} kg of the precious blood to mosquitoes! Realizing further, however, that around 100 mosquito species only incline to take human blood, and that a female will survive enough to take a minimum of four feeds in her lifetime within a period of two weeks, the total loss of human blood amounts to 196 x 10^5 kg in 15 days, 392 x 10^5 kg in 30 days, and an amazingly stupendous quantum of 4604 x 10^5 kg in an year. The significance of this enormous loss of the vital fluid can be further judged from the fact that this volume of blood equals to nearly 92.08 x 10^6 (Ninety two million) mortals at a given point of time. The enormity of this loss is further compounded by our awareness of the fact that the damage-causing process is not only recurrent but is also irreversible. Ironically, all mosquitoes do not like to feed human blood, and in fact less than 1% of all species prefer to essentially drink human blood. A larger proportion of mosquitoes, however, chance upon humans for blood feeding if the predisposed host kind is not available. This phenomenon in a way alludes towards the significance of other animals which provide larger share of blood feeding to mosquitoes. Providen-tially, in nature, realizing all kinds of biotic and abiotic stresses, less than 5% of the eggs laid by a female get to become adults, thereby reducing enormously the overall blood-feeding burden by the mosquito on vertebrate hosts, including man (Fig. 5.1).

HOST-SEEKING BEHAVIOUR IN MOSQUITOES

A female mosquito seeking human blood is faced with a very hostile quest. She may never receive cues to human presence if the cues are masked by a chemical repellant. If she is successful in sensing a host cue, she senses a very small change in the concentrations of carbon dioxide in the air and orients towards it. Upon arrival at a host, the mosquito could face such physical barriers as clothing, or bed netting to impede her ability to blood feed. Other animals have their means of defense as well. For example, elephants may avoid mosquito attack by covering themselves with mud (Edman and Scott, 1987). Even in the absence of physical barriers, a mosquito may be susceptible to the defensive action we may employ (*i.e.*, smashing, shooing, trapping, or zapping). Yet vast numbers of mosquitoes do succeed and, in addition to an irritating bite, leave us with diseases like malaria, dengue, yellow fever etc.

It has been long understood that hematophagy evolved as insects equipped with piercing mouthparts (for plant feeding) shifted their feeding preference from other insects to vertebrate hosts - a switch which provided better nourishment. It is also possible that progress toward hematophagy occurred in small steps - from organic matter and faeces of vertebrates to skin and blood (Waage, 1979). In the family Culicidae, only females have the necessary mouthparts to feed on vertebrate blood.

The mouthparts of a female mosquito are highly modified for the purpose of blood-feeding. The maxillae, for example, are sharply tapered piercing mechanisms, while the labrum forms a food channel, and a complex network of muscles aids in blood sucking. Mouthparts are highly innervated with sensory receptors. Receptors in the head region are responsible for mechanoreception, proprioception and chemoreception. Information from such receptors is sent to the central nervous system of the mosquito, where it is integrated and a behavioural response, such as the fixed action pattern for host-seeking behavior, is triggered. The mosquito, however, must be in the correct physiological state in order to receive and respond to stimuli. Female mosquitoes use sensory receptors to exploit both vertebrate host information and nectar resources for energy in between blood feedings. Male mosquitoes feed only on nectar; their mouthparts are, generally, shorter and weaker than those of the female. Strong female mouthparts are vital in reproduction of mosquito species.

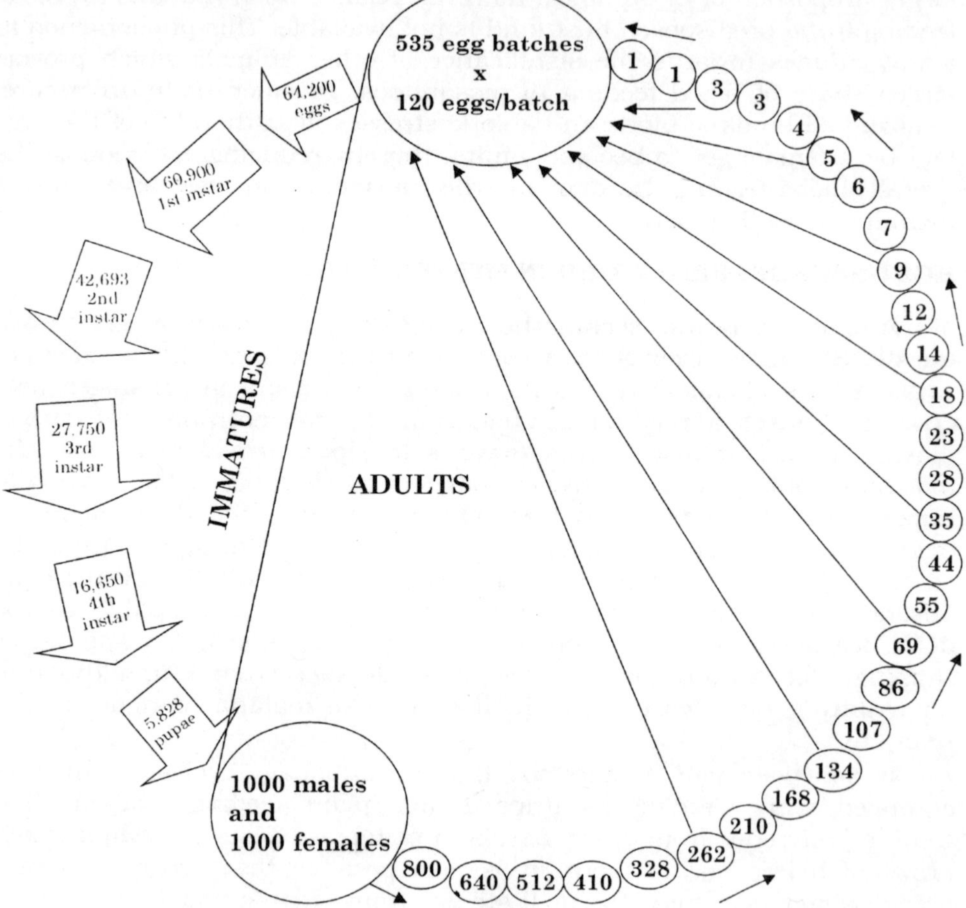

Figure 5.1. Relationship of survival and egg deposition to number of mosquitoes in different development stages.

Mosquitoes exhibit holometabolous metamorphosis and therefore, have a life cycle consisting of egg, larvae, pupa, and adult. Adult female culicids feed according to their state in the gonotrophic cycle. Following oviposition, the mosquito will seek a blood meal. Blood ingestion begins and continues until stretch receptors in the abdomen signal that the female mosquito can not hold anymore. This is referred to as Distention-induced Host-seeking Inhibition. Blood is then digested and egg maturation begins. Host-seeking is then further inhibited in Oocyte-induced Host-seeking Inhibition. Following the oviposition of this batch of eggs, the cycle begins again. Variations in the cycle occur in mosquito species which bloodfeed more than once during egg maturation. Vertebrate blood contains essential proteins necessaiv for eggs to mature. Egg production is dependent upon amino acids and does not require the presence of other constituents in the blood. It has been found that female *Aedes aegypti* fed on an artificial diet high in protein developed equal amounts of eggs to those fed on bovine blood (Kogan, 1990). Feeding is also dependent upon the circadian rhythms of the specific mosquito species.

Circadian rhythms

Circadian rhythms determine the time of day or night that a mosquito will feed. This is of great importance in determining a means of mosquito control; for example, bed netting can be used to prevent blood feeding while prospective hosts sleep. Mosquito behavior is also dictated by a species' tendency to blood feed on humans (anthropophagic) or animals (zoophilic) and whether the mosquito prefers to feed indoors (endophagic) or outdoors (exophagic). *Anopheles gambiae*, a vector of malaria in Africa and *An. culicifacies*, a vector of malaria in india, feed on sleeping hosts inside their homes. Host-seeking behaviour is dependent upon chemical and visual cues specific to a desirable host, as well as their internal species-specific cues.

Carbon dioxide attractant to mosquitoes

Carbon dioxide has long been recognized as a cue for mosquito host location. Odour bursts of human breath contain approximately 4.5% carbon dioxide, while normal atmospheric levels of carbon dioxide are between 0.03-0.04%. As with other wind dispersed odour, a plume of carbon dioxide is distributed by turbulence in the atmosphere. An insect detects the odour as bursts, for instance, *Ae. aegypti* has shown to orient to a plume of odour in a wind tunnel using upwind optomotor anemotaxis (Kennedy, 1939; Carde, 1996). This implies that a mosquito will increase its turning rate to stay in the range of a carbon dioxide plume. In contrast to gypsy moths orienting to pheremone, mosquitoes seek a potentially hostile host. The model of upwind optomotor anemotaxis by gypsy moths may not precisely fit the behavioral pattern of mosquitoes, but gives researchers a good idea of insect orientation mechanisms.

Sensory receptors for carbon dioxide are located in the palps, each containing 3 neurons to detect fluctuations in carbon dioxide levels. These receptors are sensitive enough to detect changes of 0.01% but can not detect changes above 4.0% - a percentage corresponding to that of human breath. An over-abundance of carbon dioxide is understood to have a narcotic effect on mosquitoes. Perhaps the most interesting property of carbon dioxide in this sense is as an enhancer of other cues to host location. L-lactic acid, 1-octen-3-ol, warmth, and other cues in the presence of carbon dioxide are considered stimulants in host-seeking behaviour of the mosquito.

L-lactic acid as a mosquito attractant

L-lactic acid is another chemical cue used by mosquitoes in host-seeking. A positive correlation has been found between the amount of L-lactic acid in test subjects and the attractiveness of the subjects to *Aedes aegypti*, a vector of yellow fever. L-lactic acid, however, is ineffective in the absence of carbon dioxide. The take-off rate, flight activity, landing behaviour, and probing behavior have all been found to be affected in presence of L-lactic acid and carbon dioxide concentrations. All of these behaviours, excluding landing, were generally enhanced by the presence of both substances. The presence of carbon dioxide, however, was proven to be the most important stimulant in eliciting behaviours associated with host-seeking, with L-lactic acid enhancing the behaviours.

Receptors for L-lactic acid are located on the antennae in the grooved peg sensilla and are physically separated from receptors of carbon dioxide. In addition, receptors for L-lactic acid are not stimulated by concentrations of carbon dioxide. The relation of these two stimulants must therefore occur as a higher function of the mosquito central nervous system. Cork (1996) got similar results in studies of 1-octen-3-ol. Combined with carbon dioxide, this compound may be an attractant to mosquitoes, however, the validity of this compound as a mosquito attractant is currently debatable.

Biting site selection

As a part of their host-seeking behaviour, mosquitoes select meticulously a biting site. For zoophilic mosquitoes, host-accessibility is dependent upon physical barriers to blood feeding. Many mosquito species prefer to feed on relatively hairless, or featherless areas of a host. These areas are, conveniently, more vascularized to keep the host warm. Chipmunks and squirrels are more likely to be bitten on the ear (versus the back or feet). As for relatively hairless beings, it seems unlikely that mosquitoes would prefer any one biting site over another. This is not a universal truth since it is proved otherwise. De Jong and Knols (1996) used a test subject inside a bed net, wearing only close-fitting underwear, to determine body regions most likely to be bitten by specific mosquito species. Mosquitoes were released into the bed net, and were recorded if they bit within three minutes. *Aedes aegypti* was found to prefer the shoulder and head region, while *Anopheles gambiae* preferred the feet. *An. gambiae* has also been proven to be more attracted to Limburger cheese than human breath in wind tunnel experiments (De Jong and Knols, 1995). This suggests that bacterial stimuli may also stimulate host-seeking behavior, and is an interesting possible clue to better understanding host-seeking behaviour in mosquitoes. Selection of a biting site can also be further dictated by the presence of a parasite.

Parasite enhancement of host-seeking

Host-seeking behaviour may also be enhanced by the presence of a parasite in the vector's body. To increase their dispersal, pathogens within a mosquito vector may alter the host's behaviour. It is clear that by altering vector behaviour to increase hematophagy, a pathogen will gain increased dispersal. Pathogens can affect host attractiveness, increase probing behaviour, and reduce engorgement time (Edman *et al.*, 1984), as well as alter the vector's circadian rhythms. Filarial worms, for example, develop according to the activity of the vector and are successfully dispersed.

Terminating host-seeking behaviour

After orienting toward a host according to chemical cues and using upwind optomotor anemotaxis, a mosquito has to have cues to land. Arresting stimuli, such as sweat, humidity, heat, colour, and movement are possible factors which lead to the termination of host-seeking, and the beginning of probing behaviour. However, as mentioned above, host-seeking behavior may be prevented by certain stages of the gonotrophic cycle.

Some researches have indicated that the host-seeking inhibitory responses of *Ae. aegypti* result from the neuropeptide known as Head Peptide 1. Increased amounts of this neuropeptide then act to desensitize sensory receptors to inhibit host-seeking behaviour. It has been further determined that nutritional and mating states, and other hormones released during egg maturation control the inhibition of host-seeking behaviour. There is a great deal of research yet to be done on hormonal control of host-seeking behaviour as understanding of these mechanisms can lead to genetic manipulation for vector control.

BLOOD FEEDING BEHAVIOUR OF MOSQUITOES: *A REMARKABLE EVOLUTIONARY FEAT*

In most blood-feeding insect species, the haematophagous behavior is confined to insects in the adult stage, most often females (Beaty & Margquardt, 1996; Knaus et al. 1993). For instance, only adult female mosquitoes take blood meals; they also take flower nectar. The male mosquitoes feed solely on nectar, and the larvae are aquatic filter feeders. In this case, the primary utilization of blood is for developing eggs. On the other hand, blood provides all the nutritional needs for insects such as the kissing bugs, in which both sexes, and nymphs as well as adults, bloodfeed. Insects required several adaptations in order to be able to exploit the rich resource of blood. These adaptations are believed to have evolved through one of two routes (Waage, 1979):

(i) In one route, insects that preyed on other insects and plant-feeding insects, both of which had mouthparts and digestive tracts already preadapted for piercing and consuming tissues, took a short evolutionary hop and shifted their feeding preferences to vertebrate blood. This hop may have occurred when the predaceous insects evolved a preference for vertebrates living near their previous insect prey or when insects preadapted for feeding on nectar from flowers serendipitously punctured the skin of vertebrates and gained the resulting advantages.

(ii) In the second possible route, insects that had no preadaptations for blood feeding may have gradually developed a more intimate association with potential vertebrate hosts. They would have done this by first feeding on the organic matter in nests, then moving on to dung and dead skin particles from the host, and finally consuming live skin and blood.

In the case of either route, blood feeding required morphological adaptations. These included (a) developing morphological adaptations, such as developing mouthparts able to penetrate skin, (b) physiological adaptations, such as the developing proteolytic enzymes for blood digestion, and (c) behavioural adaptations, such as developing the ability to find objects that have blood and distinguish them from objects that do not. At the same time, the mosquitoes have had to deal with unwilling blood donors (such as people) that display defensive behaviour that both discourage mosquitoes from feeding and create a source of significant insect mortality.

Blood-Feeding Techniques of Mosquitoes

Generally, there are two types of blood feeders: vessel feeders and pool feeders. Vessel feeders take blood directly from venules or small veins, while pool feeders take blood from a blood pool resulting from the laceration of blood vessels (James and Harwood, 1969) (Fig. 5.2 see also Fig. 1.5).

The process of a mosquito questing for a blood meal involves a complex, interconnected cascade of behaviours, each probably having its own cadence of cues, be they visual, thermal, or olfactory. The complexity of these questing behaviours may account for the bewildering variations in trapping efficiency noted for certain species of mosquitoes at different times.

The finely toothed maxillae of the fascicle begin sawing into the tissue of the skin with fine back-and-forth movements. A fascicle is "a bundle of muscle or tendon fibers". The fascicle is guided into the skin between the labella. As it goes into the skin the labium folds back like a hairpin and the mosquito shifts its legs closer to its body. When about half of the length of the fascicle has been inserted into the skin, the mosquito begins to draw blood. After the mosquito's abdomen is filled, she straightens her front legs to quickly withdraw the fascicle. The fascicle springs upward and forward out of the wound and is fitted back into the deep groove in the labium. Illustrations given in pages belows show the stylet bundles of mosquitoes piercing, while their sheaths (labium) bend and the stylets come out of the grooves. A stylet is "a long thin pointed instrument".

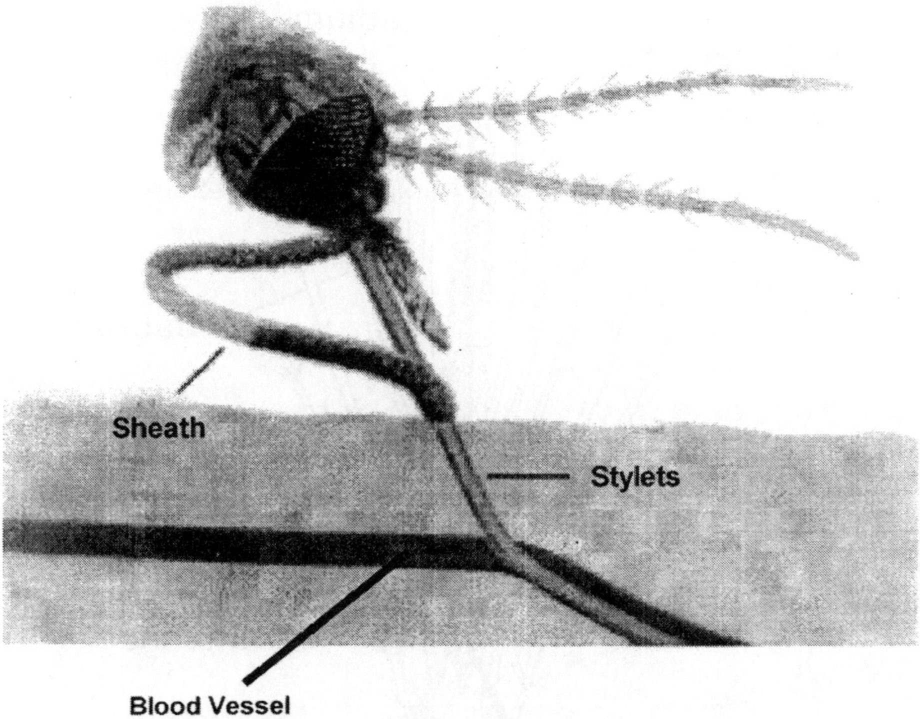

Fig. 5.2. A mosquito probing into the host. It appears that the entire proboscis goes into the host when, in reality, the outer sheath is pushed away from the stylets, and remains entirely above the skin.

To better understand the complexity of what is involved in the act of blood feeding, we need to know something about the anatomy of the mosquito's proboscis and alimentary canal.

The proboscis projects directly in front of the insect. It consists of a large, scaly outer lower lip called the labium, which terminates at the tip in two hairy lobes called the labella. The labium forms a deep trough in which is concealed a smaller bundle of long tapering pale yellow feeding stylets that are collectively called the fascicle. The largest stylet is the sharply pointed labrum, which forms an inverted gutter up which the blood is drawn.

At the side of the labrum are two long, thin mandibles. Under them are two much larger needle-like maxillae with fine saw-toothed tips. Under them in turn is a flattened stylet, the hypopharynx, down the center of which runs a single salivary canal. The fascicle acts as both a piercing mechanism and a food canal. When the mosquito is poised for the moment of piercing, it has all six legs on the surface of the victim's skin, its hairy labella touching the skin and its maxillary palps raised. It is believed that after the mosquito spreads its labella the finely toothed maxillae of the fascicle literally saw their way into the tissue of the skin with extremely rapid back-and-forth movements.

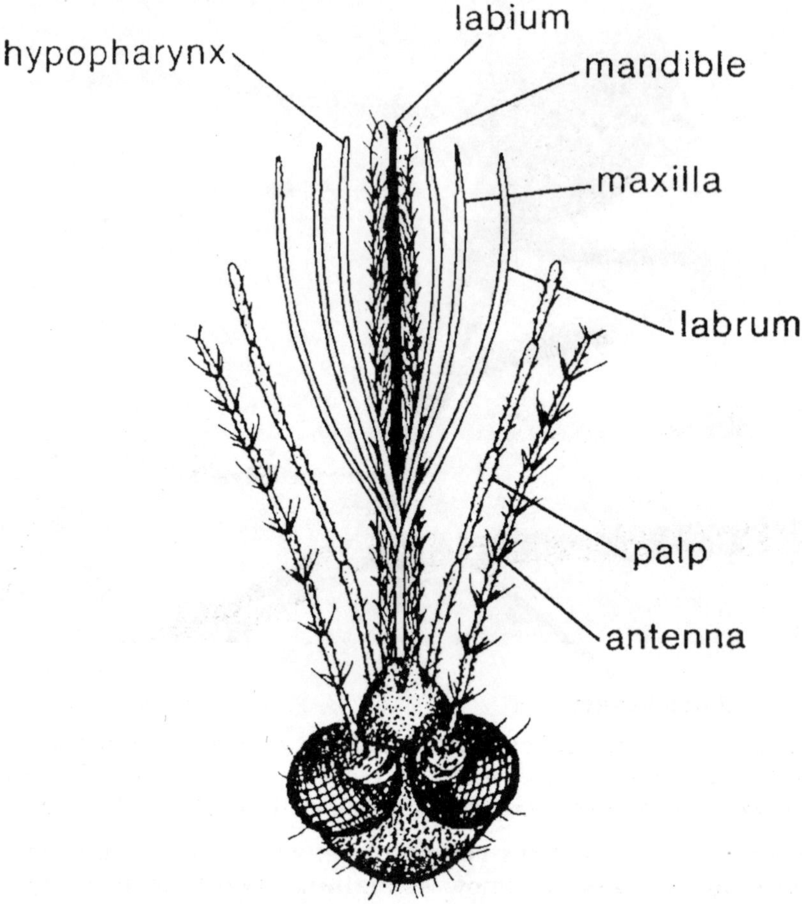

Fig. 5.3. Dorsal view of mouthparts of a female mosquito.

Next, the large labium is folded back like a hairpin. The skin is actually punctured without any obvious thrusting of the mosquito's body. About half the length of the fascicle is inserted into the skin. As it is inserted, the mosquito gradually shifts its legs closer to its body. It takes an average female mosquito about fifty seconds to insert her fascicle into the human skin, and unless she is interrupted (perhaps by being swatted by the victim!) she usually sucks blood for about two and a half minutes.

When the female mosquito has drunk her fill, she withdraws the fascicle rather quickly (in about five seconds). The procedure is to straighten her front legs, leaning slightly backward, and moving the labium from side to side. Once the fascicle has been removed from the skin, it springs upward and forward and then returns to the deep groove of the labium (Fig. 5.4).

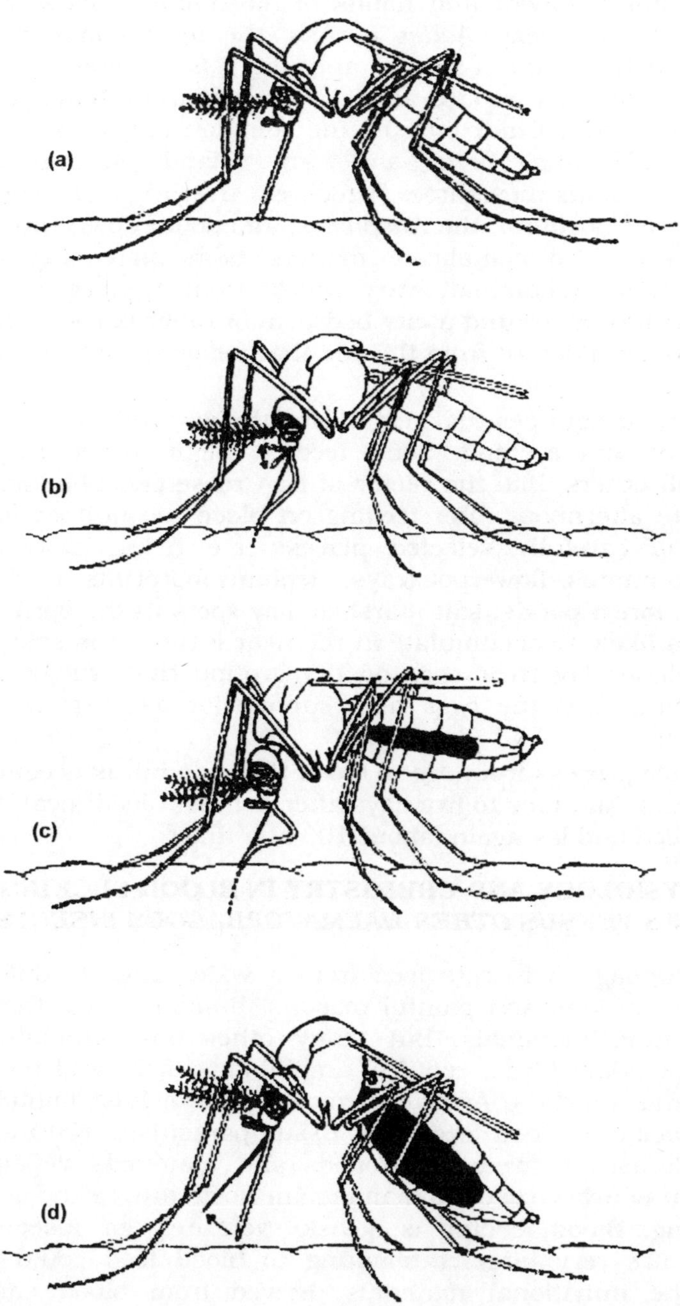

Fig. 5.4. Feeding mechanism of a mosquito: (a) A female mosquito bends and settles to probe, (b) It forces its mouthparts through the skin, (c) The mouthparts are guided deper to suck blood with the support of labial sheath sticking outside host's body, and (d) After engorgement, mouthparts are withdrawn.

The feeding behavior and habits of mosquitoes varies widely from one species to the next. *Aedes vexans*, the most common plague of Americans, both inside country and city, is different from *Aedes excrucians* of northern wooded areas, as well as from *Aedes tormentor* of the southern woods; *Culex pipiens*, the common house mosquito of the tropics; *Coquillettidia perturbans*, an inland swamp mosquito; *Toxorhynchitis rutilus* mosquitoes, which eat the larvae of *Aedes aegypti*, the yellow fever mosquito, the *Anopheles*, and so on down through 3 000 different species. To complicate matters even more, the species of mosquitoes that predominate vary widely from area to area, so that the type that hovers around a city bedroom or other parts of the home is likely to be quite different from the one that follows people around on the golf course or in the park.

There are mosquitoes that rest during the day and emerge in the dim light of dawn and evening. Some feed at night, other during bright daylight; still others, like the vector of La Crosse encephalitis, are most active in late afternoon. After feeding on blood mosquitoes lay eggs in thousands of carefully selected places: tree holes, discarded tires, decorative fountains, flowerpot trays, elephant footprints, pitcher plants, empty cans, forest ponds, salt marshes, low spots in the yard; anywhere that water is likely to accumulate in the near future. It is said that some species of *Aedes* lay their eggs on dry ground that will be subject to future flooding, and the eggs may survive for as long as two years without water.

The whole process from egg to blood-thirsty adult is accomplished in about ten days, and four to five days after her last blood meal, the female is ready to feed and lay again (about 100 at a time).

SALIVA PHYSIOLOGY AND CHEMISTRY IN BLOOD-SUCKING: *MOSQUITOES VERSUS OTHER HAEMATOPHAGOUS INSECTS*

Haematophagous insects feed from a wide range of different host animals. For obvious and painful reasons, humans know that many of them feed from mammals, but many other host animals are also exploited, including birds, reptiles, amphibians, fish, and even insects, arachnids and annelids. Arthropod modulation of host immunity likely evolved to facilitate blood feeding or tissue parasitism. Natural cycles of arboviral diseases between arthropods and mammals depend on the transmission of arboviruses to mammalian hosts during the act of vector blood feeding. Blood feeding is a risky venture for insects because vertebrates are very large; attempting to blood feed could be lethal. However, the nutritional materials derived from blood can support present and future life for the insect, and all this fascinating and yet intricate activity is accomplished by the saliva present in the salivary glands of arthropods, of which mosquito is a member (Ribeiro, 1995). The saliva of haematophagous insects contains a variety of pharmacologically active substances that counteract the normal hemostatic response to injury in vertebrate hosts.

Bioactive salivary compounds exist in the salivary glands of ticks, mosquitoes, sand flies and many other vectors. These compounds have anticoagulant, antihemostatic and immunomodulatory properties that facilitate blood feeding. Blood-feeding insects can suppress host's innate and specific acquired immune defenses. Thus, insect blood feeding should be thought of in a continuum over time instead of as an individual and sometimes painful event. The fact that insect saliva can enhance pathogen transmission leads to the possibility that vaccinating the host against the components of vector saliva will inhibit pathogen transmission. Researches all over the world are now taking into account these critical interactions between insect saliva and blood feeding to develop a more realistic representation of arthropod-borne diseases. This knowledge could lead to useful tools to develop novel control strategies aimed at reducing the impact of arthropod-borne diseases worldwide. An additional benefit of research could come from the medical community's exploration of the use of insect salivary compounds for potential pharmalogical uses (Tyagi, 2003).

Researchers now know that the flying bloodsuckers are not simply flying syringes and that administering pathogens to animals by simple injection does not mimic natural disease transmission. These blood feeders produce bioactive molecules in their salivary glands that are inoculated into host skin via saliva during feeding. The salivary bioactive molecules have antihemostatic and immunodulatory functions. Such functions allow haematophaghous insects to live with and contend with the host immune system. Vector salivary secretions have been shown to modulate host immune responses, and arthropod-borne pathogens may benefit from the effects of vector salivary factors that are primarily involved in blood feeding. These effects may influence the pathogenesis and epidemiology of arthropod-borne diseases.

Because saliva is the only thing an arthropod injects during the blood-feeding process, salivary proteins must be responsible for insects' ability to defeat mammals' blood coagulation response. Paesen *et al.* (2000) reviewed an example of a blood-feeding insect's ability to do this. They studied the tick histamine-binding proteins in tick saliva. Histamine has an important role in inflammation and allergy. Histamine can lower blood pressure, dilate capillaries and increase blood flow.

Another example of an insect's ability to prevent coagulation exists in the saliva of mosquitoes (e.g., *Aedes aegypti*), which have vasodilators in their saliva. These vasodilators are in fact tachykinins, which are peptides that have diverse pharmacological actions in the central nervous system, cardiovascular, and glandular tissues (Ribeiro, 1992). The yellow-fever mosquito, *Ae. aegypti*, secretes an active enzyme called apyrase that inhibits ADP-dependent platelet aggregation. The amino acid sequences obtained match the conceptual translation product of a cDNA clone isolated from an adult female salivary gland library. Sequence comparisons indicate similarities with a ubiquitous family of 5'-nucleotidases. The mosquito protein differs from other members of the family by lacking a carboxyl-terminal hydrophobic domain. The apparent conversion of a gene encoding an enzyme involved in a common metabolic event at the cellular level to a gene involved in the antihemostatic response of mosquitoes illustrates one way this particular insect has adapted to the challenges of bloodfeeding.

Many ectoparasitic arthropods have compounds that help them evade the effects of host grooming, a process that represents a significant threat to ectoparasitic arthropods, especially for long-term blood feeders. For example, salivary glands of the tick, *Ixodes dammini*, contain an enzyme, kininase, which is capable of destroying bradykinin, a mediator involved in the sensation of the itch (Ribeiro *et al.*, 1985). Decreasing host awareness of the insect on the skin surface increases successful parasitism. It is noteworthy that mosquitoes fall under short-term ectoparasitic arthropods.

Anti-aggregating factors, anticoagulants and vasodilator properties

Platelet aggregation, vasodilation and blood coagulation make up the hemostatic processes triggered in host skin that is lacerated by the action of feeding insects. Anti-aggregating factors, vasodilators and anticoagulants have been found in every blood-feeding arthropod that has been examined. Hematophagy appears to have evolved independently in at least 13 different families of arthropods. Thus, the conservation of these antihemostatic activities emphasizes that platelet aggregation, vasoconstriction and coagulation represent significant obstacles in hematophagy.

Anticoagulants are substances that prevent the clotting of blood. Coagulation is desirable in vertebrates when they become injured; it prevents unnecessary blood loss. Thrombin is a blood coagulation enzyme often found in insect saliva. Vasodilators can make blood vessels larger, allowing for more blood flow. Vasodilatory substances have been identified in ticks; nitric oxide-binding proteins are present in triatomine bugs, and novel vasoactive peptides exist in sand flies and mosquitoes (Champagne, 1994).

When a vector probes for blood, tissue is damaged, resulting in the activation of platelets and vasoconstriction. These hemostatic processes occur within seconds and are countered by potent salivary platelet, anti-aggregating factors and vasodilators. Stark and James (1996) used in vitro clotting assays with plasma to examine eight mosquito species for the presence of anticoagulants in salivary gland extracts. Independent parts of the coagulation cascade were assayed in the presence of salivary gland extracts to determine where in the cascade the anticoagulants exerted their inhibitory effects. Results indicated that members of the malaria transmitting *Anopheles* species produced thrombin-directed anticoagulants, while members of the Culicinae, possessed Fxa-directed anticoagulants. This finding suggests that, (i) hematophagy comes before the separation of major mosquito subfamilies and the different activities reflect adaptations to preferred hosts, or (ii) blood feeding evolved independently after the separation of the subfamilies.

Mosquito anticoagulants seem to possess different potencies, too. For instance, the filariasis vector, *Culex quinquefasciatus*, exhibited an anticoagulant activity twice as high as that of *Aedes aegypti.* New World *Culex* mosquitoes are commonly thought of as bird feeders that have expanded there host range to include humans and other vertebrates. The observed differences between the two mosquito species may reflect the fact that bird hemostasis relies more heavily on coagulation than does mammalian hemostasis.

A soft tick called *Ornithodorous moubata* has been shown to have a heat-labile, thrombin-directed anticoagulant activity in the salivary glands, gut, coxal fluid and egg extracts; this activity was present in all developmental stages. When the tick extracts were incubated for 40°C for 15 minutes, the salivary glands still exhibited an anticoagulant activity that prolonged the aPTT, but not the PT, clotting assays. Incubation inhibited FIXa by reconstituting FIX-deficient plasma that had been pre-incubated with the salivary gland extract. Physical and chemical profile analyses showed that the two anticoagulants were different molecules. Additionally, Fxa-directed anticoagulant has been isolated from whole bodies of *O. moubata* and characterized as an anticoagulant called TAP (tick anticoagulant peptide).

Horn fly saliva also targets thrombin or cooagulation action in hemostasis. The horn fly, *Hematobia irritans*, is a troublesome livestock pest because the adult stages of both sexes are aggressive blood feeders. Remarkably, even though horn fly adults feed recurrently on their host as ectoparasites, these flies lack the ADP-responsive antiplatelet aggregation and vasodilator antihemostatic systems described for other blood-feeding Diptera. Horn fly salivary gland extracts do interfere with the normal coagulation process, however, as demonstrated by the recalcification time assay (Cupp *et al.*, 2000). Thus, as compared with other zoophilic Diptera such as black flies, the horn fly inhibits coagulation of blood.

Biting midges (*Culicoides variipennis*) have evolved salivary components that counteract hemostatsis. The host forms a platelet plug within seconds after the wall of small blood vessels is disrupted. Adenosine diphosphate (ADP), however, is a potent agonist of platelet aggregation. An apyrase in biting midge salivary glands degrades ADP to AMP and inorganic phosphate (Leon & Tabachnick, 1996). This increased blood flow resulting from the action of the salivary vasodilator could facilitate blood feeding.

A salivary nitrophorin (nitric-oxide-carrying hemoprotein) helps two blood feeders to feed on their vertebrate hosts. Nitrophorins are haemoproteins found in saliva of bed bugs (*Cimex lecturlarius*) and conenose bugs (*Rhodnius prolixus).* The salivary nitric oxide is confirmed to promote vasodilatation and inhibit platelet aggregation (Valenzuela & Ribeiro, 1998).

The human body louse (*Pediculus humanus*) also has anticoagulants, but the compound and glandular source have not been determined. Mumcuoglu *et al.* (1996) characterized an anticoagulant activity in thoracic extracts containing two kinds of glands, but as yet the specific glandular source of the activity has not been determined.

Immnomodulatory activities of saliva

Immunomodulating substances alter, suppress or strengthen the body's immune system. The sand fly, *Lutzomyia longipalpis*, contains adenosine deaminase, an enzyme, which is ubiquitous in mammalian tissue. People who lack this enzyme become immunosuppressed. In the process of sequencing a subtracted cDNA library from the salivary glands of the sandfly mentioned above. Charlab et al., (2000) found cDNA with similarities to gene products of the adenosine deaminase family. They felt that the possible role for this enzyme is converting adenosine to inosine during blood feeding of the sandfly. This is one highly useful enzyme which following intensive research could lead to the discovery of useful pharmacological materials.

The black fly, *Simulium vittatum*, causes economic losses by reducing beef and milk production in cattle, reducing the efficiency of agricultural and industrial workers, and spreading disease. *Simulium vittatum* is a quick feeding insect that modulates host defenses. The specificity of the antibody response to salivary gland antigens of this fly was investigated by comparing the responses of mice bitten by flies to the responses of mice immunized with fly salivary gland extract. Mice bitten by the fly developed antibodies to fewer salivary gland molecules than did immunized mice (Cross *et al.*, 1993a). This finding indicateed that molecules in the salivary gland extract used for immunization were either absent from the saliva introduced with the bite or were poorly immunogenic in the context of total saliva. Antibodies produced by mice immunized with black fly salivary gland extract were of the IgM, IgG, and IgE classes, which implies that both circulating antibodies and IgE allergic reactions mediate immune response. Both fly-bitten and immunized animals developed IgE antibodies, which contribute to immediate allergic responses to fly bites. The salivary gland of *S. vittatum* is immunosuppressive (Cross *et al.* 1994). Mice inoculated with this extract have fewer spleen cells expressing class II major histocompatibility complex (MHC) surface antigens than do non-inoculated mice. These molecules are important for the presentation of immunogens to T lymphocytes by macrophages and other antigen presenting cells. They found that the same *S. vittatum* salivary extract inhibits the ability of B and T lymphocytes to proliferate *in vitro* in the presence of mitogens. Mitogens are substances, which cause cells, particularly lymphocytes, to undergo cell division.

Additionally, immunization of mice with sheep red blood cells combined with black fly salivary gland extract resulted in a more intense antibody response than immunization with sheep red blood cells alone

(Cross *et al.*, 1994). This suggested that black fly saliva contain factors that can enhance certains immune functions. To investigate how *S. vittatum* might affect the immune response, mice were inoculated repeatedly with salivary gland extract or saline before receiving an injection of ovalbumin. Ovalbumin is an antigen that elicits an immune response in animals with functional T lymphocytes. Results indicated that lymphocytes derived from mice inoculated with the fly salivary extract before injection with ovalbumin produced less of the cytokines interleukin–5 (IL5) and interleukin-10 (IL-10) *in vitro* than did lymphocytes from saline pre-inoculated controls. Preinoculation of mice lymphocytes with salivary gland extract did not affect the ovalbumin-elicited production of three other cytokines: interleukin-2 (IL-2), interleukin-4 (IL-4) and interferon-gamma. All these findings indicate that this quick-feeding fly of medical importance can down-regulate host defenses.

Evidence indicates that saliva protects the parasites and thus augments infectivity of the host. For instance, to induce infection with *Leishmania major* in experimental mice, investigators commonly inject millions of parasites with a syringe. The sand fly, in contrast, creates a *Leishmania major* infection by injecting only about 1-100 parasites. It is a matter of further inquiry as to why the sandly is more efficient at creating an infection compared to the mechanical syringe technique, besides, of course, causes of the low number of parasites surviving the hemostatic processes of the host. Because the sand fly injects the parasite while injecting saliva into the host, it is possible that something in the saliva helps the parasite survive the host defenses. Titus and Ribeiro (1988) attempted to substantiate this by testing mice infected with *L. major* in the presence or absence of sandfly salivary gland lysate to determine whether the presence of saliva enhances the infectivity of *Leishmania major* for mice. Lysates are phage particles released from the host and help fight off non-desirable foreign particles. They found out that in addition to enhancing lesion size, sandfly salivary gland lysate markedly enhanced the parasite burden within the lesions. Sometimes the parasite burden in saliva-treated lesions was 5580 times greater than the parasite burden found in the control or saliva-free lesions.

HOW DO MOSQUITOES DIGEST BLOOD?

Female mosquitoes require the amino acids in blood proteins to mature their eggs. It is during the process of taking blood that the mosquito can transmit diseases such as malaria and dengue fever, which are re-emerging as very significant health problems. It is interesting to know that the female mosquito is not able to digest the proteins at the moment the blood meal is ingested because she has no active trypsin in her gut. Trypsin synthesis requires a series of complex mechanisms following blood feeding. In the first four-six hours following a blood meal, the female gut produces a special type of trypsin called early trypsin. Although only small amounts of early trypsin are made, the activity of early trypsin plays a vital role in the regulation of digestive enzyme synthesis by the gut. Somehow early trypsin, acting on the meal proteins, produces a signal that induces translation of the several proteolytic enzymes required for blood meal protein digestion. The nature of this signal, which is not free amino acids, remains a secret that the mosquito has refused to share despite many attempts to identify it. The synthesis of early trypsin itself is regulated in a unique manner. Transcription of the early trypsin gene starts a few hours after the adult emerges from the pupal state. However, the early trypsin mRNA is stored in the midgut epithelium and remains untranslated until a blood meal is taken. It seems that free amino acids in the blood are the signal for translation, apparently by providing amino acids to charge tRNAs in the gut, which in the absence of a blood meal are uncharged, thereby limiting translation. Thus, early trypsin is part of a fascinating signaling system that conveys the presence of protein in the gut lumen to the epithelial cell. Once the signal is received, the epithelial cells can produce the enzymes required for protein digestion; in the absence of the signal the cell is quiescent. These regulatory mechanisms ensure that proteases are made only when needed.

Given the importance of amino acids for producing the proteins of the egg, one might suspect that the mosquito would be quite efficient in using the amino acids. In fact, the contrary is true. A female mosquito uses only 10-15% of the amino acids to make egg proteins. Some of the rest are used to make egg lipids, and some are used to build energy reserves of the female. But more than 80% of the amino acid carbon is oxidized to CO_2 to provide the energy needed for egg production. Matters related to the partitioning of the meal proteins into various products and how the nutritional state of the female affects these processes are a major subject of research now. It is probable that females who had poor larval nutrition will need more than one blood meal to mature a batch of eggs, because the first or even the second meal is needed to build up the energy reserves effecting egg production. Under such circumstances the possibility of disease transmission increases, so it becomes important to understand how the female manages the blood meal. Because the female oxidizes 80% of the amino acids, she has a very large amount of ammonia to deal with. The female mosquito seems to place the ammonia in proline, which she stores in her hemolymph (blood), until she can convert the ammonia to urea or uric acid for secretion. This is a unique process, still not understood adequately, in mosquitoes and the hemolymph proline concentration can be as high as 80 mM.

CHAPTER - 6

MOSQUITO-BORNE HUMAN DISEASES:
MOSQUITOES BITE MORE THAN ONCE

Mosquitoes are vectors for many deadly and debilitating human diseases such as malaria, yellow fever, dengue, filariasis and various types of encephalitides – just to name a few important ones. These mosquito-borne infections have in recent years been subject to both extensive and intensive research. As a result much more new information, supported by sound molecular interpretations, has been added up on the biology of both vectors and pathogens. Although, a full-scale description of these diseases is outside the scope of this book, nevertheless, a firsthand information about the natural history of these diseases will be most useful, considering the fact that many of these infections are endemically prevalent in India and there these repeatedly surface in epidemic proportions with very high morbidity and mortality, particularly so in the Thar Desert - a "future haven of all vector-borne diseases."

The role of mosquitoes as vectors of pathogens that cause disease is not a simple one. The mosquito/pathogen relationship is a complex co-evolutionary relationship that influences the ability of a pathogen to invade and develop in its definitive host. The biodiversity of mosquitoes, such as the evolution of sibling species complexes, that transmit pathogens responsible for disease are extensive and relatively poorly understood. For example, malaria in the southwest Pacific is transmitted by a closely related group of mosquitoes called the *Anopheles punctulatus* group (*An. punctulatus* and related species) which warrants designing of a control strategy for a correctly identified species among the 'complex'. This group consists of 12 species with identical or overlapping morphology making traditional light microscope identification of species totally inadequate. While in Africa, *An. gambiae* sibling complex has been a problem for decades, in India a similar situation is posed by *An. culicifacies* complex of five sibling species of which only two are efficient vectors. This has been proved beyond doubt that each of the vector species in the complex behaves differently eco-biologically in its various vector-pathogen interaction related characteristics, a subject which has been exploited by scientists for targeting vector populations for effective disease control. In recent times, dengue vectors too have obfuscated the dynamics of disease transmission, and in some places, like Kerala State, *Aedes albopictus* seems to have outraged the potential of the prime vector, *Ae. aegypti*, making the disease control programmes all the more difficult as the former is having a domain in the sylvan and mountainous region not easily accessible.

For efficient control of mosquito-borne disease there must be an understanding of the mosquito biology and disease epidemiology, of which the first step is the accurate identification of the incriminating vector species. This knowledge will then facilitate studies on the ecology and disease transmission potential of important vector species. Thus the implementation of focused mosquito control strategies may proceed without wasting time or money controlling mosquitoes of no medical importance.

1. DENGUE

Dengue, also commonly called as breakbone fever, along with its two clinical rather lethal manifestations, the dengue haemorrhagic fever (DHF) and the dengue shock syndrome (DSS), has emerged as a major public health problem (Gubler, 1998). In fact dengue fever, the commonest amongst all the arthropod-borne viral diseases, is one of the most important and fast emerging disease of the tropical and sub-tropical regions, affecting both urban and rural areas. The primary vector mosquitoes, *Aedes aegypti* and *Ae. albopictus*, have recently spread throughout the tropics and onto susceptible human populations in urban areas. The urbanization process, which has left many without adequate water, sewer systems or waste management, and created new breeding grounds for the vector, has hastened the spread of disease. Dengue is a notifiable disease.

Distribution

Dengue is endemic in 100 countries throughout the Americas, South-east Asia, the Western Pacific islands, Africa and Eastern Mediterranean. Almost 40% of the world's population is at the risk of the disease. Presently, however, an estimated 50 million infections are believed to occur annually worldwide, of which about 40,000 cases are those of DHF. It is, therefore, not very overwhelming to observe that in the South-eastern Asian region dengue (including DHF and DSS), identified for the first time as a new disease in the Philippines in 1953, has by now become a leading cause of hospitalization and deaths among children during last 10-15 years, ranking after only the diarrhoeal disease and acute respiratory infections. The estimated quantum of annual cases in the Region is between 2030 million and DHF cases about 2,00,000. The case fatality rate is higher in the endemic countries, being around 3.5%. The disease has also formed new foci in the hitherto almost a *terra incognito* for the infection, such as Kerala State. In Table 6.1, statistics pertaining to dengue is briefly tabulated.

Table 6.1. Global status of dengue

Characteristic	Quantity
At risk	2.5 – 3 billion people (40% of world's population)
Hospitalized cases	500,000/year (90% of those affected are children)
New infections	50 million annually
Deaths	24, 000 annually
Disease burden	465.000 Disability Adjusted Life Years (DALY)

So far India is concerned, dengue cases, which were confirmed for the first time in 1945 in Kolkata, have since been regularly appearing both in urban and rural areas, with increasing proportion of DHF and DSS (Banerjee, 1994). All the four serotypes of dengue virus have been isolated from different parts in the country. The first epidemic of DHF in India, triggered off by DEN-2 along with Chikungunya virus, took place in 1950s, with about 200 deaths, thousands of hospitalization and hundreds and thousands of cases of the milder disease. The disease was particularly severe in the children under the age of 15 years. During 1964-65, epidemics of dengue with haemorrhagic manifestations occurred in a number of eastern coastal cities like Visakhapatnam, Chennai, Vellore, Pondicherry, Rajamundhry, Kakinada, besides certain interior areas such as Jabalpur in 1966. One of the worst ever epidemics of dengue/DHF in the Indian history occurred between August-November, 1966 which resulted in more than 10,000 cases and 435 deaths in Delhi. Subsequently, epidemics of dengue became more frequent and occurred in varied physiographic situations within the country, such as Delhi (1982, 1991, 1992, 1995), Nasik (1993, 1995), Surat, Shajahanpur and Mandya (1992), Jammu and Kolkata (1993), Rourkela (1994), Bankura, Dharmapuri, Satara and Dhule (1995). In recent times, the year 2003 would be better forgotten for the worst spate of dengue haemorrhagic fever epidemics in several states like Delhi, Maharashtra, Andhra Pradesh, Tamil Nadu, Karnataka, and, of course, Kerala where alone 3377 cases with 66 deaths had been recorded till 10th July.

In Rajasthan State, but outside the Thar Desert, dengue epidemics had been recorded only twice. During an epidemic in Ajmer district in 1969, among 97 cases studied, only two cases of clinical encephalitis were seen. Dengue virus serotype 3 was isolated from cerebrospinal fluid of one, but no shock syndrome was seen. The second epidemic of dengue fever was recorded in Jaipur in 1971. So far as the Thar Desert is concerned, of the two epidemics the latest occurred in 2001 in Jodhpur district, while the former came about in 1985 in Jalore district (Chouhan et al., 1991). The 1985 epidemic in Jalore district was reportedly associated with the dengue haemorrhagic fever (DHF), a total of 110 cases (80 males and 30 females) were reported, of which five had died.

Involvement of central nervous sytem was found in at least 3 patients who were detected to suffer with encephalitis with tonic spasm and delirium. Recently, *Aedes aegypti* was suggested to transmit the infection through the trans-ovarial route (Joshi et al., 1996). The author has found prodigious breeding of *Ae. aegypti* in the "tanka" in Thar rural environs, and correlated this new breeding behaviour of the vector with the possible spread of the vector and the disease in the Thar desert region.

BIOLOGY OF DENGUE VIRUS

Dengue vectors

The dengue virus is transmitted through the bite of various day-biting mosquitoes belonging to subgenus *Stegomyia.* The principal vectors are *Aedes aegypti* and *Ae. albopictus*; the former being a more ubiquitous and dangerous vector mosquito transmitting infection both in the urban and rural environs, whereas the latter plays a significant role in the sylvatic environs.

Pathogen

Four viruses, dengue 1 to 4, classified in an antigenic complex of the flavivirus genus, family flaviviridae, are the aetiological agents of this entity. These spherical agents of 40-50 nm in diameter have a lipid envelope and a positive single stranded RNA. The viral genome of approximately 11 kb in length encodes three structural proteins (capsid, C, membrane protein, M, and envelope glycoprotein, E) and seven non-structural proteins (NS1, NS2a, NS2b, NS3, NS4a, NS4b, and NS5). The 5'and 3'non-coding regions are important for regulating viral replication. The main biological properties of the viruses are located in the E protein, including receptor binding, haemagglutination of erythrocytes, neutralizing antibody induction, and protective immune response (cf. Guzman & Kouri, 2002).

The four serotypes are antigenically very similar to each other but different enough to elicit only partial cross protection, after infection by one of them. Recovering victims are generally immune to future infections, but only from the strain they were infected with earlier. A person can potentially experience all four different strains of dengue. In man each of the serotypes has been associated with classical dengue and dengue haemorrhagic fever. It is generally understood that DEN-2 has a strong correlation with the onset of the dengue shock syndrome (DSS).

Transmission

Dengue viruses are transmitted to humans through the bites of infective female *Aedes* mosquitoes. Mosquitoes generally acquire the virus while feeding on the blood of an infected person. Once infected, a mosquito remains infective for entire life. Infected humans, on the other hand, circulate the virus in their blood. After virus incubation for 8-10 days, an infected mosquito is capable, during probing and blood feeding, of transmitting the virus, to susceptible individuals for the rest of its life. Humans serve as an amplifying host, though some monkeys may also serve as a source of the virus. The virus circulates in the blood of infected humans for 2-7 days, at approximately the same time as they have fever. *Aedes* mosquitoes may acquire the virus when they feed on an individual during this period. Both female and male mosquitoes can also transmit the virus vertically (*i.e.*, transovarially in case females and, by the males when they pass the virus to the females during copulation), particularly by the former, under adverse conditions such as the prolonged dry season and/or, in case of female mosquitoes, lack of availability of suitable host for feeding. The phenomenon helps pass the virus down to the next generation, thereby additionally circulating the virus in nature. The rather complicated transmission cycles of dengue viruses by various vector species in Africa, Asia and the Americas are presented in Figure 6.1.

Forest/Enzootic **Rural/Epidemic** **Urban/Endemic/Epidemic**

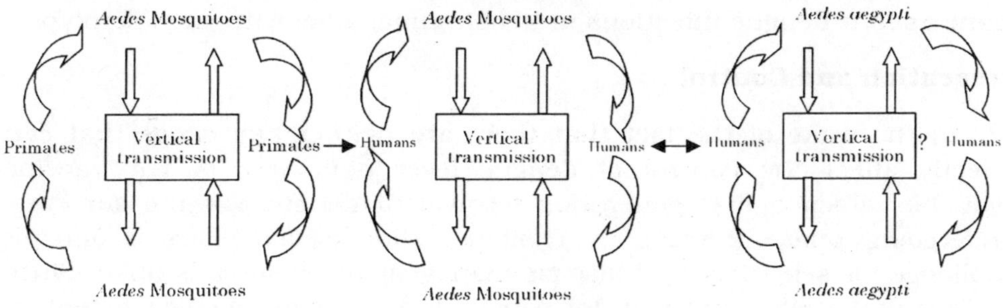

Africa = *Aedes (Diceromyia)* **Ae. aegypti** **Ae. aegypti**
= *Aedes (Stegomyia)* **Ae. albopictus**
Asia = *Aedes (Finlaya)* **Ae. polynesiensis**
= *Aedes (Stegomyia)* **Ae. mediovittatus?**
? Americas = *Aedes* sp.

Figure 6.1. Transmission cycles of dengue viruses.

Symptoms

Dengue fever, a flu-like illness with varying characteristics, is characterized with the onset of excruciating and prolonged pain particularly in joints. It is also characterized by a sudden high fever, severe headache, backache and pain in joints. Dengue, transmitted from person to person as dengue fever/dengue haemorrhagic fever (a more serious and potentially fatal disease), is a spectrum of viral syndromes ranging from non-specific viral fever with headaches, joint and muscular pains and to a severe form of haemorrhagic manifestations with a tendency to develop into dengue shock syndrome (DSS). Infants and youths often experience an undifferentiated febrile disease with rash. Older children may have a mild febrile syndrome but more typically experience the above symptoms of high fever, severe headache, pain behind the eyes, muscle and joint pains and rash.

Dengue haemorrhagic fever is characterized by high fever, haemorrhagic phenomena, enlarged liver and circulatory failure. A sudden onset of fever is the first sign of DHF, accompanied by facial flush and other symptoms of dengue fever. The fever persists for 2-7 days and can reach up to 41°C, followed by febrile convulsions and haemorrhagic phenomena. The patient may recover and symptoms abate. But, if left untreated, the patient may go into shock (DSS or dengue shock syndrome) with a rapid, weak pulse, followed by signs of circulatory failure such as cool, blotchy skin. Without proper treatment, the patient may die within 12-24 hours.

Dengue haemorrhagic fever or dengue shock syndrome occurs with high frequency in children who have experienced a previous dengue infection. The acute phase of dengue infection lasts for 5-7 days and is followed by immune response. Infection with one virus serotype confers lifelong immunity to reinfestation with that serotype. A man can have as many as four dengue infections in his life-time, one with each serotype.

Prevention and Control

In the wake of the fact that there are neither any drugs that can directly affect the control of dengue fever nor there is any vaccine available, albeit a first generation tetravalent dengue vaccine currently undergoing phase 2 trials in Thailand, the disease offers a difficult challenge for scientists and disease management personnels alike. Since there is no specific treatment for dengue fever, a close medical attention and clinical management often saves the lives of many patients. At present, however, the only method of controlling dengue and DHF is to combat the vector mosquitoes through mostly the chemical and environmental management methodologies. Community-based clean-up campaigns remove tires, bottles, cans and other items that catch retain water, eliminating potential breeding sites for vector mosquitoes. Larval habitats are also treated with insecticides. Insecticidal space sprays, using vehicle-mounted or portable machines, are used in many programmes for emergency control of adult mosquitoes.

2. MALARIA

Malaria is the most important mosquito-transmitted human disease throughout the world. Among all the vector-borne diseases, malaria continues to be one of the biggest contributors to disease burdens in terms of death and suffering, and from the public health point of view it is undoubtedly the most devastating parasitic disease in the tropics. By undermining the health and capacity to work of hundreds of millions of people, it is closely linked to poverty and contributes significantly to stunting social and economic development.

Malaria is defined as a febrile disease which is caused by the four species of human malaria parasites belonging to the genus *Plasmodium* and naturally transmitted to people by the bite of an infected female mosquito of the genus Anopheles. Early symptoms of malaria include fever, shivering, aches and pains in the joints and headache. In *P. falciparum* malaria, infected red cells can obstruct the blood vessels of the brain, causing cerebral malaria, which is often lethal. Other vital organs can also be damaged, with fatal consequences. The various types of malaria are acute or chronic diseases.

About four decades ago there were hopes that malaria could be eradicated in a fairly short time, mainly through the intensive use of insecticides and antimalarials. However, malaria still remains a serious threat in endemic countries. Those at greater risk of dying from the disease are children under age 5 in malaria-endemic areas, pregnant women, people moving from non-malarious to malarious zones for reasons of work, migration, refuge, war or tourism, and travelers who visit endemic countries and return home with the disease. In many parts of the world, malaria is becoming even a greater problem than it has been before. Epidemics are recurring in areas where transmission has been interrupted, and are generally associated with deteriorating social and economic conditions.

Distribution

Of all the tropical diseases malaria is the most important infection, remaining widespread throughout the tropics but also occurring in many temperate regions. Malaria exists in 100 countries but is mainly confined to poorer tropical areas of Africa, Asia and Latin America. *Plasmodium falciparum* is the main cause of severe clinical malaria and death. More than 40% of the world population is exposed to varying levels of risk for malaria in 91 countries. Of a total world population of approximately 5.3 billion, 2.1 billion (39.6%) live in areas where endemic malaria has been considerably reduced or even eradicated, but where transmission has begun again and in which the situation is now unstable or deteriorating. In some areas, severe malaria problems developed following major ecological or social changes, particularly the agricultural or the exploitation of forests or socio-political interests. Each year 300-500 million cases of malaria, 90% of them in Africa, and between 1.5 and 2.7 million deaths are reported worldwide. Severe malaria morbidity and mortality are linked with areas endemic for *P. falciparum*, the predominant species in Africa. According to a report malaria kills a child somewhere in the world every 30 seconds (Alnwick, 2000), whereas it has been estimated that 800,000 children die each year of malaria in Africa alone. Mortality is lower but nevertheless important in the Americas and Southeast Asia, especially where *P. falciparum* drug resistance is widespread. There occur approximately 3.54 million cases in the Southeast Asia region every year. A succinct account of various statistics about malaria worldwide is presented in Table 6.2.

Table 6.2. Global status of malaria

Characteristic	Quantity
At risk	Over 40% of the global population (>2.1 billion people)
Cases	273 million clinical cases annually
Deaths	1.09 million annually (due to malaria alone or in combination with other diseases – most deaths occur in children under the age of 5 yrs)
Disease burden	44.99 million Disability Adjusted Life Years (DALYs)

As far as India is concerned, malaria situation has always been very grave, particularly before the national-wide long term malaria control agency was established in the early 1950s, exacting a heavy toll of illness and death. India also earned the reputation of being the cradle of the most significant and epoch-making discovery of malaria being transmitted by the anopheline mosquitoes (Ross, 1898a) which stimulated a great amount of scientific interest researching into various aspects of the disease parasitology, epidemiology and control in the country. Before malaria control programme was launched in India in 1953, nearly 100 million people suffered from the disease annually and close to 0.1 million people died every year (Sharma & Mehrotra, 1986; Sharma, 1999; Tyagi, 2002). The latter figure clearly appears to be a gross underestimate since during frequent epidemics so common in the country those days the overall number would have been much higher (Sinton, 1935), while in 1931, the number of deaths indirectly or directly resulted from malaria had exceeded 2 millions. Besides, malaria imposed the national as well as individual development as economic losses of the order of Rs. 2936 million per year were forced on India. Soon after the country-wide national control/eradication programme was launched, malaria cases were reduced to less than 0.1 million in 1965-66, without single death, and it appeared that malaria would be wiped off from India before very long. It is noteworthy here that decline in cases was in case of *P. vivax* and not in *P. falciparum*.

Within the next ten years malaria epidemics were widespread and the disease, recording the highest ever number of malaria cases (6.43 million) by 1978, invaded even those areas which were earlier not known for malaria transmission or were once freed from its scourges. Subsequently, during 1980s and early 1990s, following a series of critical reviews of the malaria eradication programmes, the malaria incidence was brought down to less than 2 million, but soon increased to nearly 3 million. In the mid-1990s, as reported by the National malaria Eradication Programme (NMEP). Ironically, the malaria incidence and deaths reported by NMEP were based on the parasitologically proved cases, thus excluded a large number of asymptomatic carriers and/or those who approached other medicinal systems for cure of disease or even resorted to self-medication. As earlier emphasized by the in-depth evaluation of the Modified Plan of Operation (MPO) (NMEP, 1985), both morbidity and mortality were apparently underestimated. While the NMEP reported 2.5 to 3 million malaria cases and about 1000 malarial deaths annually, the WHO South East Asia Regional Office put the estimates to 15 million malaria cases (nearly 6 times) and 19,500 deaths due to malaria (Sharma, 1999). Therefore, it appears safer to assume that the malaria incidence reported by the Indian national malaria control agency offers at most the disease trend and not its true incidence, knowledge of which, for logical reasons, is inevitably indispensable for effective implementation of malaria control in the country (For details on malaria economics in India, in particular, see Chapter 14).

All states in India are affected with malaria today, whereas the disease has been continuously advancing to form new foci or increasing intensity of prevalence in areas which were either hypoendemic or not harbouring malarial parasites, particularly *P. falciparum*. As glaring examples, the Kalahandi-Koraput districts in Orissa, most of the seven-sister states in the north-east and the currently malariologically towering inferno of the Thar Desert in Rajasthan State (Tyagi, 2002) offer sufficient evidence to this fact.

BIOLOGY OF MALARIAL DISEASE

Causative agent

Malaria parasite is a single-celled protozoan belonging to the genus *Plasmodium*. Four species infect humans by entering the bloodstream.

(1) *P. falciparum* Welch, 1897: It is found throughout tropical Africa, Asia and Latin America.

(2) *P. vivax* (Grassi & Feletti, 1890): This malaria parasite is worldwide and is the most prevalent one amongst all the types of human plasmodia.

(3) *P. malariae* (Laveran, 1881): Although the distribution of this parasite is worldwide, nevertheless it is quite patchy.

(4) *P. ovale* Stephens, 1922: Mainly it occurs in tropical west Africa, but has also been reported from Vietnan. With regard to India, an instance of *P. ovale* infection was reported from the Koraput/Kalahandi area of Orissa.

In India, only *P. vivax* and *P. falciparum* are considered important due either to their high incidence or causing heavy morbidity and mortality. *Plasmodium falciparum* has developed resistance against chloroquine in many countries in South and South-East Asia, Africa and Central and South America.

Mosquito vectors

A little over 60 species are vectors for malaria in the world, some of the most important being *Anopheles gambiae*, *An. atroparvus*, *An. macculippenis*, *An culicifacies* and *An. stephensi*. In India only 9 species (6 primary and 3 secondary vectors) are important from the viewpoint of human malaria transmission (Nagpal & Sharma, 1995, Tyagi, 2002). Table 6.3 below indicates the total anopheline fauna occurring in India during fast one century.

Table 6.3. Summary of anophelines of India (with major inventories so far)

Year	Number of species described			Authors
	India	Rajasthan	Thar Desert	
1912	3	-	-	Anonymous, 1912
1916	3	-	-	Anonymous, 1916
1931	38	4	4	Covell, 1931
1933	43 (excluding 8 varieties)	4	4	Christophers, 1933
1955	38	12	5	Puri, 1955
1960	39	13	5	Puri, 1960
1981	51	-	-	Ramachandra Rao, 1981
1995	58	-	-	Nagpal & Sharma, 1995
2002	-	18	12	Tyagi, 2002

Transmission

The disease is transmitted by the bite of an infective female anopheline mosquito. When inoculated the parasites are transferred from man to man via the vector mosquito. Male mosquitoes do not bite or bloodfeed. To effect the transmission the malaria parasite undergoes two distinct cycles of development: (a) The human cycle (Schizogony): Parasites multiply exponentially, through asexual reproduction, in the liver and then in infected red blood cells. Man represents the 'intermediate host', and (b) the mosquito cycle (Sporogony): Mosquitoes ingest parasites with a bloodmeal, in whose gut the parasites undergo another reproductive phase, through sexual reproduction, before being passed on to another human. Mosquito acts as the 'definitive host'. These are defined in brief as follows:

(1) ***The human cycle***: The asexual cycle of parasites in human is initiated when an infected mosquito bites a person and injects sporozoites. These tiny, motile forms of the parasite circulate in the blood stream for less than an hour before disappearing from the peripheral blood. While some of the sporozoites are able to make it to the liver cells (hepatocytes), many are destroyed through phagocytosis by the white blood cells. In the liver, following 1-2 weeks of development (depending upon a particular malarial parasite species), they develop to form hepatic schizonts. On rupture the schizonts release thousands of very tiny merozoites (a single schizont of *P. falciparum* may result into 40,000 merozoites, much 3-20 times higher than in other species). The human liver offers a safe sanctuary for malaria parasites. The schizonts burst out to release merozoites almost synchronously in *P. falciparum*, without

leaving behind any persistent tissue phase or the exoerythrocytic phase as found in all other species as a result of unsimultaneous rupturing of the hepatic schizonts which later develop into exoerythrocytic schizonts after undergoing pre-erythrocytic schizogony. This phenomenon carries tremendous pathological significance as they release merozoites into the blood stream causing relapses of these infections (in case of *P. vivax* and *P. ovale* these relapses may last up to 2-3 years, while *P. malariae* may persist up to 10-20 years or more!). Relapses in *P. malariae* are said to be due to its persistence in blood rather than in liver. However, once the parasites enter the RBC, they do not reinvade the liver at all. This phase of development of the malarial parasite associated with the liver is called as the 'hepatic phase'.

Again many merozoites become victim to the waiting phagocytes, but still quite many are able to infest the RBC by attaching to specific receptor sites. At this junction, following a very intricate and interesting mechanism involving dissolution of the cell wall of the RBC, the merozoites penetrate the host RBC. Once inside the merozoites develop to transform into various different forms called mainly the trophozoite and schizont. The latter, under pressure of growth and development of enclosed merozoites, burst to liberate the tiny merozoites to infect afresh the RBCs. This phase of development of the malarial parasite associated with the red blood cells is called as the "erothrocytic phase". The cycle is repeated over and over again until it is slowed down by the immune response of the host or the drugs or the death of the host or parasites. Each species of malarial parasite is characterized by a definite duration of the erythrocytic cycle; *P. falciparum*, *P. vivax* and *P. ovale* have 48 hours, while *P. malariae* has 72 hours.

Having undergone erythrocytic schizogony for sometime, some of the merozoites, instead of dividing and further developing into trophozoites/ schizonts, actually become male and female gametocytes - the sexual forms of the parasite which are infective to mosquito. This phase of development of parasite into gametocytes is called as "gametogony". Peripheral blood reveals only the mature gametocytes. Gametogony is completed in four days and, therefore, a person carrying these gametocytes is often referred as the 'carrier' because by bloodfeeding on such a carrier only a mosquito becomes infected.

(2) ***The mosquito cycle***: Also called as sexual cycle, or sporogony, this cycle begins with the ingestion of the gametocytes with the bloodmeal by the vector mosquito. The gametocytes continue further development in the body of mosquito. The gametogony, with formation of gametocytes, starts about 10 days, on an average, after the malarial paroxysm has started in man. The male and female gametocytes are called microgametocyte and macrogametocyte, respectively. The presence of at least 12 gametocytes, i.e., both mature male and female sexual parasites, with more macrogametocytes than their counterparts, per cubic millimeter of blood is absolutely imperative for further development of the parasite's sexual cycle in the mosquitoes. The first event to take place in the gut of the mosquito is 'exflagellation' of the microgame-

tocytes, whereby 4-8 thread-like processes called microgametes are formed. This phenomenon of exflagellation can also be observed outside the mosquito host in a moist chamber preparation of peripheral blood. The macrogametocyte, on the other hand, undergoes a simple process of maturation to transform into a female gamete or macrogamete. Through a process of chemotaxis, microgametes which, after detachment from the parent cell, actively wander around, are attracted towards the female gamete. As a rule, only one microgamete penetrates and resultantly fertilizes the female gamete, forming a zygote. The whole process of development is completed within 20 minutes, depending on ambient temperature etc. The resultant zygote is at first a motionless body but soon acquires a change in its form to become oblong 'ookinete' and begins moving within 18-24 hours. The ookinete dissolves the inner wall of the stomach and penetrates through to reach the outer surface of the stomach where it develops into oocyst. Each oocyst measures about 6-12 μm in diameter containing a single vesicular nucleus and pigment granules of the macrogamete. With further development, oocyst increases to a size of about 60 μm in diameter by the successive nuclear division containing several hundreds or thousands of sickle shaped sporozoites. The number of oocysts formed on a gut may vary from 20-30. When mature, the oocyst burts and liberates sporozoites into the body cavity (haemocoel) of the mosquito. Many of the sporozoites migrate to the salivary glands of the mosquito, and the mosquito is now infective to man. More than one malarial parasite can simultaneously develop in the mosquito body, thereby giving ample possibilities of mixed infections, such as often observed in case of mixed *P. vivax-P. falciparum* infection. The period of time required for the development of the parasite from the gametocyte to sporozoite in the body of the mosquito is about 8-21 days (*P. vivax* 8-10 days at 25°C, *P. ovale* 14-18 days at 25°C, *P. falciparum* 10-12 days at 30°C and *P. malariae* 18-21 days at 30°C), depending of course on favourable conditions of atmospheric temperature and humidity. This period is also referred to as the "extrinsic incubation period".

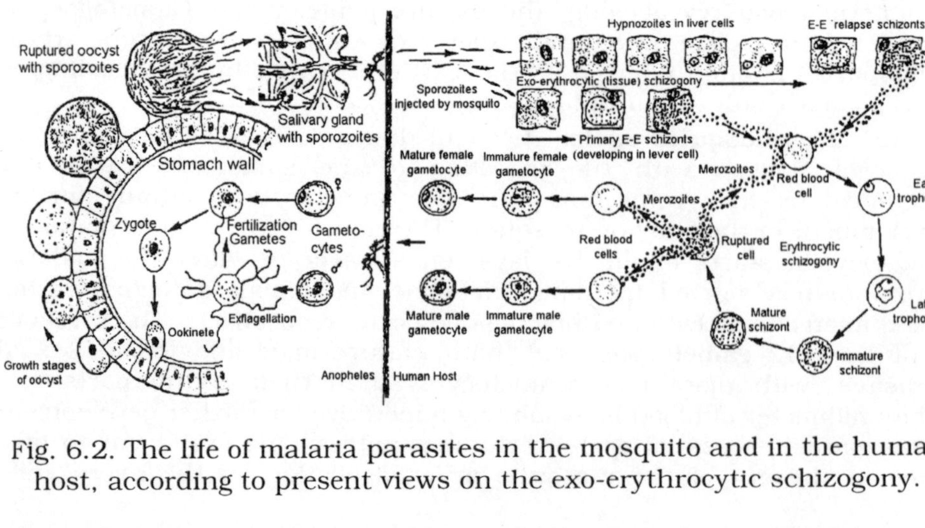

Fig. 6.2. The life of malaria parasites in the mosquito and in the human host, according to present views on the exo-erythrocytic schizogony.

Symptoms

Malaria is a complex infection and, though there are certain clearly identifiable clinical characteristics, it presents a host of features under different situations of infection. In general, nevertheless, the primary malarial fever is marked by paroxysms, corresponding to the development of parasites and their antigenic activities in the RBCs. The fever is highest on those occasions when the broods of merozoites are liberated into the blood stream. Traditionally, a typical malarial attack is to follow a sequence of three distinct stages:

(1) *Cold stage*: Initiation of this stage is marked with lassitude, headache, nausea and chilly sensation followed in an hour or so by rigors. The temperature rises sharply to 39-41°C. During this stage, the patient feels severe headache, associated with vomiting. There are certain integumental changes also, as the skin feels cold first and then hot during the initial stage of the fever. Malarial parasites appear in the peripheral blood and their presence can be demonstrated. The pulse is rapid but weak. This stage persists for 15 minutes to 1 hour.

(2) *Hot stage*: During this stage, as the term means, the patient feels very hot, feeling to doff all his clothes. The skin too feels hot and dry. The patient continues to have excruciating headache, although nausea appears to diminish. The pulse, in contrast to what was felt in the cold stage, is full. The patient has a rapid respiration. This stage lasts for 2-6 hours.

(3) *Sweating stage*: With the onset of this stage, the fever declines fast with profuse sweating, so much so that all inner clothes are virtually drenched. The temperature falls swiftly to normal and the skin is cool and moist under the impact of sweating. The pulse rate becomes slower, and the patient feels relieved and often falls asleep. This stage persists for 2-4 hours.

These stages are followed by an afebrile period during which the patient feels greatly relieved, clinically though. The febrile paroxysms recur with definite intermittent periodicity every 3rd or 4th day, depending on the type of malarial parasite involved in the infection. Based on this feature only the terms tertiary (48 hour cycle of the onset of fever as seen in *P. vivax, P. falciparum* and *P. ovale*) and quartan malaria (72 hour cycle as seen only in *P. malariae*) are assigned to different types of malarial fevers.

Malaria has a tendency to relapse (save for *P. falciparum*) and is characterized by the enlargement of the spleen and secondary infections.

Diagnosis of malaria depends on demonstration of parasites in the blood through preparation of conventional "thick" and "thin" slides and their examination under microscope. In recent times, some other methodologies have also been tried and occasionally brought in mass scale use. One of these and dependable techniques is the simple and rapid dipstick (antigen capture) assay for detection of *P. falciparum*. However, albeit the fact that on trial basis this technique has been successfully tested in some hyperendemic places in India, nevertheless the costs involved seem to be prohibitive here as well as in other developing and/or underdeveloped nations, at present at least!.

PREVENTION AND CONTROL

Malaria is a serious disease which if not treated in time may cause death of the patient. Luckily malaria is treatable, although different countries have slightly varying menu of treating the disease, which is but in due consideration of the type of malaria and malaria parasite they are encountering, *i.e.*, whether the parasite is resistant to certain drugs or drug combinations and at what level, and whether the malaria is endemic or imported *etc.* Although chloroquine continues to be the drug of choice but because of incidence of emergence of drug resistance in malaria parasites in several countries, including India, newer drugs are being employed, generally in combinations. The Chinese drug quinghao-usu containing artemisinine and its various derivatives including artemether have been found to save lives against advanced cases of *P. falciparum* infection. In India, the treatment guidelines are issued by the National Anti-Malaria Propgramme (NAMP). However, malaria can be largely prevented if certain precautions are taken. The vast subject of preventive and control measures in malaria is outside the scope of this book. However, for general understanding of various vector control methodologies available, readers are adviced to see Chapter 12.

3. FILARIASIS

Filariasis (Filum=thread-like) is a group of various different kinds of infections caused by any of the filarial worms, but traditionally the term mainly refers to lymphatic filariasis. Common human filarial infections transmitted through bites of arthropods, or, to be precise, the insects, is tabulated below (Table 6.4).

Distribution

Filariasis is widespread through many tropical and subtropical parts of the world, and, in fact, it has earned a reputation of global problem. Even though the death is rarely associated with lymphatic filariasis infection, nevertheless the disease is a major social and economic scourge in large areas of warmer Africa, Asia, Western Pacific and parts of the Americas. While 1.1 billion people in 80 countries live in areas where there is a risk of contracting infection, over 120 million people living in 73 countries are estimated to be infected with or without clinical manifestations all over the world. In south-eastern Asian region alone, according to an estimate, about 600 million (60%) people in eight countries are constantly at the risk of contracting infection; India being the most endemic and holding 70% of the Asian burden. In India, lymphatic filariasis, particularly bancroftian type of filariasis, is a major public health problem. At present, it is estimated that about 420 million (109 million urban and 311 million rural) people are living in fialriasis endemic areas. While nearly 45 million people are presently detected to have one or more chronic filarial lesions, it is estimated that there occur at least 6 million attacks of acute filarial disease annually. As part of a global effort India is seeking to put lymphatic filariasis on the list of eliminated diseases before 20 years from now, or much sooner, if possible, as declared recently in the second meeting of the Global Alliance for the Elimination of Lymphatic Filariasis (GAELF) held under the aegis of World Health Organization.

Table 6.4. Common filarial infections caused by mosquitoes.

Filarial species	Arthropod-vector type	Type of disease exacted	Zoogeo-graphy	Present in which part of India
*Wuchereria bancrofti**	Mosquito	Lymphatic filariasis	Ubiquitous; mainly tropical and sub-tropical countries	Mostly prevalent along the Coramandel coastline of western peninsular India; sporadic cases prevailing in eastern, central & northern parts of India as well.
*Brugia malayi**	Mosquito	Lymphatic filariasis	Asia	Kerala State in principle.
*B. timori**	Mosquito	Lymphatic filariasis	Asia (Indone sia & nearby islands)	-

*microfilarae are detected in the blood.

BIOLOGY OF LYMPHATIC FILARIASIS WORM

Vectors

Many mosquitoes belonging to all the three major genera (e.g., *Culex quinquefasciatus*, *Cx. pipiens*, *Aedes polynesiensis*, *Anopheles gambiae* and *An. minimus*) transmit the human lymphatic filariasis. In India, it is *Cx. quinquefasciatus* in urban agglomerations whereas, mostly within the State of Kerala, certain *Mansonia* spp. (*Ma. annulifera*, *Ma. uniformis*) transmit the pathogen; the former carrying *W. bancrofti* and the latter species the *Brugia malayi* parasites.

Mansonia spp. have an interesting breeding behaviour. The adult female lays eggs in association with certain hydrophytes. The larvae on emergence take oxygen from the bubbles attached to the submerged parts of the plant. Thus the larvae are very difficult to be controlled, while adults pose different constraints in control programmes. It has been shown in Kerala state that larval control programme with the involvement of community and government machinery can yield very effective results in both vector abatement and the disease reduction. The simple methods followed in reducing larval population in the ponds was release of larvivorous fishes, which were associated with generating income of the peoples involved, and by cleaning the breeding sites of all hydrophytic weeds that supported the larval growth.

Causative agent

Lymphatic filariasis, as occurring in humans, is a parasitic disease caused by the nematodes, *Wuchereria bancrofti*, *Brugia malayi* and *B. timori*. While man is the definitive host, mosquito is the intermediate host of both the bancroftian and brugian types of filariasis. In general, all the three parasites exhibit a similar general plan of life cycles in humans.

Transmission

In man, the definitive host, both male (40 mm) and female (50-100 mm) adult worms are usually found in the lymphatic system. Their life span is expected close to 15 years, or more. Female adults are viviparous and give birth directly to tiny worms, called microfilaria or simply *mf*, to the tune of 50,000 per day. These journey to the blood circulation through the lymphatics. Microfilaria may survive up to one year, but occasionally for more years (Fig. 6.3).

During this time when the microfilaria are circulating in the blood if a vector mosquito bites an infected person, the mosquito picks up the microfilaria along with the bloodmeal. Although the minimum number of *mf* in blood needed to make a mosquito infected is not yet known with any degree of certitude, nevertheless, it is believed, based on investigations, that at least 15 *mf* should be present in each drop of blood to infect the mosquito. Microfilaria undergo several morphological changes in the mosquito body before reaching the final infective stage. First of all, within 1-2 hours of ingestion, the microfilria come out of the sheath which encompases the tiny worm. This process, called exsheathing, takes place in the gut of the mosquito. The exsheathing gives birth to the *first stage larva* of the worm, with a spiky tail, measuring 125-250 μm x 10-15 μm in size. This, in next 6-12 hours, first penetrates through the gut wall and then moves to the thoracic muscles where it grows and develops into sausage-like, short and thick form. The first stage larva moults and increases in length to transform into a little more elongated but thick form, with an alimentary canal formed within, called the *second stage larva* (225-325 μm x 15-30 μm in size). This stage is, however, nearly immotile. The second moulting takes place and the second stage larva assumes the shape of a long and *thin third stage filariform larva* which is subject to availability in any part of the mosquito body. In case of *Wuchereria bancrofti*, this filariform ready-to-be infective stage possesses three distinct caudal papillae. This stage is highly motile and moves to the proboscis of the mosquito from where it is ready to be transported to the human host whenever the infective mosquito has a chance to bloodfeed. The duration of the mosquito cycle or the extrinsic incubation period is between 10 and 14 days, provided optimum ambient conditions particularly temperature and humidity prevail (Fig. 6.4).

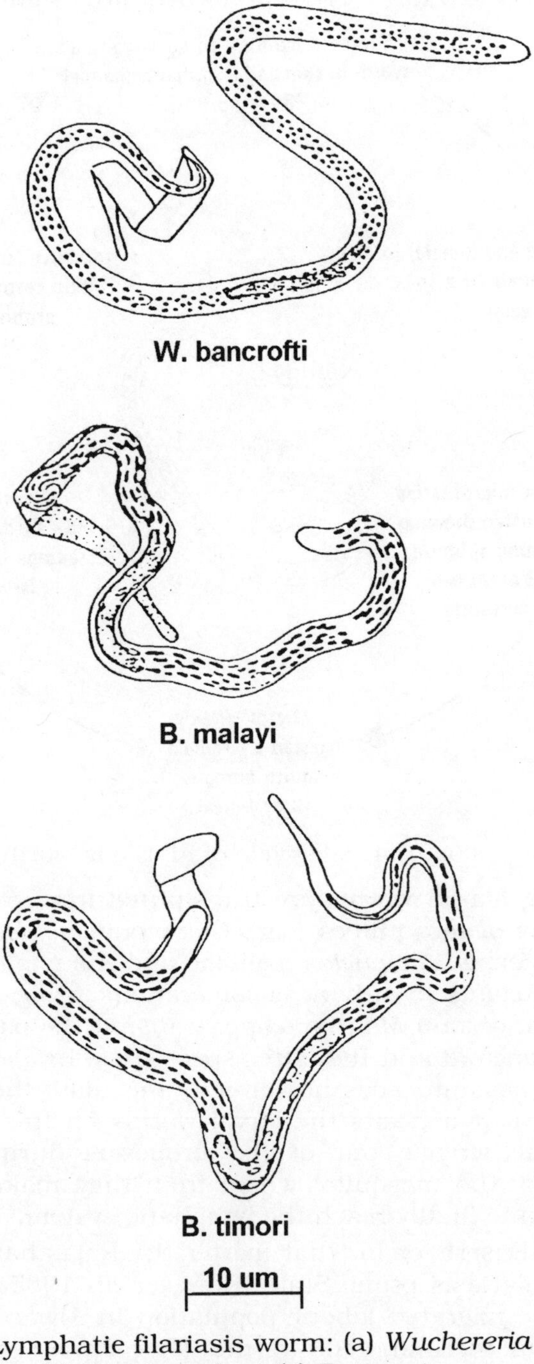

Fig. 6.3. Lymphatie filariasis worm: (a) *Wuchereria bancrofti.* (b) *Brugia malayi,* (c) *B. timori.*

In sharp contrast to the human bancroftian filariasis, which is generally not a zoonosis, not certainly in India, the other form of filariasis, the brugian type, does have zoonosis in animals like monkeys, dogs and cats, which however seem to carry little significance.

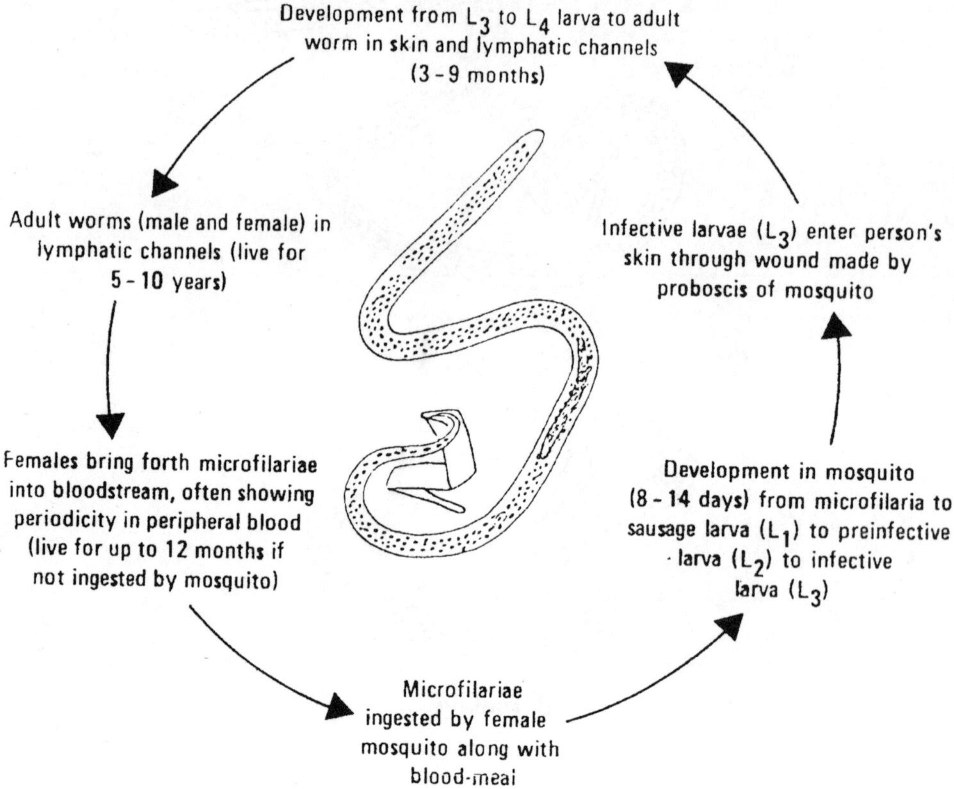

Fig. 6.4. Life cycle of filariasis worm

The young filarial worms are transmitted from person to person by various species of mosquitoes (e.g., *Culex quinquefasciatus*, *Cx. pipiens*, *Aedes polynesiensis*, *Anopheles gambiae* and *An. minimus*). In India, it is *Cx. quinquefasciatus* in urban agglomerations while, mostly within the State of Kerala, certain *Mansonia* spp. transmit the pathogen; the former carrying *W. bancrofti* and the latter species the *Brugia malayi* parasites. The infective mosquito does not directly inoculate the parasite into the host's body, but it deposits the larval worms on the site of the bite (in fact the larvae wriggle out of the proboscis during the process of bloodfeeding by the mosquito) where from they make their way in the punctured skin to finally reach the lymphatic system.

The Thar Desert, or for that matter the Rajasthan State, was never considered a filariasis prone State. However, in 1985, a few cases were detected in the migrated labour population in Alwar district, without a mosquito being incriminated. In the Thar Desert region, the author conducted a short-term entomo-epidemiological investigation in both mobile (such as those in the Indian Army) and residual populations (such as those who did not ever go to hyperendemic States in the country), in the early 1990s, and had found several cases within the Army personnels and their families suffering with the infection of lymphatic filariasis, while the known vector species, *Cx. quinquefasciatus* was also incriminated.

Symptoms

The adult nematode worms live in various parts of the lymphatic system and most people may simply harbour them without manifesting any visible symptoms of infection (asymptomatic carriers) or the filarial worms may cause inflammation as well as other complications of serious nature which in certain forms may be debilitating, such as lymphagenitis, lymphadenitis, elephantiasis of various parts of the body or as a hypersensitivity state like the tropical pulmonary eosinophilia or as an atypical form such as filarial arthritis. The microfilariae, on the other hand, circulate in peripheral blood and are available to infect vector mosquitoes when they bloodfeed. Prolonged or repeated infections may cause extreme enlargement of external genitalia, breasts, or legs in some people and is often referred to as elephantiasis when swollen and elephantoid legs occur. Though not considered as a fatal disease, it leads to a state of great suffering stigmatized by physical deformities and disabilities as well as fears of social disrespect and isolation.

Prevention and control

Diethyl carbamazine citrate (DEC) oral administration, at the rate of 6 mg/kg body weight for 12 days, is advocated to cure the infection, although the drug is basically a microfilaricidal in nature. Another drug, ivermectin, is also said to be quite effective. Vector control is essentially made by larviciding *Cx. quinquefasciatus* immatures, but also adulticiding with residual compounds at their indoor resting sites (see Chapter 12).

CHAPTER - 7

MOSQUITOES AND ZOONOTIC DISEASES OF HUMANS: *BIODIVERSITY AT ITS BEST*

Vertebrate non-human animals often provide a sentinel system for people because many of these serve as natural reservoirs of several disease agents (viral, bacterial, protozoal) that are transmitted to humans by blood-sucking arthropod vectors. When animals contract a disease, it's often a warning to people that the illness is moving into the area, and it's time to take measures to prevent its spread. This is true with insect-borne diseases such as West Nile virus, Lyme disease, Rocky Mountain spotted fever and Ehrlichiosis (a long-recognized canine disease). In dogs, Ehrlichiosis can cause chronic weight loss, paralysis, blindness and brain damage. In humans, it's an emerging zoonotic disease that causes acute respiratory distress and is potentially fatal – but no evidence exists that dogs can transfer it directly to people. What makes zoonotics so fascinating is that each disease has its own dynamics in the host, in the infected individual, in the manner in which it's transmitted, how it's maintained, and how it is prevented. Both people and domestic animals can avoid most zoonotic diseases through good hygiene, elimination of disease-carrying insects, elimination of areas that can harbour bacteria and parasites, and proper veterinary care for animals. Most of these diseases can be successfully treated if diagnosed early. There are more than 200 diseases that can be transferred from animals to humans, and even man's best friend can be unfriendly when it comes to staying healthy. Most of those 200 or so diseases that can be passed to humans are fairly rare and don't occur much in India. More so, only fewer of these are transmitted by mosquitoes, though they can be killers.

Mosquitoes are the carriers, or vectors, of West Nile virus, Eastern Equine Virus, Western Equine Virus, Yellow Fever etc., while ticks carry infections such as Lyme disease, Rocky Mountain spotted fever and Ehrlichiosis. Vectors bite an infected host, then transmit the bacteria or virus to another animal or person that may contract the disease. It can take as long as two years for the tick to complete the cycle that allows the illness to transfer to a new host. Horses can develop Equine Encephalitis and the same mosquitoes can pass it to their owners. Although horses and humans can be victims of mosquito transmitted Encephalitis, horses do not directly transmit the disease to humans.

BIODIVERSITY IN VECTOR-BORNE ZOONOSIS

The importance of species diversity or biodiversity (*i.e.*, the number and relative abundance of different species within an ecological community) in the performance of ecological functions, such as primary production (as in the vector body) and resource extraction (as in the zoonotic animal) is very well understood with reference to ecology of infectious diseases, particularly vector-borne zoonoses - diseases in humans in which the disease agent resides predominantly in a non-human animal host and is transmitted among hosts (including humans) via the bite of a vector, typically an arthropod, such as the mosquito. It is being widely agreed to that greater diversity in the functional roles of species leads to greater efficiency in the use of scarce resources. Similarly, high diversity within an ecological community may interfere with the ability of competitively superior species to become numerically dominant. Thus, diversity begets diversity. On one hand, high species diversity in vertebrate hosts of vectors may play a beneficial role by impeding dominance by particular species that act as key reservoirs of the pathogen, while, on the other, high diversity of vertebrate hosts may provide generalist vectors or pathogens with a hedge against local extinction that would accompany the extirpation of a primary host, and, therefore, may play a role in increasing the disease risk to humans.

Although comprehensive assessments of the degree to which zoonotic vector species are selective of hosts are rare, it appears that extreme specialists are few. For example, some species of mosquitoes feed largely on birds and others largely on mammals, but most or all the mosquitoes feed on at least several species within each vertebrate class. The widespread use of domestic chickens and livestock as sentinels to detect mosquito-borne diseases such as malaria, dengue fever, and various mosquito-borne encephalitis viruses capitalizes on the tendency of vector mosquitoes to feed on both human and non-human hosts. Many pathogens have multiple routes of transmission to vectors, including transovarial, venereal and oral. For some vector-borne zoonoses, especially mosquito-borne viral diseases (Table 7.1), transovarial transmission is relatively efficient, leading to moderate or high infection prevalences in newly hatched larvae produced by infected mothers (Ostfeld & Keesing, 2000).

Table 7.1. Selected zoonotic diseases for which vertical (i.e, transovarial) transmission of the pathogen is highly inefficient (<5%) or more efficient (>5%)

Degree of transmission	Disease	Vector
Vertical transmission <5%	California group encephalitis virus (LaCrosse virus)	*Aedes albopictus, Culex fatigans*
	Chikungunya virus	*Aedes aegypti, Ae. albopictus, Ae. furcifer*
	Dengue fever virus	*Aedes aegypti, Ae. albopictus,*
	Eastern equine encephalitis virus	*Culiseta melanura*
	Murray Valley encephalitis	*Aedes aegypti*
	Rift Valley fever	*Aedes juppi*
	St. Luis encephalitis	*Culex pipiens*
	Vasicular stomatitis	*Aedes albopictus, Culex fatigans, Lutzomyia longipalpus*
Vertical transmission >5%	California group encephalitis virus	*Aedes dorsalis, Ae. melanimon*
		Aedes atlanticus
	(a) Keystone strain	*Aedes atropalpus*
	(b) LaCrosse virus	*Ae. triseriatus, Culiseta inornata*
		Aedes triseriatus.
	(c) Snowshoe hare virus	*Culiseta inornata*
	Jamestown canyon virus	*Aedes provocans, Ae. stimulans*
	Japanese encephalitis virus	*Aedes alcasidi Ae. vexans, Armigeres flavus, Ar. subalbatus, Culex pipiens, Cx. quinquefasciatus, Cx. tritaeniorhynchus*
	Rift valley fever	*Aedes lineatopennis*
	Rio Grande virus	*Lutzomyia anthophora*

1. YELLOW FEVER

The yellow fever virus is a monkey Flavivirus transmitted by mosquitoes. Yellow fever is usually an acute and highly fatal viral disease, although it is occasionally only a mild infection. It appears to have originated in Africa and was carried to other parts of world, including the New World, possibly through slaves smuggled via sea passage. The disease is prevalent in two epidemiological forms, *viz.*, urban and jungle yellow fever. Though the same virus causes both types, yet the mosquito vectors and vertebrate hosts are quite different. *Aedes aegypti* is the most common vector in urban agglomerations, but *Ae. albopictus* is also a potential vector in areas where it has become established. A vaccine is available against Yellow fever, and it is mandatory for all peoples to get vaccinated before entering a yellow fever endemic country. India is so far free from this disease.

DISTRIBUTION

Yellow fever is prevalent only in Africa and the Central and the South America, although its main mosquito vector, *Aedes aegypti*, has a distribution over most parts of the Tropics and the Subtropics. While in Africa yellow fever is prevalent endemically in a total of 29 countries, defined by the presence of neutralizing antibodies, between 15 °N and 10 °S, in the Americas the disease is mostly enzootically present in tropical forests between 10°N and 40°S. Rural populations are at greater risk. Most cases occur among the young adult males who work in forest, thereby coming in contact with the virus through the mosquito bites. Ironically, in Africa these are the children which get often infected with the disease. According to a recent WHO estimate there may be about 200,000 cases and 20,000 deaths annually.

Strangely, though also providentially enough, yellow fever has never been reported from Asia, which is currently reeling under the severity of heavy morbidity and deaths caused by dengue, a closely related infection. It is however known that yellow fever mmunization does not protect against dengue and, therefore, whether the reverse is not the case is yet to be understood.

TRANSMISSION CYCLE OF YELLOW FEVER

Disease Agent

The causative agent for yellow fever is *Flavivirus fibricus*, belonging to the togavirus family. Antigenically the yellow fever virus is close to West Nile virus and dengue virus. The virus is pantropic under natural conditions, but following sustained culture in tissues (e.g., chick embryo), it gives up entirely its pathogenic properties, without losing its antigenicity, though.

Disease vectors

Various mosquito species have been confirmed to play a role in the transmission of yellow fever, such as those belonging to the genera Aedes (e.g., *Ae. africanus*, *Ae. simpsoni*) and Haemagogus.

Mode of transmission

In forest areas the reservoir of infection is mainly maintained in monkeys and mosquitoes, while in urban areas man is the reservoir in forms of clinical and subclinical cases, along with the *Aedes* mosquitoes. The first 3-4 days of illness (the incubation period in man is 3-6 days) are important because it is during this period that the blood is infective and during which time if a vector mosquito bites an infected person then it may also pick up the virus. In mosquitoes, virus develops extrinsically for 8-12 days, to make the mosquito infective. Once infected, the mosquito continues to retain the virus throughout its life.

The transmission cycles, along with vector species involved, vary from Western to Eastern Africa and in South America. There are two types of endemic areas where the virus is maintained: humid forests and emerging zones, and areas where intermediary and urban epidemics occur. On both continents, the virus circulates in forest areas by moving epizootes from one region to another among non-immune monkeys. The virus can also be maintained in an enzootic or epidemic zone by vertical transmission from a female to her offspring. This type of transmission was proven experimentally as early as 1903 by Marchoux and Simond for *Aedes aegypti*, the urban vector of yellow fever, but was observed in nature only in 1995. However, the isolation of the virus in male *Ae. furcifer taylori* in 1977 in Senegal had already suggested that this kind of transmission took place in nature. The vertical transmission of the yellow fever virus means that it can survive, from one rainy season to the next, in *Aedes* eggs, and thus increases the number of contaminating meals and the proportion of females apt to transmit since they infect at a younger age.

Prevention and control of Yellow fever

The live attenuated yellow fever vaccine 17D was found as early as 1936 by M. Theiler of the Rockfeller Foundation. This strain of yellow fever is still the only one used today. The experience acquired with this vaccine has led to various changes in its composition: (i) Use of the seed lot system (in 1941) following the accidents observed in Brazil, (ii) Elimination of human plasma as establishing agent because of hepatitis B transmission (1942), (iii) Preparation of a vaccine free of avian leukemia, (iv) Perfection of a thermostable vaccine (1984). These various successive improvements resulted in one of the most effective vaccines. Over the past years, different ways of improving the vaccine have been envisaged: change of cellular substrate, purifications, development of a new vaccine through genetical engineering. It has been recently suggested that an infected cDNA clone from the 17D strain be used as a yellow fever vaccine or as a gene-vector for other flaviviruses. This most promising approach, however, raises questions notably ones of security.

2. JAPANESE ENCEPHALITIS

Japanese encephalitis belongs to a group of important arthropod-borne viral infections that attack the central nervous system and causes inflammation of the brain.

There are several types of encephalitides, such as the Japanese encephalitis, St. Luis encephalitis etc. which are normally transmitted between animal populations and occasionally to man. In Asia, including India, it is Japanese encephalitis which is a major concern since it affects mostly children below 12 years of age.

It is a zoonotic disease caused by a group B virus belonging to the family Flaviviridae, and transmitted mainly by culicine mosquitoes. The disease acquires its name from the fact that it repeatedly occurred in the form of severe epidemics in Japan between 1873 and 1968. The disease is presently widely spread over particularly in the Southeast Asia with an annual incidence of about 45,000 cases, mostly in children under five years (as in Singapore) or 12 years (as in India). It is fatal in 7-28% cases, while 50% of the survivors tend to develop neurological handicaps. Japanese encephalitis is now a vaccine preventable disease.

Japanese encephalitis, an arbovirus (=**ar**thropod **bo**rne **virus**) infection, which, though like other virus infections, can be transmitted by blood taken for transfusion in the period of viraemia, is nevertheless particularly carried in Southeast Asia by the culicine mosquitoes belonging to the *Culex vishnui* group, particularly *Culex tritaeniorhynchus* breeding in rice fields. The incubation period is 6-8 days and the onset may be acute or gradual, with high fever, rigidity and exaggerated reflexes, headache dizziness, convulsions or even delusions and later coma. Meningeal symptoms may predominate. Many vertebrates including pigs, a sparrow, herons, certain lizards and possibly bats are naturally infected and maintain a continuous chain of infection.

DISTRIBUTION

Japanese encephalitis (JE) is prevalent in most of the Southeast Asian countries. In India Japanese encephalitis was first recorded in 1955, although the first serologically confirmed epidemic of the disease had occurred in West Bengal in 1973 when 763 cases including 325 deaths were reported. Since then the disease had been occurring now and then in many States particularly southern and northeastern States in the country. As far as the Thar Desert is considered, Sri Ganganagar district, which is also the most irrigated desert district and supercedes all other districts of the State in the production of paddy as well, is reported to breed JE vectors in abundance. Even so, Japanese encephalitis has never been reported from the Thar desert, although extensive surveys conducted by the National Insitute of Virology, Pune in India in recent years have indicated presence of specific neutralizing antibodies to JE virus in Rajasthan.

EPIDEMIOLOGICAL CONSIDERATIONS

Pathogenic agent

The Japanese encephalitis virus is antigenically related to yellow fever virus. It measures 40-50 nm, and is enclosed in a lipid layer. The virus is generally isolated from mosquito pools in areas of high transmission.

Human Host

Man is the incidental and dead end of the Japanese encephalitis virus infection, and the age group mostly observed for attack by JE is 3-6 years. The incidence in males has been observed to be higher than in females.

Vertebrate Animal Host

The list of vertebrates hosting the JE virus in nature is long, but the pigs have been incriminated as an important vertebrate amplifying host for the virus. Because of the development of antibodies in the pigs, the infected ones do not generally manifest overt symptoms of illness, although they remain viremic for several days so that the mosquitoes which feed on them become infected. Pigs are called "amplifying hosts", or simply "amplifiers", because they cause to transmit the virus to many others through mosquito bites to maintain the transmission in nature.

The Ardeid birds like herons and egrets are reported to play a significant role in the transmission of the disease in nature. Besides, pigeons and sparrows which develop viremia can also infect the mosquitoes.

In addition to above, vertebrates like bovines, horses, mules and donkeys as well as goats and sheep have also been found to harbour virus and/or antibodies.

Environment

Even though the JE transmission occurs throughout the year, it appears to have an association with the cooler part of the year. Therefore, in several instances the upsurge of JE cases is abruptly noticed following the rainfall when the density of the vector mosquitoes, particularly *Cx. tritaeniorhynchus*, invariably increases manifold.

Transmission

Japanese encephalitis virus naturally propagates in wild or domestic vertebrates and bloodsucking mosquitoes. Man is infected by the bite of an infected mosquito. The disease occurs as focal outbreaks in man under certain sets of ecological situations conducive for transmission. The transmission cycle ends in man due to lower titre of viremia and relative preference of vector mosquitoes for animals over man.

Several mosquitoes belonging particularly to the genus Culex play an important part during the outbreaks of JE, but the most prominent species are *Cx. tritaeniorhynchus, Cx. gelidus, Cx. vishnui and Cx. pseudovishnui.* Apart from Culex mosquitoes, species belonging to Anopheles, Aedes and Mansonia have also been implicated. Altogether about 30 species have been found to harbour JE virus. These vector mosquitoes generally breed in groundwater habitats particularly in rice fields and major components of irrigation systems, ground pools and shallow ditches. The breeding preference of the vector species, *Culex tritaeniorhynchus*, and paddy cultivation have been adequately advocated. They incline to feed extradomestically on the vertebrate hosts, other than on man, and also rest outdoors. The female mosquito usually transmits the infection after an extrinsic incubation period of 9-12 days.

The JE virus has also been found to be transmitted transovarially by some culicide mosquitoes.

The important factors governing the spill-over of the disease to man are the relative abundance of the vectors, the density and absolute number of infected mosquitoes, longevity of vectors, adequate man-mosquito contact and the availability of the amplifying hosts.

The incubation period of JE in human being is 5-15 days. An infected person may remain asymptomatic or develop symptoms of the disease, although in a community where people are at risk of infection the number of asymptomatic carriers is generally very high to that of symptomatic carriers (300:1), that is, for each case of JE detected in the community there must be at least 300 persons suffering from asymptomatic JE infection. Symptomatic carriers may develop meningo-encephalitis. The prodromal stage (i.e., signs before the involvement of the central nervous system) may last for 1-3 days with fever, headache and vomiting, followed by alteration in sensorium and generalized convulsions. Patients with this stage of disease may die by as much as 25%. Of the survivors, as many as 50% may develop serious neurological and psychiatric sequellae including paralysis, ataxia, personality changes, memory loss, aphasia and severe emotional instability. The virus may also infect the foetus transplacentally and in pregnant females may lead to an abortion.

SYMPTOMS AND DIAGNOSIS

The diagnosis of Japanese encephalitis from only clinical consideration is nearly improbable since many other kinds of infections overlapping in the area of JE epidemic may confuse with encephalitis, particularly in regions where many persons have positive blood smears for malaria. Japanese encephalitis can only be diagnosed with any degree of certitude through the laboratory tests. The virus can be detected in autopsied brain specimens by demonstration of the viral antigen by immunoperoxidase techniques or by fluorescent antibody technique in infected mosquitoes. Isolation of the virus can also be carried out from CSF and, only occasionally from, peripheral blood during early stages of illness. Serological diagnosis is generally achieved by the demonstration of JE virus specific IgM in CSF or serum or of rising titres of JE specific antibodies in the serum. The JE virus specific IgM can be detected in CSF or serum of patients as early as 3-4 days following onset of illness, through ELISA test; a single positive test is confirmatory. Likewise there are available some more modern and reliable diagnostic tests for JE virus.

PREVENTION AND CONTROL

No specific treatment is known. Supportive care should be made available to the patients. CDRI, Lucknow (India) has developed 6-MFA, fungal interferon inducer and found it effective in proven JE cases. At present two types of killed vaccines are in use for immunization against Japanese encephalitis: (i) One is prepared from the brain of mice inoculated with the Nakayama JE strain, and (ii) the second type is prepared in primary cell cultures derived from hamster kidney cells. Vaccination is not advisable in persons suffering from ill health. Vector Control is important and is carried out to bring down the density of the vector species with the help of fast acting insecticides.

3. WEST NILE FEVER (WNF)

ETIOLOGY

The WNF virus belongs to the Togaviridae family; it is part of the complex of the encephalitis viruses of San Luis and of Japanese encephalitis. West Nile Fever (WNF) is a mosquito-borne disease that can cause encephalitis, or inflammation of the brain. It is caused by the West Nile virus, named after the West Nile district of Uganda where the virus was first isolated in 1937. The viruses that cause West Nile Fever and St. Louis Encephalitis come from the same family flaviviridae, and cause diseases that are similar to one another. West Nile Virus (WNV) infection may affect horses, humans, dogs and some domestic poultry. The main reservoir for the virus is birds of the crow family - crows, ravens, magpies, blue and gray jays. The WNV is spread from bird to bird, and to other animals, by the bites of mosquitoes. The horse is a dead-end host for WNV. The virus does not spread by person-to-person contact, or directly from birds to persons. A mosquito becomes infected by biting a source of infection, mostly birds, which carries the virus. In past couple of years, West Nile Virus has created havoc in United States, killing several persons. It's invation in USA was sudden, "In 1999, a disease never before seen in the United States began killing birds and then humans in New York City. The virus that causes West Nile encephalitis was carried by that most common of seasonal pests, the mosquito, and suddenly the backyard barbecue took on an ominous image as a hunting ground for mosquitoes."

The infection infrequently occurs sporadically in India, too, mostly along with Japanese encephalitis and/or dengue, but is not considered a serious one.

DISTRIBUTION

The disease has been prevalent in Egypt, Asia, Israel, South Africa, parts of Europe and for some years has also been regularly reported from the Western Hemisphere. Possible explanations for the WN virus spread across the Atlantic sea on to the Americas is the introduction through an infected bird or by mosquitoes that were brought into this area from a region where West Nile Fever is known to occur. The West Nile Virus first appeared in North America in 1999 and has spread steadily across the country over the past five years. In India it is present throughout the mainland.

Transmission cycle

The infection confuses with other closely related infections. The virus has been isolated from several species of mosquitoes (particularly *Culex* spp.), bats and man. The West Nile virus is sympatric with Japanese encephalitis virus and immunization against WN reduces the severity of the JE infection, and *vice versa.* The WN virus is not transmitted from person to person and there is no evidence that a person can get infected by handling live or dead infected birds. But, to add a further level of safety, if birds or other potentially infected animals must be handled, a protective barrier (*e.g.*, gloves, inverted plastic bags) should be used. The WN virus in nature is maintained through a bird-mosquito-bird cycle (Fig. 7.1).

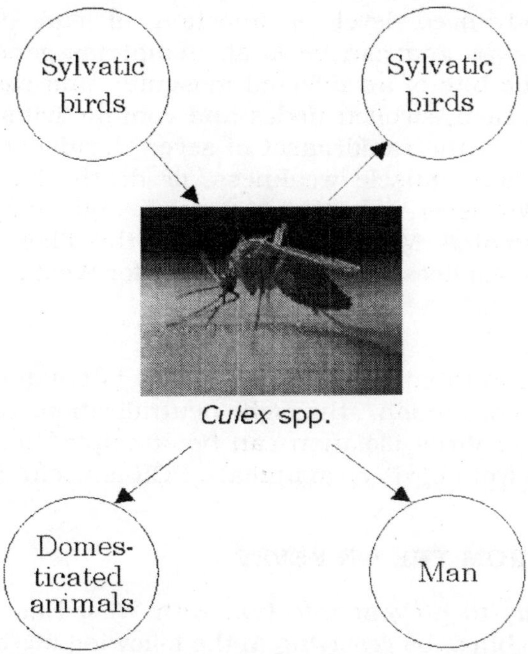

Figure 7.1 Transmission Cycle of west Nile Virus

The WN virus infects a large number of vertebrate hosts, among them human, domestic animals and several species of birds. Only the birds have the condition to act as reservoir, since they have a high and prolonged viraemia. Furthermore, in the areas of viral distribution there is a large number of birds that reproduce with an adequate rate for providing a significant number of susceptible birds, maintaining in this way the infection cycle. The virus has been isolated from pigeons, crows, blue jays, sparrows, chicken, and pheasants. However it does not kill all infected birds. American crows have shown more susceptibility to die. Horses are susceptible to WNV infection and about 40% of equine cases result in death.

SYMPTOMS

The symptoms of West Nile virus infection are very similar to those of St. Louis Encephalitis, Japanese Encephalitis and dengue, but inflammation of lymph nodes is often more marked, and there may be a meningeal syndrome and myocarditis. Less than 1% of infected people develop more severe illness that includes meningitis or encephalitis. Mortality is insignificant. Of the few people that develop encephalitis, a small proportion die but, overall, this is estimated to occur in less than 1 out of 1000 infections. Elderly and immune deficient individuals are more susceptible to die. Most people who are infected have no symptoms or may experience mild illness such as fever and headache before fully recovering. In some individuals, particularly the elderly, West Nile Fever can be a serious disease that affect the central nervous system. At its most serious advanced level of infection, it can cause permanent neurological damage and can be fatal. Symptoms generally occur 5-15 days following the bite of an infected mosquito, and range from a slight fever, headache, rash, swollen nodes and conjunctivitis (irritation of the lining of the eye) to the rapid onset of severe headache, high fever, stiff neck, disorientation, muscle weakness, or death. There is no specific treatment for WN fever, yet the symptoms and complications of the disease can be treated. Most people who get this illness recover from it. As of now, there also does not exist a vaccine for West Nile Fever

DIAGNOSIS

Laboratory confirmation of human cases is obtained upon verifying the serological conversion, through neutralization tests or modified ELISA techniques. Virus isolation can be attempted from mosquitoes or brain specimens from birds or mammals. PCR is useful for viral detection in mosquitoes.

PREVENTION FROM THE WN FEVER

The best way to prevent infection with West Nile Virus is to avoid getting mosquito bites, by resorting to the following methods:

(1) Use insect repellants.

(2) Wear protective clothing outdoors.

(3) Empty water from outside objects such as bird baths, old tires and any containers in which water accumulates and where mosquitoes may breed.

(4) Make all the house doors and windows properly screened.

(5) Dead birds in the area might be an indicator of West Nile Virus (WNV) activity in the area. Many birds are susceptible

to infection. However crows and jays become ill and can die. The public can collaborate in the surveillance reporting sick or dead birds to state or local health departments.

4. OTHER ENCEPHALITIDES

An encephalitis - inflammation of the brain - is most commonly caused by a viral infection, which can be contracted through insect bites, food, drink or skin contact. Once the virus is in the bloodstream it localizes in the brain, which is swamped with white blood cells attempting to fight the infection. This causes the swelling and can lead to nerve and brain damage, and bleeding within the brain. Encephalitides are a group of important arthropod-borne viral infections that attack the central nervous system and causes inflammation of the brain. These diseases are normally transmitted between animal populations and occasionally to man. There are several types of encephalitides, such as the Japanese encephalitis, St. Luis encephalitis, western equine encephalitis (WEE), eastern equine encephalitis (EEE), Sinbis (SIN) virus, Semliki Forest (SF) virus, Barmah Forest (BF) virus, Rocio (ROC) virus, Wesselsbron (WSL) virus, Murray Valley encephalitis (MVE) virus, Omsk haemorhagic fever (OMSK) virus, Louping ill (LI) virus, Powassan (POW) virus, Bunyamwera (BUN) virus *etc.* In Asia, including India, it is Japanese encephalitis which is a major concern since it affects mostly children below 12 years of age (*vide supra*). Some of the most important encephalitides, including the JE and those which are prevalent outside India, are described below:

Table 7.2. Percentage threat-to-life coverage of various kinds of Encephalitides spread by mosquitoes

Type of encephalitis	% Mortality rate
Australian and Californian encephalitis, West Nile fever (along with Colorado tick fever)	10
Western Equine encephalitis	<10
St. Luis encephalitis (under 65 yrs)	15
Japanese encephalitis	30
St. Luis encephalitis (over 65 yrs)	<40
Eastern Equine encephalitis	50-70

GEOGRAPHICAL DISTRIBUTION OF ALPHAVIRUSES

Most of the above viruses are alphaviruses, amongst which EEE/WEE was the first to be found, as a filterable viral agent, in October 1930. Alpha-viruses are typically transmitted by mosquitos, and they have been isolated worldwide; New World, Old World and Oceania as well as from the Tropics to near the Arctic Circle. Their chronological distribution is presented in Fig. 7.2.

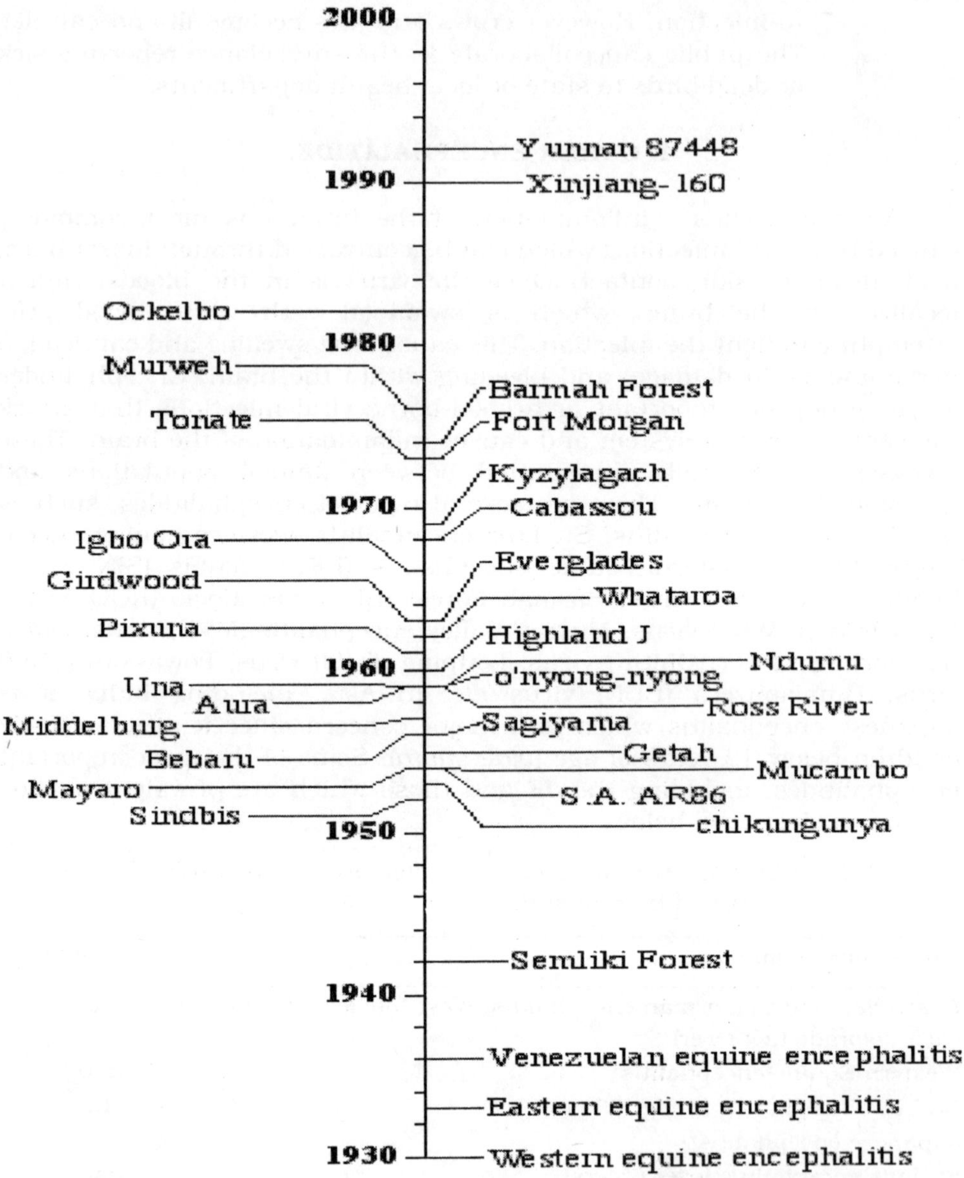

Figure 7.2. Chronological disribution of discoveries of alphaviruses in time-scale.

(i) Rift Valley Fever (RVF)

Rift Valley Fever (RVF), is a zoonosis (a disease which primarily affects animals, but occasionally causes disease in humans). It may cause severe disease in both animals and humans leading to high morbidity and mortality. The death of RVF-infected livestock often leads to substantial economic losses. Rift valley fever (RVF) is an extremely serious mosquito-borne virus disease of ruminant animals, particularly of sheep. It is endemic in parts of Africa, but causes periodic major epidemics when climatic factors favour spread of the insect vector and virus. It causes heavy losses through abortions of pregnant animals and a very high death rate in young animals. It is also an important zoonosis, causing a severe influenza-like disease in humans and sometimes a fatal illness. RVF has the potential to move rapidly to new countries and continents.

Since 1930, when the virus was first isolated during an investigation into an epidemic amongst sheep on a farm in the Rift Valley of Kenya, there have been outbreaks in sub-Saharan and North Africa. In 1997-98, there was a major outbreak in Kenya and Somalia. In September, 2000, RVF was for the first time reported outside of the African Continent. Cases were confirmed in Saudi Arabia and Yemen. This virgin-soil epidemic in the Arabian Peninsula raises the threat of expansion into other parts of Asia and Europe.

Many different species of mosquitoes are vectors for the RVF virus. There is, therefore, a potential for epizootics (epidemics amongst animals) and associated human epidemics following the introduction of the virus into a new area where these vectors are present. This has been demonstrated in the past and remains a concern.

RVF TRANSMISSION

RVF Virus

The virus, which causes RVF, is a member of the *Phlebovirus* genus, one of the five genera in the family Bunyaviridae.

Vectors

RVF virus is primarily spread amongst animals by the bite of infected mosquitoes. A wide variety of mosquito species may act as the vector for transmission of the RVF virus; in different regions a different species of mosquito may prove to be the predominant vector. In addition, the various vector species play differing roles in sustaining transmission of the virus. *Aedes* mosquitoes, for example, may acquire the virus from feeding on infected animals, and are capable of transovarial transmission (transmission of the virus from infected female mosquitoes to offspring via eggs), so new generations of infected mosquitoes may hatch from their eggs. This provides a durable mechanism for maintaining the virus in nature, as the eggs of these mosquitoes may survive for periods of up to several years in dry conditions. During periods of inundation of larval habitats by rainfall, for example, in the rainy season, the eggs will hatch, and the mosquito population will increase and spread the virus to the animals on which they feed.

Previously uninfected *Aedes* and other species of mosquitoes will feed on infected, viraemic (virus circulating in the bloodstream) animals and thus amplify and perpetuate the outbreak by transmitting the virus to the animals on which they subsequently feed.

RVF Virus in Non-human Hosts

Many types of animals may be infected with RVF, and disease may be severe in many domesticated animals including cattle, sheep, camels and goats. Sheep appear to be more susceptible than cattle and goats are less susceptible. Exotic breeds, which have been recently introduced into an endemic area, fare worse than breeds long adapted to local conditions. Animals of different ages also differ in their susceptibility to severe illness: over 90% of lambs infected with RVF die, whereas mortality amongst adult sheep can be as low as 10%.

The abortion rate amongst pregnant, infected ewes is almost 100%. An epizootic (epidemic animal disease) of RVF is usually first manifested as a wave of unexplained abortions amongst livestock. This may signal the start of an epidemic.

Transmission to Humans

During epizootics, people may become infected with RVF either by being bitten by infected mosquitoes, or through contact with the blood, other body fluids or organs of infected animals. Such contact may occur during the care or slaughtering of infected animals, or possibly from the ingestion of raw milk. The virus may infect humans through inoculation (e.g., if the skin is broken, or through a wound from an infected knife), or through inhalation as an aerosol. The aerosol mode of transmission has also led to infection in laboratory workers.

Symptoms

The incubation period (interval from infection to onset of symptoms) of RVF varies from two to six days. There then follows an influenza-like illness, with sudden onset of fever, headache, myalgia (muscle pain) and backache. Some patients also develop neck stiffness, photophobia (the patient finds exposure to light uncomfortable) and vomiting; in these patients the disease, in the early stages, may be mistaken for meningitis.

The symptoms of RVF usually last from four to seven days, after which time the immune response to infection becomes detectable with the appearance of IgM and IgG antibodies, and the disappearance of circulating virus from the bloodstream.

CLINICAL FEATURES OF SEVERE CASES

While most human cases are relatively mild, a small proportion of patients develops a much more severe disease. This generally appears as one of several recognizable syndromes: eye disease, meningoencephalitis (inflammation of the brain and surrounding tissue) or haemorrhagic fever. The proportion of patients developing these three types of complications is about 0.5-2% for eye disease, and less than 1% for meningoencephalitis and haemorrhagic fever syndrome. The fever and other clinical symptoms may appear in association with eye disease, which characteristically manifests itself in retinal lesions. The onset of eye disease is usually one to three weeks after the first symptoms appear. When the lesions are in the macula, some degree of permanent visual loss will result. Death in patients with only ocular disease is uncommon.

Another syndrome which manifests itself with acute neurological disease is meningo-encephalitis. The onset of this syndrome is also usually one to three weeks after the first symptoms appear. Death in patients with only meningoencephalitis is uncommon.

RVF may also manifest itself as haemorrhagic fever. Two to four days after the onset of illness, the patient shows evidence of severe liver disease, with jaundice and haemorrhagic phenomena, such as vomiting blood, passing blood in the faeces, developing a purpuric rash (a rash caused by bleeding in the skin), and bleeding from the gums. Patients with the RVF-haemorrhagic fever syndrome may remain viraemic for up to 10 days. The case-fatality rate for patients developing haemorrhagic disease is high at approximately 50%.

Most fatalities occur in patients who have developed haemorrhagic fever. The total case fatality rate has varied widely in the various documented epidemics, but overall is less than 1%.

DIAGNOSIS

Several approaches may be used in diagnosing acute RVF. Serological tests such as enzyme-linked immunoassay (the "ELISA" or "EIA" methods) may demonstrate the presence of specific IgM antibodies to the virus. The virus itself may be detected in blood during the viremia phase of illness or post-mortem tissues by a variety of techniques including virus propagation (in cell cultures or inoculated animals), antigen detection tests, and PCR, a molecular method for detecting the viral genome.

TREATMENT

The antiviral drug ribavirin has been shown to inhibit viral growth in experimental systems, but has not been evaluated in the clinical setting. Most human cases of RVF are relatively mild and of short duration, so will not require any specific treatment. For the more severe cases, the mainstay of treatment is general supportive therapy.

PREVENTION AND CONTROL

RVF can be prevented by a sustained program of animal vaccination. Both live, attenuated, and killed vaccines have been developed for veterinary use. The live vaccine requires only one dose and produces long-lived immunity, but the presently-available vaccine may cause abortion if given to pregnant animals. The killed vaccines do not cause these unwanted effects, but multiple doses must be given to produce protective immunity. This may prove problematic in endemic areas.

An inactivated vaccine has been developed for human use. This vaccine is not licensed and is not commercially available, but has been used experimentally to protect veterinary and laboratory personnel at high risk of exposure to RVF. Other candidate vaccines are under investigation.

The risk of transmission from infected blood or tissues exists for people working with infected animals or people during an outbreak. Gloves and other appropriate protective clothing should be worn, and care taken when handling sick animals or their tissues. Healthcare workers looking after patients with suspected or confirmed RVF should employ universal precautions when taking and processing specimens from patients. Hospitalized patients should be nursed using barrier techniques. As noted above, laboratory workers are at risk, so samples taken for diagnosis from suspected human and animal cases of RVF should be handled by trained staff and processed in suitably equipped laboratories.

Other approaches to the control of disease involve protection from and control of the mosquito vectors. Personal protection is important and effective. Where appropriate, individuals should wear protective clothing, such as long shirts and trousers, use bednets and insect repellent, and avoid outdoor activity at peak biting times of the vector species. Measures to control mosquitoes during outbreaks, e.g., use of insecticides, are effective if conditions allow access to mosquito breeding sites.

(ii) St Louis Encephalitis

Named after the town where it was first classified in 1933, St Louis encephalitis (SLE) is the most common viral encephalitis in the US. It is thought to occur naturally in birds and is spread by mosquitoes. It produces flu-like symptoms such as fever, headache and lethargy, and in severe cases high fever and convulsions. The elderly are most at risk from the virus, with 30% of those infected dying as a result. However, most people who pick it up show no symptoms.

(iii) Australian Encephalitis

Symptoms of Australian Encephalitis - also known as Murray Valley Encephalitis - vary from mild to severe with permanent brain damage or sometimes death. First reported in 1917, it has been found across the Australian mainland and is passed from water birds to humans via mosquitoes. Only one person in 500 infected becomes noticeably ill, although if they do the common symptoms are fever, anorexia and headache sometimes along with vomiting, nausea, diarrhoea and dizziness. In severe cases, brain function may be impaired after a few days and lethargy, irritability, drowsiness and confusion may set in. Convulsions and fits come next leading to the possibility of coma and death.

(iv) California Encephalitis

Covers a group of viruses carried by mosquitoes and some mammals, but primarily the La Crosse Encephalitis which accounts for 8% to 30% of all cases of encephalitis in the US. It is most likely to affect children and many cases produce no symptoms. In those cases where they become apparent, symptoms are likely to be mild. They start with a slight fever or headache but fast develop into severe headache, high fever, muscle aches, stiffness in the back of the neck and disorientation. There is a risk of permanent brain damage in severe cases. As with other insect-borne diseases, outbreaks are most likely to occur during mosquito breeding season.

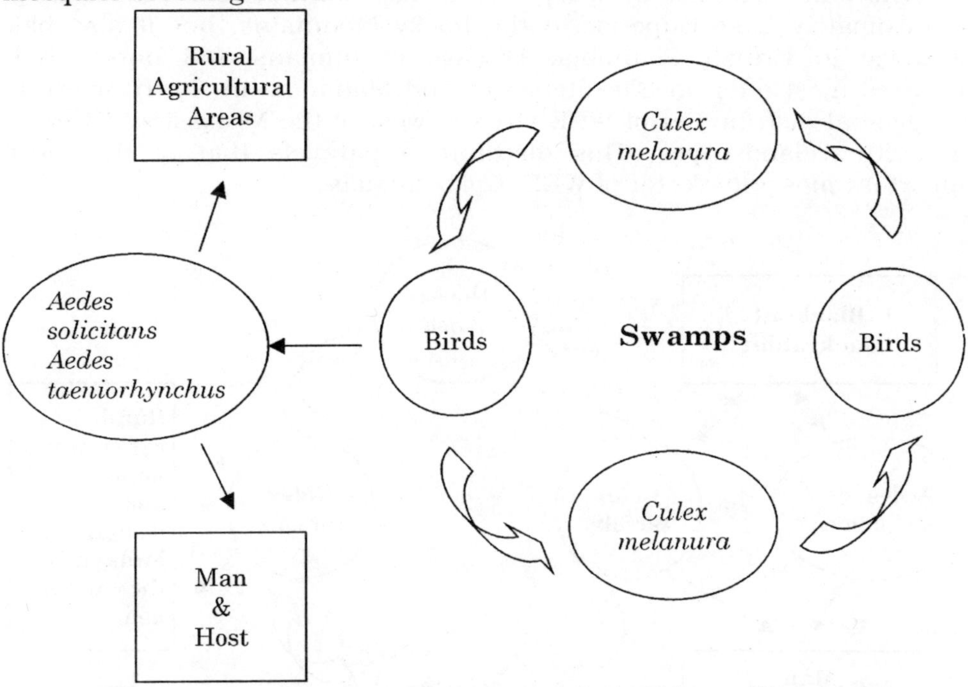

Fig. 7.3. Life cycle of eastern equine encephalomyelitis virus.

(v) Eastern Equine Encephalitis

A rare but frequently deadly virus, Eastern Equine Encephalitis too is spread from birds to humans and horses via mosquitoes (Fig. 7.3). It is found mainly on the east coast of the US and the eastern Gulf coast, and while most people show no symptoms, those who do will suffer one of two fates. Some get symptoms similar to a mild case of flu-fever, headache, and sore throat, while others will suffer infection of the central nervous system, which leads to severe headache, seizures and coma. About half of those in the second category will die, while those who survive may have permanent brain damage. Symptoms usually appear within four to 10 days of infection, and there is no cure. Young children and people over 55 are most at risk.

(vi) Western Equine Encephalitis

Western Equine Encephalitis (WEE) virus is a member of the Alphavirus genus within the family Togaviridae. The virus was first recognized in a horse in California in 1930. It is an enveloped, single-stranded RNA virus. WEE virus is one of a group of 6 closely related viruses that once were thought to represent a single WEE virus. North American members of this closely-related group of viruses include WEE itself, Highlands J virus and Fort Morgan virus. The group also includes Sinbis virus (Africa, Europe, Asia), Aura virus (South America) and Y62-33 virus (former USSR).

The WEE virus has been reported across south-central Canada from approximately Lake Superior to the Rocky Mountains, but it also has occurred in British Columbia. Disease in humans and horses has occurred most often in Saskatchewan and Manitoba. In North America, the general distribution of WEE virus is west of the Mississippi River in agricultural landscapes. This distribution parallels that of the most important mosquito vector of WEE, *Culex tarsalis*.

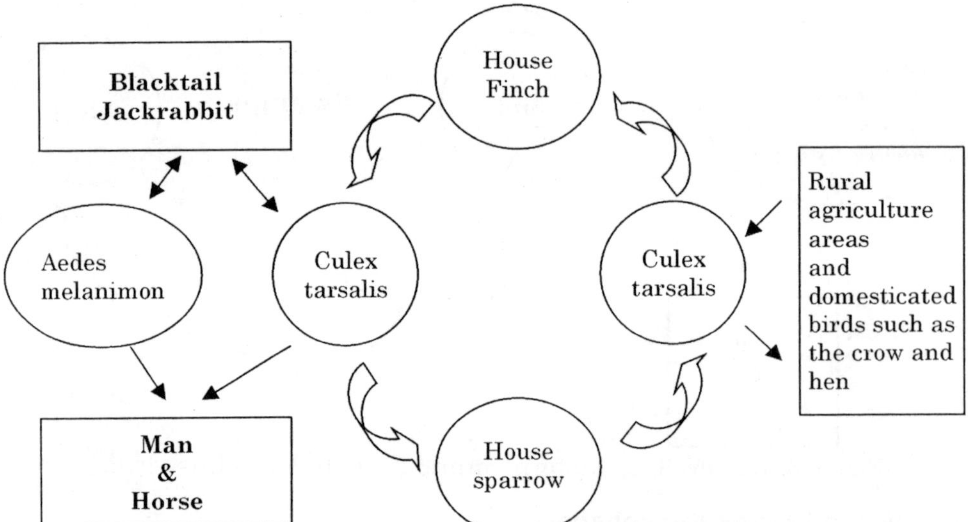

Fig. 7.4. Life cycle of western equine encepalomyelitis virus in western North America.

The WEE virus normally cycles among wild birds and mammals, and mosquitoes. In western Canada, *Culex tarsalis* is the mosquito host of greatest importance, but the virus has been found in a variety of mosquito species of five genera: *Aedes, Anopheles, Coquillettidia, Culex* and *Culesita* (Fig. 7.4). Birds are thought to be the most important vertebrate hosts for WEE. The virus replicates very effectively in birds, and birds thus act as amplifiers of the virus and important sources of infection for mosquitoes. In Western Canada, the House Sparrow and the

House Finch appear to be particularly important amplifying hosts. WEE virus, or antibodies against the virus, have been found in a wide range of wild bird species. Mammals also may play roles in the ecology of WEE. Virus has been isolated from Richardson's Ground Squirrel (*Spermophilus richardsoni*) and Snowshoe hares (*Lepus americanus*). The virus also has been isolated from Garter Snakes (*Thamnophis* sp.) and from Leopard Frogs (*Rana pipiens*).

Western Equine Encephalitis virus goes through an annual cycle of amplification and dissemination. Nestling birds are infected in the spring by virus-infected mosquitoes. The virus replicates readily in these birds and other mosquitoes that feed on the birds become infected. They, in turn, can infect more birds, and the amplification cycle is repeated. Over the summer, the prevalence of infection in vector species of mosquitoes increases. Cold weather in the fall ends these cycles as mosquitoes die or become dormant. It is not known precisely how WEE virus is maintained during winter. Mammals and reptiles or amphibians may play roles in over-winter survival and also in summer amplification. Outbreaks of disease in people and horses tend to occur in years when, due to a complex of ecological factors, the prevalence of infection in mosquitoes is particularly high. Infection in mosquitoes is highest at the end of summer - in August and September - and infections in people and horses are most common during these months.

Disease in animals

Wildlife: It is not known if WEE is an important pathogen to its normal vertebrate or mosquito hosts, but in general this appears not to be the case. WEE virus has been isolated from nestling house sparrows that were clinically ill or dead and from some rodents suspected of having rabies. However, it is not certain that WEE was the cause of disease in these animals. It seems likely that WEE may cause disease in wild animals from time to time, but there is no evidence to date that it is an important wild animal pathogen.

Domestic Animals: Horses can suffer severe, often fatal disease after infection with WEE. In 1937-38 in the prairie provinces, 52,500 horses suffered clinical disease and 15,000 died. Clinical disease develops 1-3 weeks after infection and ranges from a general fever followed by recovery to severe infection and inflammation of the brain leading to deranged behaviour and death. In reported outbreaks of WEE, from 10% to 50% of horses that became clinically ill died of the disease. One disease outbreak in turkeys in the United States was thought to have been caused by WEE.

Human disease

Of the arboviruses that occur in Canada, WEE has been the most important cause of human disease. Sporadic outbreaks have been recognized in Canada each decade since the 1930. An epidemic occurred in the prairie provinces in 1941 in which 1094 human cases were recorded, which is most of the total of cases (1500 or so) recorded in Canada up to 1990. However, most people infected with WEE suffer only mild disease or no disease at all. Children are more likely to suffer disease than are adults; approximately half of children under one year of age will become ill if infected while slightly less than one in 1,000 people aged 14 years or older will become ill if infected. The mortality rate from WEE is reported to range from 8 to 15% in people who develop clinical disease after infection.

Because of its annual cycle of amplification, risk of disease in people and horses varies greatly from year to year, depending on the prevalence of infected mosquitoes of species likely to feed on both birds and mammals. Surveillance programs of various kinds are in place or have been used in the past to monitor the prevalence of infected mosquitoes. When a high prevalence is detected, education programs to urge people to avoid exposure to mosquitoes and mosquito control programs can be undertaken. Earlier assessment of the magnitude of the amplification cycle is possible by surveys of nestling birds or use of sentinel bird flocks which are monitored for exposure to WEE virus. Since mosquito control generally is not very effective when undertaken late in the summer, early detection can trigger control efforts earlier in the season which are more likely to succeed. Television and air conditioning appear to have reduced the incidence of disease from WEE in people by modifying human behaviour in such a way as to reduce contact with infected mosquitoes.

Effective vaccines are available for use in horses but none is available for general use in humans. The small number of human cases of WEE is insufficient to justify a public vaccination program.

Western Equine Encephalitis is fatal less often than its eastern namesake and is found in the western and central US. The effects of infection range from no symptoms to fatal disease. Milder effects consist of a headache with, perhaps, a fever while the severe form causes sudden high fever, headache, drowsiness, irritability, nausea, and vomiting. This is followed by confusion, weakness, and coma or, in young infants, seizures. Symptoms usually appear within five to 10 days of infection, there is no cure and people of all ages are at equal risk.

(vii) Highlands J (HJ) virus

Aetiologically, as the HJV is belonging to the western equine encephalomyelitis complex, it is closely related pathologically to the Western Equine Encephalitis. The HJ virus is prevalent in the same ecological environs and transmission cycle as the Eastern Equine Encephalitis virus. Though, a major infection transmitted by *Culiseta melanura* in songbirds in the eastern half of North America and in other domestic or exotic birds elsewhere, it also infects mammals with low pathogenecity. Rarely, however, the HJ virus is associated with human disease.

(viii) Venezuelan equine encephalomyelitis (VEE) virus

The VEE virus, transmitted mainly in equines through biting of a *Culex* mosquito belonging to subgenus *Melanoconion*, is a complex of several enzootic and epizootic serotypes and varieties, all of which are associated with human illness, mostly the children, albeit rare involvement of encephalitis, central nervous system and death. The symptoms of the disease in humans simulate influenza, although other routes like handling of infected rodents or high concentrations of virus in laboratory experimentations are also documented. Death of infected equines occurs in 5-14 days. Enzootic VEE varieties are capable to exist throughout the year. Besides rodents, in some locations birds are also involved in the transmission cycles.

(ix) Chikungunya (CHIK) virus

Chikungunya, having origin in Tanzania, and disease distribution in Africa and Asia, produces an acute illness in humans within two days of exposure that lasts 3-5 days. It is characterized by sudden onset of fever, chills, headache, photophobia, arthralgia and rash, although symptoms are often confused with those of dengue, as generalized myalgia with pain in the back and shoulders is common. Most patients recover in 5-7 days, but some severe cases can require months of recovery and occasionally a few patients may never recover from articular dysfunction. In Africa and Asia the CHIK virus is transmitted by *Aedes aegypti*, maintaining human to human cycle. Additionally, however, in Africa non-human primates (*e.g.*, Baboons and *Cercopethicus* monkeys) are also involved in transmission cycles. Mosquitoes responsible for enzootic transmission belong to two subgenera of the genus *Aedes* (*Ae. africanus, Ae. lutecephalus* and *Ae. opok* under subgenus Stegomyia, and *Ae. cordellieri, Ae. furcifer* and *Ae. taylori* under subgenus Diceromyia). In certain situations, even *Mansonia africana* is also implicated in enzootic transmission.

(x) O'nyong-nyong (ONN) virus

Also known as O'Nyong-Nyong virus disease, ONN, it is synonymous to Epidemic Polyarthritis and rash. The disease is found in Africa. and infects humans. It belongs to Togaviridae (formerly group A arboviruses), genus Alphavirus, and is spherical, and made of enveloped virions 60 nm in diameter, single-stranded, positive-sense RNA genome. O'nyong-nyong virus is a mosquito-borne alphavirus that causes a febrile disease characterized by generalized, debilitating joint pains and often the disease is accompanied by a maculopapular skin rash and lymphadenopathy. Fatalities are rare, but morbidity often is significant characterized by arthralgia or arthritis typically in the knee, ankle and small joints of the extremities, followed by a maculopapular rash in 60-70% of cases; buccal and palatal enanthema can occur. It's incubation period is greater than 8 days. However, the disease is largely self limiting and there is no evidence that the disease can be spread person-to-person. The typical epidemic mosquito vectors are *Anopheles funestus* and *An. gambiae*, two of the region's major malaria vectors. During the second half of 1996, an epidemic of o'nyong-nyong fever was documented in rural south-central Uganda. This represents only the second recognized epidemic of this disease since its initial description in 1962.

(xi) Ross River (RR) virus and Barmah Forest (BF) disease

Traditionally known as Epidemic Polyarthritis, both Ross River (RR) and Barmah Forest (BF) disease, are caused by viruses which are transmitted to humans through the bite of mosquitoes.

Ross River virus (RRV) is a fascinating, important arbovirus that is endemic and enzootic, and is the most commonly transmitted mosquito-borne viral disease to humans in Australia, besides Papua New Guinea. It was also epidemic in the South Pacific in 1979 and 1980. Infection with RRV may cause disease in humans, typically presenting as peripheral polyarthralgia or arthritis, sometimes with fever and rash. The first well-described outbreak occurred in 1928. During World War II there were more outbreaks, and the name epidemic polyarthritis was applied. During a 1956 outbreak, epidemic polyarthritis was linked serologically to a group A arbovirus (Alphavirus). The virus was subsequently isolated from *Aedes vigilax* mosquitoes in 1963 and then from epidemic polyarthritis patients. The number of cases has averaged >5,000 per annum during 1991-1997. The virus appears to be endemic in most rural areas, and there has been an increasing incidence near major cities. The BF disease is less common, but the number of cases appears to be increasing annually, with several outbreaks occurring during the1990s. For most of Australia, peak incidence of the diseases is through the summer and autumn months, particularly from January through to March, when the mosquito vectors are most abundant. However, in southwestern Australia and eastern Victoria, RR activity often begins in the spring months and peaks in early summer. Areas under intensive irrigation and localities close to saltmarshes are most productive for mosquito populations and hence tend to result in the highest number of human cases of disease. Outbreaks occur when local conditions of rainfall, tides and temperature promote vector abundance.

Serological studies and laboratory investigations have indicated that native mammals, most likely kangaroos and wallabies, are natural hosts for RR virus but little is known about the hosts of Barmah Forest virus. RR virus transmission from human to mosquito to human (thus occurring without the involvement of an animal) has been proposed, and there is now little doubt that such a cycle involving only humans and mosquitoes occurs during periods of intense virus activity.

The RR and BF viruses have been isolated from many mosquito species, indicating wide susceptibility among mosquitoes. In inland regions, the major vector is *Culex annulirostris* which breeds in freshwater habitats, especially in irrigated areas. Along coastal regions, saltmarsh mosquitoes represent the major threat, including *Aedes vigilax* and *Ae. camptorhynchus* in northern and southern coastal regions respectively. There is some evidence that 'floodwater' *Aedes* species such as *Ae. normanensis* play an important role in transmission in inland regions following heavy rains or floods, and *Coquillettidia linealis* is a secondary vector in areas with established wetlands. In the domestic urban situation, there is evidence to suggest that *Ae. notoscriptus* may be a vector, while *Cx. quinquefasciatus* is not.

Human infection with RR virus or BF virus, may result in the clinical condition known as polyarthritis. The effects range from a symptomless condition, through a transient rash and mild illness with fever, to polyarthritis affecting chiefly the ankles, fingers, knees, and wrists, but other joints may be affected. The disease is not fatal. For RR virus, symptoms become evident from 3-21 days (average 9 days) after infection, and mild cases may recover in less than one month but many persist for months to years. Recent studies have indicated that the rash may be more florid with BF virus infections but that the arthritic symptoms are greater with RR virus infection. People of working age are most likely to be afflicted with the diseases, whilst symptoms are rare in children.

A variety of blood tests are used to demonstrate the presence of specific antibodies to RR and BF virus. Blood samples should be taken during the acute and convalescent phases of the illness, and a fourfold rise in antibody levels will confirm the clinical diagnosis. Specific therapies do not exist to treat the disease, rather it is the symptoms that are alleviated. This includes various analgesics to reduce the pain and fevers, and anti-inflammatory agents for the arthritic symptoms.

Prevention of mosquito-borne viral diseases is mainly accomplished through reducing the threat of bites from mosquitoes. This can be achieved either through undertaking active mosquito control or by the use of personal protective measures. A variety of active mosquito control measures are available including habitat modification in order to reduce water availability for breeding of the larval stage, through to the use of appropriate insecticides for controlling the larval or adult stage. These are large scale control measures which can only be undertaken by government bodies. On a small scale, house-holders can ensure that their own backyard does not contain water holding containers which can provide suitable mosquito larval habitats (e.g. undrained pot plants, blocked gutters, disused bottles, old tyres, etc). Personal protective measures include: avoiding known mosquito infested areas, especially at dawn and dusk when mosquitoes are most active; ensuring that houses are adequately screened; using insect repellents and reapplying it regularly; and wearing long sleeved shirts and pants. Other preventive measures include government based programs that undertake mosquito monitoring and virus surveillance from mosquitoes. These programmes aim to act as an early warning system for virus activity by monitoring mosquito populations, viruses such as Ross River or Barmah Forest and weather patterns.

CHAPTER - 8

MOSQUITO AND THE HIV TRANSMISSION:
IS THE DEVIL STILL HAUNTING THE BEAST?

Over three decades have passed following the epoch-making discovery of the virus called human immunodeficiency virus (HIV) that causes the much dreaded Acquired Immunity Deficiency Syndrome, or AIDS, in the humans. Even though the modes of HIV transmission are well documented by now (*e.g.*, drug abuse, unprotected sex etc.) nevertheless, a large number of people, both professional and amateur alike, keep questioning about the possible role of a mosquito, or for that matter any haematophagous insect, in effecting the transmission if conducive situation prevails! That even the scientific communities have constantly confabulated the subject and pointed out to draw a conclusion with caution, is substantiated with two recent events; (i) First during the sixteenth National Congress of Parasitology (Jodhpur; October 31-November 2, 2001) a question, "Whether mosquitoes can transmit AIDS virus or not?", was asked to Dr. S. Sabesan from Vector Control Research Centre, Pondicherry, and an eminent entomologist himself and, (ii) Secondly, a rather broad spectrum question on the same theme "Whether blood-sucking insects can spread the AIDS virus or not?", was asked to me during the National Symposium on Recent Trends in Zoological Sciences (Patiala; March 12-13, 2003). The two scientists offered their views to explain why mosquitoes could not carry this job though one could feel skepticism in the eyes of at least a few acknowledged participants and their desire to debate more on the subject. Ironically, in recent years, the piercing and sucking mouthparts of the mosquito had been paralleled with a drug abusers' hypodermic syringe. It is noteworthy that such drug abusers resorting to repeated usage of the hypodermic syringe are a potentially risk group of HIV infection. This comparison lends further credential to the people's imagination implicating mosquitoes in the vicious disease transmission cycle. A widely acknowledged viewpoint is that mosquitoes, or so to say any haematophagous insect, are unable to transmit the virus, though the arguments given in support of this conclusion, however, differ widely.

In the following pages, therefore, an attempt is made to assimilate all the possible information available on the subject, and it is expected that even the faintest glimmer signifying insects' role, in whatsoever way, would continue to arouse interest in the scientific community, notwithstanding our definite knowledge about the other well documented means of transmission of the disease (Miike, 1987).

MOSQUITO FACTOR IN DISEASE TRANSMISSION: *THEORY & PRACTICE*

Different blood feeding insects involve varied mechanisms in effecting the transmission of disease pathogen (see Chapter 5). Mosquitoes, for instance, can transmit a virus from one infected person to another (uninfected) person during probing for blood feeding on different hosts and by inoculating virus into the blood of an uninfected host. When the virus-mixed blood is sucked into the crop, the pathogen has to penetrate through the gut wall to reach the haemocoel from where they travel to salivary gland by using some path- and site-identifying markers. Once lodged in the salivary glands, they are ready to be injected into the blood stream of a person whenever the infected mosquito finds an opportunity to feed. These and other various conditions have to be full-filled by a virus in order to make the mosquito infective. In principle, there are three theoretical mechanisms which would allow blood feeding insects such as mosquitoes to transmit HIV.

(1) ***Mechanism One:*** A vector mosquito should blood-feed on an HIV-infected person and ingests the virus particles along with the blood. In the gut of the mosquito, where several physical and chemical barriers naturally exist (e.g., pharyngeal comb, acid, enzymes etc.), the virus has to overcome all these stumbling blocks, attach to site and penetrate through gut wall, and migrate to the mosquito's salivary glands. Then, the infective mosquito should subsequently feed on blood of an uninfected person so that the virus are passed on to its new host. Most mosquito-borne infections like malaria (a protozoal infection), yellow fever, dengue and Japanese encephalitis (all viral infections) involve this kind of mechanism of disease transmission.

(2) ***Mechanism Two:*** A vector mosquito begins to blood-feed on an HIV-infected person, but is interrupted in her feeding and forced to give up this host after its successful initial feeding (i.e., adequate blood was drawn along with virus particles to get access to the suction canal in mosquito's mouthparts) for another host, preferably/incidentally an uninfected person. Because the mosquito still did not fully engorge, the urge for full blood meal draws her to take on to the rest from the uninfected host(s). In the process, the virus particles which were left behind stuck on the mouthparts during the 'interrupted' previous blood-feeding get inoculated into the body of the next (uninfected) host as soon as the mosquito's mouthparts penetrate through the latter's skin.

This type of mechanism is rather uncommon in mosquitoes, but equine infectious anemia is transmitted to horses by biting flies in this manner.

(3) ***Mechanism Three:*** In this a mosquito initiates the cycle by beginning to feed on an HIV carrier (in a manner given above), but on interruption, resumes the partial blood-feeding on a different (uninfected) host. Here, the mosquito is knowingly or unknowingly squashed or smeared to death on a wound site on the host's body into where the virus particles can make way in. Theoretically, any mosquito can be subjected to this type of mechanism of pathogen transmission, but the new uninfected host may pick up infection provided the inoculum of contaminated blood from crushed mosquito's gut had adequate virus particles.

VECTOR-DIGESTED PATHOGENS CAN NOT CAUSE INFECTION

A pathogen has to keep alive in the mosquito's body if it has to cause infection to another (human) host on transfer following a subsequent blood meal. On the other hand, if the mosquito digests the parasite/ pathogen, the transmission cycle is terminated and the parasite/ pathogen can not be passed on to the next host. Several mosquito-borne parasites/pathogens exercise varied ways to avoid being treated as food for the mosquitoes. While, some parasites/pathogens are simply refractory enough to the digestive enzymes inside the mosquito's stomach and pass out intact, others target the receptive gut wall and bore their way out of the stomach as fast as possible to evade the deleterious effects of digestive enzymes that would otherwise decimate their very existence. For example, malaria parasites survive inside mosquitoes for 9-12 days and actually go through a series of necessary life stages during that period. Encephalitis virus particles survive for 10-25 days inside a mosquito and replicate enormously during the incubation period. Similarly, a lymphatic filariasis worm also takes a little over one week's time to stay in the body of mosquito before getting ready to be ejected for a transfer to a (human) host. In contrast to these, many a studies have shown that the HIV virus seemingly serve as a 'food' to the mosquito and is digested along with the blood meal. Therefore, as soon as the mosquito vector has fed on the blood of an HIV-infected person, the ingested blood is digested within 1-2 days, destroying completely all viruses which otherwise could potentially produce a new infection if inoculated undamaged. Even intrathoracic inoculation of mosquitoes was carried out with an objective to bypass any gut barriers, but the virus still did not show any sign of reproduction (Webb *et al.*, 1989). So, as there exist no virus particles that could multiply in mosquito's body and invade the salivary glands, they obviously stand no chance to effect any transmission from one (mosquito) host to another (man).

MOSQUITOES DO NOT INGEST ENOUGH HIV PARTCLES TO TRANSMIT AIDS BY CONTAMINATION

Certain insect-borne disease agents which have the potential to be transferred from one individual to the next via contaminated mouthparts must circulate at very high levels in the bloodstream of their host. Transfer by mouthpart contamination requires sufficient infectious particles to initiate a new infection, though the exact quantum may vary from one disease to another. The fact that AIDS virus circulates at very low levels in the blood; far below than those of any known mosquito-borne disease. The infected individuals rarely circulate more than 10 units of HIV, and 70-80% of HIV-infected persons have undetectable levels of virus particles in their blood. Statistically, if a mosquito is interrupted while feeding on an HIV-infected person, circulating 1000 units of HIV, has a 1:10 million probability of inoculating a single unit of HIV from contaminated mosquito mouthparts. In the same way, if a fully fed mosquito containing AIDS positive blood will also not approach the levels needed to initiate infection. It appears, therefore, that simply mechanical transmission of AIDS by HIV-contaminated mosquitoes is but a far outcry!

MOSQUITOES' MOUTHPARTS VERSUS HYPODERMIC NEEDLES

Rhetorically enough the mouthparts of vector mosquitoes have recently been called the tiny "flying needles", comparable to the hypodermic needles employed by the drug abusers in injecting some kind of narcotic stuff and in the process interchange between those are HIV infected and others who are initially not. The argument here is that if the hypodermic needles can successfully transmit HIV from one individual to another then why can the mosquitoes not be able to do the same. Although it is already broached above about very low circulation level of HIV in the blood of the infected person, but for theoretical reason if for a moment it is presumed that, instead, a high level of virus particles circulate, even then mosquitoes blood-feeding mechanism can not be paralleled with the methods that are employed in used syringes or hypodermic needles. Mosquito's feeding apparatus is an extremely complicated structure that is completely different in its physiology than the crude single-bore syringe. In mosquitoes, both the food canal and salivary canal are separate passages. Unlike a syringe, the mosquito regurgitates saliva before initiating feeding through one passage (and never the remnants of its last blood meal, as one might wrongly infer!) and draws blood up another passageway. As a consequence, the food canal is not flushed out like a used syringe or 'hypodermic needle' and the blood flow is always unidirectional. In fine, it is inappropriate to compare mosquito's mouthparts with the hypodermic needles as mechanics involved in mosquito feeding are totally different from those employed by the drug users with the syringes (Fig. 8.1 & 8.2).

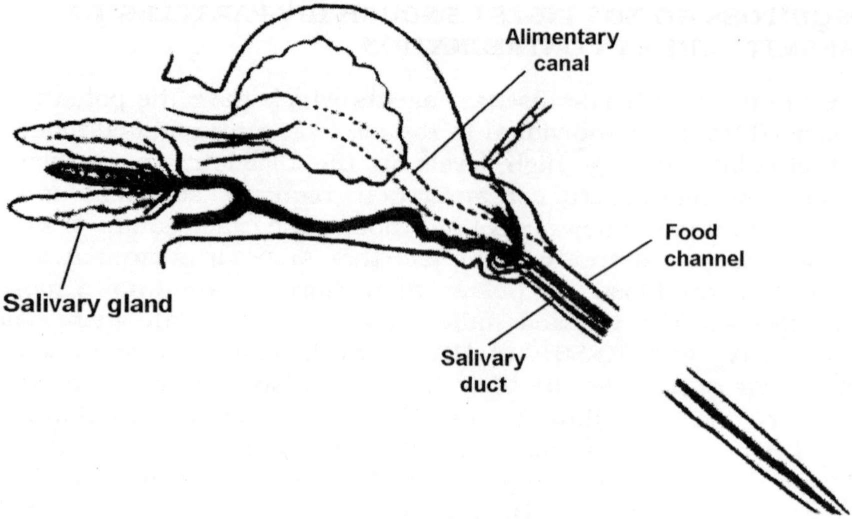

Fig. 8.1. Anatomy of a mosquito showing separate passageways of each food channel and salivary duct.

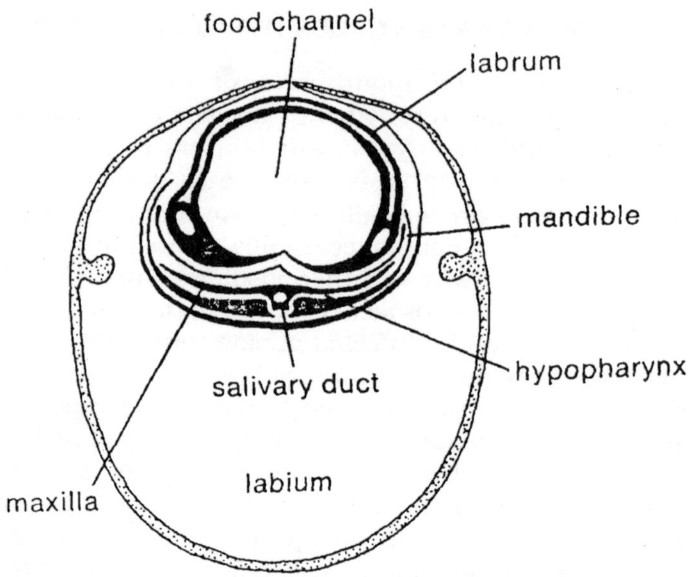

Fig. 8.2. Section through a mosquito's mouthparts showing position of different structures, inclduing food channel and salivary duct.

CAN OTHER INSECTS TRANSMIT HIV/AIDS?

Besides mosquitoes who have been quite elaborately investigated in relation to AIDS transmission, certain other insects like stable flies, bed bugs and ticks (actually an arachnid) have been systematically studied both in laboratory and field.

Webb et al. (1989) determined that the AIDS virus could survive in bedbug for 8 days, whereas Humphrey-Smith et al. (1993) demonstrated, based on *in vitro* artificial fed meals to ticks with high levels of virus ($\geq 10^5$ tissue culture infective doses per milliliter (TCID/ml), that the HIV virus could stand survival for 10 days in ticks, without any inkling of multiplication though. Intra-abdominal inoculation of bedbugs was likewise used to avoid any encounter with gut barriers, but again the virus did not grow in number. What is more, even when *in vitro* cultures of HIV were developed with a number of arthropod cell lines, it became clear that HIV was incapable of multiplication in these systems. The possibility of mechanical transmission of HIV by bedbug was tested on the basis of estimation of blood meal residue on mouthparts (7×10^{-5}ml). When 50 bedbugs, interrupted during their blood-feeding and containing 1.3×10^5 TCID/ml HIV, failed to contaminate the uninfected blood on which they finished feeding or even the mouse skin membrane through which they refed (*cf.* Goddard, 2000). Brandner et al. (1992) have demonstrated through laboratory studies that stable flies had regurgitated 0.2 μl of fluid containing an estimated 10 TCID, as soon as they were fed blood with 5×10^4 TCID of HIV, alluding that, though highly unlikely, the HIV was transferable, at least theoretically.

CONCLUSION

Mosquitoes are unlikely to transmit the HIV infection. Mosquito's feeding mechanism is highly evolved and suitable to her physical and safety requirements. Her food canal and salivary canal are separate passageways in the mosquito. Mosquito's feeding apparatus is an extremely complicated structure that is totally unlike the crude single-bore syringe. Unlike a syringe, the mosquito delivers salivary fluid through one passage and draws blood up another. As a result, the food canal is not flushed out like a used needle, and blood flow is always unidirectional. The mechanics involved in mosquito feeding are totally unlike the mechanisms employed by the drug user's needles. In short, mosquitoes are not flying hypodermic needles and a mosquito that disgorges saliva into your body is not flushing out the remnants of its last blood meal. Another reason attributable to specific mosquito feeding is, of course, that the virus responsible for the AIDS infection is regarded as food to the mosquito and is digested along with the blood meal. Mosquitoes do not ingest enough HIV particles to transmit AIDS by contamination. An AIDS-free individual would have to be bitten by 10 million mosquitoes that had begun feeding on an AIDS carrier to receive a single unit of HIV from contaminated mosquito mouthparts.

Various investigations conducted have so far indicated that mosquitoes, or for that matter, hamatophagous insects are unable to transmit the HIV/AIDS virus, although there is a school of thought who, based on laboratory studies and epidemiological surveys, believe that the transmission is but very remotely possible. Biological mode of transmission seems improbable, but mechanical transmission would most likely occur, even though this too depends on several viraemic characteristics both in the infected host and on the mouthparts as well as the host's wound contamination through mosquito's regurgitation of the partially ingested blood. In some tropical countries this view has got much strength particularly where there are large populations of biting insects and a high prevalence of HIV infection (*cf.* Goddard, 2000).

In view of the above evidences, it appears almost certain that mosquitoes, or other blood-sucking arthropods, do not have potential to transmit the HIV/AIDS infection – in a practical sense, at least! The debate may, however, go on for some more time with more scientific experimentations performed, and who knows what may turn out eventually; after all "No research is ever complete" is the underlying fundamental rule to yearn for knowing more and novel, in science. The subject of transmission of the disease by a "mosquito" is still very much "alive".

CHAPTER - 9

THE MOSQUITO 'THEORY', 'DAY' & 'SAMPLER':
AN INVENTION IS ALWAYS GREAT TO ITS ORIGINATOR, A PERSONAL AFFRONT TO SOME OTHERS!

Although for almost 2,500 years man had known mosquitoes being associated with disturbance in sleep, public nuisance and loss of general tranquility by virtue of their painful and irritating biting habits, yet it was only in the third quarter of the 18th century that a serious attention was paid to their possible association with transmission of certain pathogens like protozoans, viruses, worms *etc.* The idea that mosquito could have been involved in some way in spreading certain communicable diseases had been born out even a century ago but these were mostly ill-defined conjectures and assumptions without any material display. It eventually occurred in the early 1880s when Dr. Patrick Manson, a surgeon officer with the British army in China, first provided an irrefutable and scientifically demonstrable evidence in support of his 'mosquito theory of human lymphatic filariasis', thus also sowing the seed for the birth of that fascinating and adventurous discipline called 'Medical Entomology'. The subsequent mosquito theories, particularly 'the mosquito theory of malaria' proposed by Dr. Ronald Ross, in 1897, on the urging of Patrick Manson only, have been rife with much claims and counterclaims, and often with a pinch of bitterness! Like many other hair-raising stories in the annals of scientific pursuits by many great scientists of yore which were not without a tinge of battle for supremacy or primacy, the so called mosquito theories, too, were full of dramas that saw much ups and downs, making the science ever so interesting.

But, before reaching a conclusion on who claimed a given mosquito theory with respect to a particular infection, it is inevitably indispensable to comprehend beforehand, in truly lexiconic sense, as to how the meaning of 'theory' differs from a mere conjecture or 'hypothesis'. According to The New Pocket Oxford Dictionary (2002), the word 'Theory' is defined as *'an idea or set of ideas that is intended to explain something or a set of principles on which an activity is based'*, whereas the word 'Hypothesis' is spelt out as *'a proposed explanation based on limited evidence, used as a starting point for further investigation*'. Thus, it is comprehensible that a theory is differentiable from a hypothesis in its inexorability of demonstration of an idea or act.

MOSQUITO THEORIES

1. The Mosquito Theory of filariasis

Fewer tropical diseases appear to frighten as fiercely as possibly the lymphatic filariasis. Some of the chronically affected grotesque cases of elephantiasis, a form of filariasis, leave an indelible imprint on mind of the people who have discerned a case to such a great extent that they simply can not avoid being affected in some way by its intensity of severity and incapacitation to the affected. During the British Raj, a self-acclaimed period of global governance by the British, particularly in the South-east Asia, one of the major concern of the British government was to save its people from the deadly and/or debilitating mosquito-borne diseases amongst which, after malaria, lymphatic filariasis threatened them most, exacting enormous costs, even though the disease is hardly ever fatal.

Dr. Patrick Manson, then in his late twenties, and working as a British medical officer in Taiwan (at that time known as Formosa, annexed with China), saw a challenge in deciphering the cause responsible to carry transmission from one human to another. He witnessed a large number of cases of elephantiasis, on several of which he would work to drain and surgically remove the bloated flesh resultant to this disease as no antidote was invented to treat elephantiasis. For over a decade, he worked assiduously with an unparalleled devotion, often secretly performing autopsies, to encounter the animal associated with the disease and to track down its life cycle outside the human body. Manson drew the opinion that the elephantiasis disease in patients had been caused by the victim's immune system reacting to the presence of the copulating worms (Spielman & D'Antonio, 2001). All along this venture, Manson understood that filarial worm could not lay trapped to their dead end in the body's lymph nodes and appreciated nature's ingenuity in devising means and ways to unlock and set free the elephantiasis-causing worms from their human abode so as to be able to pass on to another human through possibly an organism that fed on the blood of the infected one.

Mosquito, for several reasons, was the first probable culprit to be thought of, and he began to work meticulously on this hypothesis which was already aired more than nearly two decades ago in cases of malaria and yellow fever, though, by Beauperthuy and Nott in the Americas. Still more, the idea had actually trickled down over about 100 years when Lancisi, an Italian physician, conjectured a vague association between insect-infested swamps and the resultant human illnesses. Finally, pursuing rigorous and painstaking path of systematic human and mosquito dissections, Manson, in early 1880s, was able to demonstrate that the filarial worm lived and developed in both man and mosquito, although his part of mosquito theory was not totally flawless. But, even these follies only show how monumental was Manson's single-handed accomplishment in opening an altogether new chapter in furthering our understanding of this socially and physically the most devastating infection.

2. The Mosquito Theory of malaria

Even though Charles Laveran, a French Army doctor on an assignment to Algeria to check epidemics that plagued the French outposts, had discovered the first of the four human malaria parasite species on November 6, 1880, and many assumptions implicating the mosquito for playing some role in the transmission of this and other diseases were already making rounds, yet it was inarguably the epoch-making finding of Dr. Patrick Manson, a British medical officer, who, following his posting in 1870s to Taiwan (then a part of China and called Formosa), demonstrated with certitude in 1880s the involvement of the mosquito, *Culex pipiens quinquefasciatus*, in playing the inevitably indispensable role of an intermediate host, the vector, before executing the transfer of the lymphatic filarial worms to an another, possibly, uninfected human. A little more time ago, the Cuban scientist Dr. Carlos Finlay had practically proved the inexorable role of the mosquito, *Stegomyia fasciata* (which we now know as *Aedes aegypti*) in transmitting yellow fever in Havana, on October 13, 1880, albeit strong repudiation for about two decades till the mosquito's prowess as a vector of malaria and filariasis was finally confirmed by Ross and the Italians (later, on October 23, 1900, a USA giant of a scientist Dr. Walter Reed had proclaimed having deciphered the yellow fever enigma by charging *Ae. aegypti* for carrying the disease pathogen!). These prolonged scientific feats had had to leave an indelible imprint on the young mind of Ronald Ross, who, though was an unlikely, if not reluctant, scientist, later rose to write the history by solving the malaria puzzle almost single-handedly.

Ronald Ross, eldest of a British general's ten children, was born in the foothills of the Himalaya, at Almora, India, on May 13, 1857. A mediocre in the school, he showed nothing much remarkable, save for his strong inclination and a distinct talent for subjects such as art, music and literature. However, since his father was determined that he should become a doctor, young Ross reluctantly joined the Medical School in London where he managed to pass. For joining the Indian Medical Service, he took the examination a second time, and was selected for serving in the Madras Presidency.

Ross served in the Indian Army from 1881 to 1899 before he returned to England. During this period of his highly troubled service in India he studied malaria almost by chance and due to his unending desire to do something great. Amidst the pressure of his frequent movements due to transfers, Ross began studying malaria without even having seen a parasite until much later time of Laveran's discovery of

Plasmodium malariae, in 1880, due mostly to his initial contretemps to this new finding. In addition to this, he was also unable to identify mosquitoes which compounded his problems manifold (ironically, Ross could not have identified the mosquito, he called "dabbled" one, which he had been successful to get infected in his attempt to prove *Anopheles* mosquito's role in malaria transmission and was for long confused about only a few anopheline species' power to carry the parasite).

Ronald Ross became a medical scientist almost by accident. Although, always yearning to do something remarkable, he was utterly unsuccessful in sighting malaria parasites, despite his obtaining further training in public health, bacteriology and microscopy during his first furlough to England in 1888. Soon back to India, this time married, he, a truly polymath, had plenty of time to spend playing golf and fulfilling his deep interests in poetry and music. He published a romance novel in London and composed poetry, but even with this initial success he was bored and yearned intensely for achievement. During his second furlough in 1894, he returned to London and met Patrick Manson, who had by then already demonstrated mosquitoes role in transmitting lymphatic filariasis and, rightly so, earned the reputation of the Father of Tropical Medicine. At the Seaman's Hospital, where Manson, a true giant of medical science, had taken a position after leaving China, he showed to Ross malaria parasites for the first time, and demonstrated to him how to find the parasites in blood smears. Manson had inferred, based on his earlier experiences in China in respect of filarial worm, that mosquitoes could also be responsible for transmitting malaria, a theory then circulating, although Ross felt skeptical about this. Yet, Ross took note of his advice quite seriously, and returned to India fully determined to prove Manson's "Grand Induction" by observing malaria's growth and development in mosquitoes. Ross looked in Manson a father figure and a mentor – a bond endorsed by the 173 letters, rich in both scientific and human terms as well as remain one of the celebrated scientific correspondences, that the two exchanged between 1895 and 1899. These letters record, in great detail, Ross's research in India elucidating the role of mosquitoes in the transmission of malaria and worked to firmly cement their intimacy. These are testimony to not only Ross's excitement in his pursuit to make the discovery, but also his immense frustration and vexation he had had to endure on many fronts – his army bosses' annoyance, day-to-day cumbersome duties, frequent outstation postings and constant failures in his endeavours to incriminate a mosquito with a malaria parasite. This massive correspondence is, ironically, also a testimony to the rise and fall of a relationship, as Ross's preoccupation with his place in the history of malariology led to a breach between the two men. That Ross was completely obsessed with garnering every possible laurel in the wake of his "mosquito malaria theory", it ultimately spoiled his relationship with his mentor. Themes of priority, nationalism, and personal vanity punctuate this latter correspondence, which also reveals new insights about the golden years of tropical medicine. This happened primarily because Manson thought to credit Grassi, who was a Professor of Zoology with extensive mosquito/malaria-related research experience and was considered, rightly so, by many scientists to be the pre-eminent medical zoologist of his time, for his commendable contributions, but Ross turned cool toward his former teacher, mentor and father figure. Ross continued to critically despise Manson, ignoring latter's immense assistance and guidance in his pursuing malaria research as evidenced from scores of affectionate letters that passed between them. This bitterness culminated to its zenith when Ross, recuperating from a heart stroke by the end of his life, wrote a highly unpleasant book charging that Manson even hindered, rather than helped, his malaria work!

Ross worked relentlessly about finding the mosquito's role in malaria transmission. Though unaware of mosquito identification and dissection techniques, he burnt oil all through night dissecting all the mosquitoes thought of carrying the pathogen. Alas, success still eluded him. Ross all the while burning with that unquenchable thirst for achieving something very great, continued to search for the parasite in the mosquito body. The most significant step forward occurred on 20th August, 1897, when he got the first sight of 'pigmented cysts' in the stomach of a mosquito. Ross celebrated this historical event as "Mosquito Day", and the jubilation showed up in his memorable poem that he wrote soon after the discovery was made (see Chapter 3).

3. The Mosquito Theory of yellow fever

Yellow fever had been the scourge of the Caribbean and the coastal and southern United States for centuries. In the 18th and 19th centuries, devastating epidemics of yellow fever had claimed thousands of lives. In 1792, yellow fever claimed 10 percent of the population of Philadelphia (4,000 out of 40,000). New Orleans was swept by two epidemics: in 1853 almost 8,000 died and 4,000 more in 1878. That same year over 5,000 died in Memphis. Many thousands more died in other American cities and numerous Caribbean islands. The sheer terror that the epidemics brought was often overwhelming. Despite years of efforts by the best medical experts, no one knew how it was spread, why some were stricken and other family, friends or neighbors simply passed over. Equally frightening, there seemed to be no way to protect yourself or your loved ones. It appeared that no defense, no barrier, no precaution could stop its deadly progression. The epidemic in Memphis was so volcanic that it overpowered normal human behaviour: the frightened men, women and children poured out of the city by every avenue of escape, by every possible conveyance such as the hacks, carriages, buggies, wagons, furniture vans, street drays, bateaux, or by anything that could float on the river, and by the railroads. The ordinary courtesies of life were ignored, politeness gave way to selfishness, and the desire for personal safety broke through all social amenities.

"How could Yellow Fever, the dreaded killer, be stopped ?", was the daunting task of the Yellow Fever Commission that was established in Cuba some 100 years ago. It consisted of Walter Reed, James Carroll, A. Agramonte and Lazear. In the first few years, unsuccessful attempts were made to prove a relationship between yellow fever and the bacteria, *bacillus icteriodes*. It failed, and the Commission members focused their efforts determining how yellow fever was spread. They had found mounting evidence against the commonly held fomite theory of transmission. Among the several other theories, Dr. Jesse Lazear was most intrigued by the possibility of a mosquito vector. The mosquito theory of transmission of yellow fever had been around for many years. Possibly the first to discuss it was Dr. Josiah Nott of Mobile, Ala., in 1848. It had been proposed again by Dr. Carlos Finlay of Havana in 1881. But try as he might, Finlay had been unable to prove his theory in almost 100 experiments on human subjects. Finlay's mosquito theory remained in abeyance until attention was again drawn to it by the demonstration in recent years of the part played by these insects, the mosquitoes, in the causation of other tropical diseases. Many in the medical field belittled him because of his passion for his mosquito theory, but the discovery in 1897 by Dr. Ronald Ross of the British Indian Medical Service that malaria was transmitted by the mosquito (*vide supra*) gave added weight to this possibility. Lazear, who had previous experience with mosquitoes during his malaria research, was most interested in the theory. Finlay was convinced that only the female *Culex fasciatus* mosquito (now called *Aedes aegypti*) transmitted the disease. Lazear carefully hatched the mosquitoes and allowed himself, along with another Army doctor, A.S. Pinto, to be bitten by mosquitoes that had fed on yellow fever patients on the fifth day of illness which had been 12 days earlier. They remained well. Seven other volunteers were bitten; they also remained well. The researchers were beginning to doubt there was anything to this mosquito theory.

On Aug. 27, 1900, Lt. James Carroll and Lazear were discussing their lack of results when Carroll suggested that they try one on the mosquitoes on him. Carroll's mosquito had bitten a yellow fever patient on the second day of illness, 12 days before. Two days after the bite, Carroll was not feeling well and on the third day he was clearly ill with fever. A blood test for malaria was negative and as Carroll became sicker, it was obvious he had yellow fever. Lazear and Aristides Argamonte were very concerned about Carroll's health as he was somewhat frail and well past 40 years of age. However, they were excited that the mosquito bite could have caused his yellow fever. This would be impossible to prove conclusively, as Carroll had been in contact with other yellow fever patients at the hospital in Havana and in the autopsy room. He could have gotten it in any of these places; no one could know for sure. On Aug. 31, the day Carroll was diagnosed with yellow fever, Lazear was in the post hospital at Columbia Barracks. Lazear and Argamonte were much relieved several days later when it became apparent that Carroll was going to survive.

There had been no cases of yellow fever on the post or in the hospital. Pvt. William H. Dean of the Seventh Cavalry was about to be discharged from the hospital for a minor condition when Lazear asked him if he would be willing to submit to an experiment involving his mosquitoes. Dean was newly arrived in Cuba and had never been off post. He agreed and was experimentally bitten that same day. Being careful not revel in someone's illness, they must have been elated when Dean became ill and was diagnosed with yellow fever. Dean's was the first experimental case of yellow fever transmitted by the bite of a mosquito that was irrefutable as to its origin.

This discovery was, however, subjected to much harsh treatment and was always in the center of stormy debates. Even celebrated men of science became locked in fierce disagreement. These differences in opinion gave rise to wildly different approaches to prevention and treatment. Around 1793, in Washington, D.C., Dr. Albert King hypothesized that mosquitoes were the real culprits, but he was ridiculed and repudiated publically for his ideas to protect his nation from such infectious mosquito bites. Another hundred years would pass before scientists realized that King was on the right track. In 1881, a Cuban physician, Dr. Carlos Finlay, made a discovery that would revolutionize disease control and ultimately make possible the construction of the Panama Canal. He identified the *Aedes aegypti* mosquito as the carrier of yellow fever.

Finlay's breakthrough had glitches, however. He correctly had surmised that the mosquito was spreading the disease by biting people previously infected with the disease and then transmitting the disease agent to other people through its bite. But when he attempted to prove his theory by having infected mosquitoes inoculate healthy volunteers with the disease, he failed. What Finlay didn't know was that the virus needed to incubate for approximately 12 days inside the insect in order to develop. He was attempting to inoculate his volunteers with the virus using mosquitoes that had only been infected a few hours. More years would pass before these missing pieces were added to complete the puzzle. Finlay continued his quest to identify the agent causing yellow fever and reveal how the mosquito transmitted it. At the same time, building on the work of Sir Ronald Ross, a British physician and Nobel laureate (1902) who identified a malaria plasmodium in the stomach of an *Anopheles* mosquito like the one Laveran had found in human blood, Finlay formulated a detailed plan for combating both yellow fever and malaria.

In 1898, the United States battleship Maine blew up in Havana's harbor. The U.S., already at odds with Spain over its harsh colonial policies in Cuba, declared war. During the hostilities, it soon became apparent to the U.S. military that yellow fever was claiming more lives than the actual war itself. In response, the government dispatched a medical commission, whose members included Dr. Walter Reed, to find a way of halting the epidemic. Under the leadership of Dr. William C.

Gorgas, the commission mounted a vigorous campaign to rout the disease, but since neither Gorgas nor Reed put much stock in Finlay's mosquito theory, efforts focused initially on ridding the Cuban capital of the filth and contamination presumed to cause the disease. But when the yellow fever death toll rose despite a massive cleanup effort, Reed decided to test Finlay's theory.

Reed constructed two small buildings through which air could circulate freely to first show that the disease was not airborne. In one building he had volunteers sleep for 20 nights in the foul bed-sheets and garments of yellow fever patients. In the other building, volunteers slept next to infected mosquitoes, but were separated by screens. In this experiment, Reed did not use recently infected mosquitoes like Finlay had, but rather insects that had been infected for some time. Then one volunteer was exposed to infected mosquitoes for three successive nights. This volunteer was the only one to develop yellow fever. Proof at last was conclusive: *Aedes aegypti* was the culprit. Reed had yielded the knowledge that would free Havana from yellow fever, although even now certain geniuses of the time, like Gorgas, remained unconvinced of Finlay's theory but resorted to eliminate the mosquito and its breeding places, besides quarantining the patients to prevent them from the remaining infective mosquitoes. The results were spectacular as within less than a year yellow fever had been virtually eradicated.

Final conclusion certainly is that it was Dr Carlos Finlay of Havana, who in 1881, propounded the theory that mosquitoes were the carriers of Yellow fever.

4. The Mosquito Theory of dengue fever

Incidentally, dengue is amongst the last of the main mosquito-borne infections to be causing great concern to both the people who are at risk of infection with its deadly forms, the dengue haemorrhagic fever (DHF) and dengue shock syndrome (DSS), and the health planners all over the world, but particularly in the tropics, due to disease's tendency to spread to non-endemic areas in all the world's five continents. Owing to the growing concern associated with its causing enormous morbidity and mortality, and the subsequent sequellae to health retardation at various levels, the Tropical Disease Research of the World Health Organization has in recent past incorporated dengue within its fold of priority research.

As it does even today, the disease was often mistaken for either yellow fever or other similar illnesses in Asia, Africa and America. Dengue came to be recognized as a separate entity only during first part of the nineteenth century when a severe epidemic occurred in the Caribbean country. Finally, the mosquito *Aedes aegypti* was firmly confirmed to be playing a vector's role in transmitting dengue fever virus in 1906. Subsequently, other species were also identified to carry the transmission, but much less seriously, and the most notable such a species is the Asian tiger mosquito, *Ae. albopictus*. The latter species, from whom dengue virus was detected in India as late as 1988, has in recent years gained tremendous significance due to its implication in disease transmission in Kerala State, hitherto largely a *terra incognito* for epidemic dengue (Tyagi et al., *pers. comm.*). In Malaysia a simian zoonosis has been identified which has rendered an altogether new dimension to the dengue epidemiology.

5. The Mosquito Theory of encephalitis

Almost each of the encephalitide-mosquito theories, all of which have been floored during the 20th century, is as full of scintillating and stimulating scientific zeal, charm and obsession as were characterized any of their predecessors with respect to other mosquito-borne infections. Yet, nearly all these certainly went on to be accepted in the concerned research community without an unreasonable confabulation or consternation. The Japanese encephalitis' mosquito theory dates back to 1930, while that of the infamous St. Louis encephalitis is credited to 1942.

WHO'S WHO ABOUT THE 'MOSQUITO DAY'

Dr. Ronald Ross, following his epoch-making discovery of Anopheles mosquito transmitting malaria in 1897, designated August 20th as Mosquito Day, which he religiously celebrated himself all along his life; the last one he celebrated was in August 1931, a year before his death in 1932. On this day, while being posted in Secunderabad, India, he dissected anopheline mosquitoes which had fed on a malarious patient called Husein Khan on August 16th. From August 17th to 21st, Ross dissected the mosquitoes, and on August 19th he identified some 'peculiar vacuolated cells' on the stomach of the mosquito. The next day (August 20th) he dissected another mosquito, and found many of these cells on the stomach wall of the mosquito. He concluded that these (oocysts) were the malaria parasite stages in the mosquito.

The significance of this finding is that until then no one had any idea of how parasites in the blood of malarious patients were transmitted via mosquitoes. There were many theories; for example, that infected water carried the parasite from drowned mosquitoes into people when they drank it. Further experiments by Ross showed that the oocysts contained sporozoites, which later on appeared in the salivary glands, and that these salivary gland parasites were able to produce malaria infection (in this case, of birds). It is for this work that Ross was awarded the Nobel Prize for Medicine in 1902. There is a plaque commemorating the discovery in Secunderabad.

It is an irony that in India, where Ronald Ross made such a great discovery on the anopheline mosquito, the 'first' Mosquito Day could be celebrated only recently on August 20, 2003, in Mumbai, when the famous mosquito repellant cream manufacturer, Godrej Sara Lee Pvt. Ltd., took the initiative on their own. The author of this book was invited in this celebration, which actually pleasantly took the shape of a scientific colloquium.

MOSQUITO SAMPLER

Sampling is amongst the most important research activities warranted to determine a lot of various different bio-ecological characteristics about the mosquitoes, especially ones acting as vector species for communi-cable disease pathogens. In mosquito sciences, adult sampling, leading to per-man hour density parameter, has for long been in use, and in India it is still practiced through the use of a mouth-operated aspirator. This aspirator-device, comprising a plastic or glass tube connected to one end with a rubber tube, and separated with latter by a thin mosquito plastic or metallic netting at the joint (so that mosquitoes sucked are held there at the end of the tube), was made available through the World Health Organization since nearly the beginning of the malaria control programme in the country. Even though this device has undergone enormous modification and, in some developed countries, like USA and UK, has also been successfully replaced by certain motorized sampling devices, in India however the same mouth-operated device had more or less been exclusively employed in the exercise (except in some institutes where, too, the motorized samplers were obtained by purchase from abroad!) – totally unaware of the fact that the aspirator was a health hazard, besides of course not foolproof to basic entomological vagaries in field adult sampling. This mouth-operated aspirator, though simple in appearance, could be operated up to nearly two feet above the erect hand (meaning a sampling person could collect adult mosquitoes everywhere but the roof, which in most urban set up is about 10ft. above the ground but with the sampler held aloft he could barely access 8.5 ft, considering average human height of 5ft 6in), only a part of the adults actually present indoors could be sampled during a specified period of time (meaning insufficient and misleading sampling), and, above all, the risk of contracting throat infection by the sampling human (the inhaled mosquito carries a lot of dust fouled with fungi, bacteria and or similar unwanted things).

When I joined the mosquito sciences, through the way of malariology, in 1981, on a Tropical Disease Research (WHO) project in the southern Gujarat state, the mouth-operated aspirator was an integral companion of an entomologist and a replacement could not be even thought of in India. The tool was obsolete, cumbersome and hazardous but still an inseparable part of an entomologist's sampling kit. One incident, however, forced me to dwell on the subject; one of my staff members, Govind Ram, a laboratory technician, who regularly joined me on routine entomological outings, showed first signs of severe and uncontrollable coughing whenever he resorted to sampling with the mouth-operated sampler. His deteriorating health soon made him totally incompatible to adult sampling with the aspirator, and he finally left the job, too. I clearly understood a need to find a replacement to the aspirator in vogue.

Some years later, during 1984-88, having joined the ICMR mainstream through VCRC, Pondicherry, and participated in numerous field-oriented entomological programmes in south and eastern Indian states, where the mouth-operated aspirator was in exclusive use for adult sampling indoors, I witnessed several of my accomplices resorting to spitting soon after each inhalation and, even suffering with occasional heavy breathing,dyspnoea. This scenario fired in me the thought the seed of which was first sown on my mantle in 1981, and I drafted my first drawing of a substitutive sampler, in my mind, though.

Come summer of 1988 and I was moved on transfer to another ICMR institute (DMRC, Jodhpur) in the hostile Thar desert. The mouth-operated aspirator continued to be witnessed in all adult sampling ventures here, too. On March 22, 1990 I developed my invention of a mechanical mosquito sampler, which the ICMR subsequently named as the "Tyagi Sampler" and filed application for seeking patent with the National Research Development Corporation, Delhi, in 1996 (*vide infra*). This possibly was the first invention in the ICMR's over eighty years history of existence that was of direct significance in field-based entomological investigations.

I have had numerous opportunities to read about the all-time great inventers and discoverers, right from Aristotle's times (500 BC), and was quite aware of the various difficult moments each of these scientists had had to go through to prove the worth of his or her research. They were criticized, ridiculed and faced fears of ostracism, yet, as they were, braved through all thick and thin by the dint of their ceaseless perseverance, undaunted determination and focused application of wisdom. Interestingly, often repudiations and innuendoes were hurled on these innovative scientists by their own contemporaries within research community due either to scientific rivalry for claiming credit on the discovery or, at times, out of jealousy, simply to create a foul atmosphere to discredit the inventor from the recognition due to him/her or, as occurred in case of my mosquito sampler, as a negative and destructive scientific approach or attitude, even procrastinate the process of realization of the significance and public utilization of the invention. The patent is currently at the last stage of award, in 2004!!

WHY IS A DISCOVERER OR AN INVENTOR FIRST CRITICIZED ONLY?

By nature a discovery or an invention is a daring effort of deviation from the conventional principles in work. Peoples particularly those who are in the same business of scientific research, if not honest to scientific research or not sensitive enough to patiently hear new ideas, lack the sense of appreciation to others' novel thoughts and thus have a tendency to remain firmly bonded with obsolete and obscure dogmas of any act. They find a discovery or an invention quite hard to appreciate. It is most unfortunate, therefore, that while a discovery is applauded by some people, most others consider it a personal affront. Some people, more or less always within the known scientific community, consider a discovery by a scientist as a signal of insecurity to their own survival or development. In other instances, some comrades, both within and outside the scientific fraternity, feel being lagged behind in the race to stardom resulting into raising intense jealousy and thereby lobbying against the discoverer when the time comes to reward the originator. In extreme cases, a discoverer or an inventor is inhumanly treated, ridiculed and even expatriated because the novel ideas underlying the discovery are far beyond the comprehension of the rude, haughty and idiotic conventional minds. Yet, if the discoverer or inventor has based his novelty on sound and demonstrable principles and is resolved to prove his worth then in course of time peoples of all sections in the society come forward to brace the new development and accept it finally. The road to success, however, is not easy and is like treading on a razor's edge. Often the road is full of thorns (i.e., all kinds of personal and professional hurdles, criticisms, inhuman treatment etc.) and has to be completed solitarily by the inventor. In this often long and difficult journey through all thick and thin only his unflinching confidence, firm determination and sound knowledge of the subject and/or matter of discovery are the originator's accomplices. The discoverer wins eventually, but not without passing through a fire-test.

No one, however great by background of position or professional attainments, escaped the criticism though it varied from person to person. While the early guessers like Lancisi, Beauperthuy and Rasori seemed to have had a low-keyed repudiation for their offbeat ideas, great 18th century scientists like Laveran and Koch faced stern opposition and criticism pertaining to their initial discoveries. Crawford was unfortunate being maltreated and ostracized, while Finlay's 'mosquito yellow fever theory' was almost hijacked by another genius in the field. Manson was a rather exception, though not without a trace of fault-finding, if not direct criticism, in both his 'mosquito filariasis theory' and the undemonstrated 'mosquito malaria theory'. The Grassi-Ross classical bickering, full of mud-slinging and work assassination for each other's contributions in deciphering man-mosquito-malaria mystery, is well known worldwide. A hundred years later, even today the scenario has not changed a bit and the originator of a discovery is made to face the oft-repeated worthless criticism.

CHAPTER - 10

"THE FIRST 'MOSQUITO' WORLD WAR" : *MAN VERSUS MAN, AND MOSQUITO - BATTLES FOR SUPREMACY AND PRIORITY!*

It was a world war with a difference: the enemy, a tiny and feeble-bodied insect - the mosquito, was still not quite known for its role in bringing deadly infections to man but was always looked upon with draconian suspicion and great scientific interest alike. For this tiny, feeble-bodied mosquito was soon going to occupy the central position of all the world's strongest minds' attention and discourses unmatched in the annals of science of any other field. There have been many a great mosquito-related battle but none was fought like the battle of 'malaria mosquito', in the last quarter of nineteenth century! The tiny mosquito assumed such a high importance that the conflict engulfed more countries into its bitter folds than probably the real World War One nearly a quarter century later! But, even this 'great malaria mosquito war' was fought in a military style. This was mankind's first war against another species. This too was an intra-specific human war, with mosquito playing the pivotal role in the entire tumultuous saga of primacy and priority over the discovery of malaria-mosquito connection.

PRELUDE TO THE 'GREAT MALARIA MOSQUITO WAR'

The 'Great Malaria Mosquito War' (which I have discussed here as the "First Mosquito World War") erupted with the first glimmer of proclamation on 20th August, 1897, by Dr. Ronald Ross, whom many claim as an unlikely, if not reluctant, scientist or malariologist, that mosquitoes transmit malaria (amazingly, and still interestingly, Ross did not yet know identifying one type of mosquito from another, leave aside one genus against another!). However, based on colour patterns, he classified them for his convenience as grey or bar-backed mosquitoes (one of which was perhaps *Culex quinquefasciatus*, the common house mosquito), brindled mosquitoes (the two mosquitoes which he perhaps used included *Aedes aegypti*, the yellow fever mosquito) and the big brown mosquito with striped (actually spotted) wings which he later called as the dapple-wing mosquito (which was in fact an anopheline mosquito belonging to the genus *Anopheles*, members of which are capable to transmit human malaria parasites and pass them to people; whether it was *An. stephensi* can only be conjectured). For this reason only, Dr. Grassi, considered amongst medical zoology fraternity a preeminent scientist with enormous experience in malaria/mosquito-related subjects, regarded Ross, who was completely unknown to mosquito taxonomy, as an amateur and an interloper. In the face of the background of the announcement of establishment of the prestigious Nobel Prize an year before in 1896, endowing in prize money for scientific discoveries the largest monetary payments ever, the first seeds of 'The great malaria mosquito war' or the war of giants were seemed to have been sown, which mushroomed in a more vicious way in years to come, for the rest of their lives, both Ross and Grassi, the former in particular, were horn-locked hurling insinuations of all types on each other with a view to secure priority or supremacy in several discoveries!

THE NOBEL PRIZE FACTOR

During the late 1870s and early 1880s, mosquito had begun to attract much attention of tropical disease researchers and Manson's assertion of mosquito's role in transmitting human lymphatic filariasis worked to further ignite subject-ebullience at a fevered pace all over the world. All the three serious diseases of the day, *i.e.*, malaria, filariasis and yellow fever, were being looked into with a microscopic precision. Still, the attraction which malaria commissioned was unparalleled in the history of vector-borne tropical diseases around the globe. Malaria always asserted some claim to be special. Widespread and debilitating, malaria – the most common of all tropical diseases – was of all time great interest, as scientists in France, Italy, Germany, United States, England and India vied in their independent and isolated pursuit with a rather nationalist fervor for research excellence and discovery claims. The struggle was no less intense than the sprint to discover the New World or the golden bird that India had once been, and it foreshadowed future scientific contests over the polio or small-pox vaccine and the atom (Spielman & D'Antonio, 2001). Most of the names in circulation were those of much acclaimed and seasoned scientists; the Italian genius Professor Giovanni Battista Grassi, commanding a team of as dedicated and qualified researchers as Dr. Amico Bignami, Dr. Marchiafava and Dr. Camillo Golgi, was already a hugely respected scientist with a unique knowledge of both medicine and zoology, and a background of several years of mosquito-malaria research; Dr. Patrick Manson, truly a genius and creative mind always perceiving connections and resemblances between disparate phenomena through his sound and logical imaginations, was an undisputable authority on tropical diseases; Dr. Charles Luis Alphonse Laveran, though unimportantly dispatched to Algeria, in 1878, on an official assignment to help save the lives of French troops from some kind of arcane infection scientifically unknown to the mankind till then and providentially pitched into investigating malaria, was later credited for discovering the first human malaria parasite *Plasmodium malariae.* In contrast to all these and many others who worked on malaria and mosquitoes, Dr. Ronald Ross, though egoistical and ambitious, like Laveran, and often regarded as an unlikely, if not reluctant, scientist, on one hand, and an inexhaustible and obsessed malariologist and a polymath, on the other, was in fact a genius *par excellence* who was ultimately credited for unraveling the long time baffling secrets of malaria parasite's journey via mosquito! All these scientists regardless of their academic and professional background had one thing in common: they were endowed with the seemingly limitless energy, patience and optimism that are indispensable traits of discoverers.

In this different kind of war game the enemy was unknown and the actors played their part of roles in all directions with a phlebotomist's cautious precision, hoping to strike the bull's eye first amongst all. While Manson's pursuit of the filarial worm between late 1870s and early 1880s cleared all doubts and dark points pertaining to mosquito's role in effecting transmission of the filarial worm and ignited equally subtle minds around the globe, scientists particularly in Europe and North America competed with nostalgic resolution and pace to win the ultimate race to scientific discovery, which was certainly like hitting the jackpot. All kinds of means were adopted to press in the claim of primacy; advance reports of 'successful' experimentations were regularly advertised in the premier dailies, scientists indulged in verbal and written duels to forward first their 'original' claim to secure credit, some scientists, if not all, lobbied not only in the world of science but in the public realm as well to muster in support to win the all-important victorious vote in the event of a contest for priority over discovery or when it mattered most, and, worst of all, a few scientists clandestinely spied over others' progress in work! Skepticism and doubt loomed large even though scientists, in order to understand the enigmatic matters of malaria a bit more better, logically involved in exchanging written, published or unpublished records, including microscopic slides. Patriotic appeals for increased funding in research became intense and nationalistic sense of competition took the uppermost tone in expressing solidarity to the home economy and health of the countrymen. Notwithstanding all this precarious environment, the scientific fraternity, comradeship and cooperation amongst scientists of various different countries too continued, though not always unscathed, as illustrated through the scores of correspondence made by Manson, Grassi and Ross in the midst of the year of discovery -1897.

Destined, as if it were, to effervesce an already ebullient quest for malaria mystery through mosquito, the announcement of setting up of the Nobel Prize in Medicine and Physiology (as a part of a series of world prizes instituted in 1896 following the death will of the dynamite king Alfred Nobel, a Swede who had a great benevolent heart and appreciation for scientists' remarkable contribution in making a better world to live in), with the largest monetary payments ever for scientific discoveries, apparently ignited every mind around genuinely engaged in deciphering malaria enigma and to proclaim first to reach the ever awaited conclusion! The contest eventually narrowed down to two scientists of contrasting characters, Grassi and Ross. Professor Grassi was a zoologist of international repute and he had been seriously toiling to prove a malaria-mosquito connection for several years. In contrast, Dr. Ross, in the words of Grassi, was an interloper and an amateur who could not distinguish one mosquito from another. For the 1902 Nobel Prize, the prize committee had initially recommended a joint award for both Grassi and Ross, in recognition of their independent and highly valuable malaria research. But, one of the members of the jury, Dr. Robert Koch, a German who also wielded tremendous scientific influence and had still not forgotten the ill treatment he had received at the hands of Grassi several years ago during an invited visit to Italy, got an opportunity to settle a personal score and, using his influence, maneuvered to vote against Grassi in the final round. Professor Grassi was most unfortunately dropped, and Dr. Ross emerged a victor.

The 'Mosquito World War' which had already begun around early 1890s now reached the climax – a point of no return. Henceforth both Grassi and Ross battled on virtually every issue of priority in several discoveries in malaria research they encountered. Influenced by the widespread British Empire and, in part, Ross's own genius in effecting timely scientific publications and/or exposition of his achievements in established scientific forums, mostly through generosity of the ever helping Dr. Manson, in Great Britain, Ross was destined to be more widely recognized and rewarded. Ross was said to be so insolently obsessed with squeezing all the possible honours that he eventually marred his highly affectionate and long-time relationship with Dr. Manson whom Ross had been calling his mentor and teacher. When Manson wished to highlight the malaria research contributions by Grassi, Ross harshly discounted Manson's science and even suggested that Manson hindered rather than helped his malaria work. Thus, the Great Malaria Mosquito War had set in with full-blooded attacks and counterattacks over the primacy of mother of all discoveries, that is an *Anopheles* mosquito obligatorily transmits the malaria parasite from man to man.

THE ITALIAN FOLLY

Scientists in Italy had a long tradition of working on insect-borne illnesses, including 'intermittent fevers' allegedly resulting from the widespread 'mal-air emanating swamps, dangerously prolific with mosquito productions. The scientific community in Italy was blessed with a string of geniuses like Grassi, Celli, Golgi, Marchiafava, Guarnieri, Dionisi, Bastianelli and many others. Yet, in spite of all the treasure of data generated through long years of dedicated perseverance, they would best be known to have won in the last lap of the medley by pinpointing irrefutably the role of particular *Anopheles* species in mediating malaria transmission as well as the malaria cycle, in human. In a parallel to what they earlier did following Laveran's discoveries, they repeated history by completing, rounding out and refining Ross's discoveries. And, therefore, even with a massive scientific background they lagged behind a solitary medical scientist in India.

In working out a connection between malaria and mosquitoes, at least two follies were significantly discernible in their endeavours. First, they '*procrastinated*' publication of their findings. If only intensity, dedication and volume of work done is considered then Prof. Grassi, the most eminent living zoologist then, would certainly have finished first in the race to pin claims to the discovery of the transmission of malaria by the mosquito. But no, he reserved his well studied judgement. He busied himself in arguing that the parasite, if it was a parasite, had to have a nucleus and none had yet been found. He did not cease at it and further argued that it had to eat and no one had definitely shown that it did. So, little realizing about Ross's amazing writing prowess and an insatiable hunger for seeing his manuscripts published at the earliest, Grassi '*waited*' for the last tittle of evidence which was accomplished only in 1889 after several scientists' researches. Later on, picking up the old threads, the Italians began work in the middle of July, 1898. By that time Ross's proof was complete and partly published. By the end of summer, Ross was able to see his results made public both by circulating his official report to all those working on the subject of 'mosquito malaria connection', including the Italians, and the lecture that Manson presented in Edinburgh, printed in two reputed journals, the *British Medical Journal* and the *Journal of Tropical Medicine*, in 1898. Secondly, none, even the preeminent Grassi, stuck to the 'mosquito malaria connection' on sustained and long term basis, and there were many gaps in resuming researches. The kind of unfazed zeal, focused dedication and resolution, and maddening obsession that Ross exhibited in every action, verbal or written, was nowhere seen at any stage of 'mosquito malaria connection' research by the Italians. Black sheeps and scientific charlatans, who existed everywhere looking for suitable opportunities, lost no time to impugn scientific credibility of those toiling hard to serve the humanity by unraveling malaria mystery. Consequently, the duel Grassi-Ross war turned more ugly with the introduction of great German bacteriologist, Dr. Robert Koch, who too was tempted to have claimed as *first* the 'mosquito-malaria connection' discovery as early as 1883, without putting forth much proof, though! Finally, the rivalry took a triangular tangle, where Grassi was sandwiched between the giant Koch and the flamboyant Ross. This equation had its own very unfortunate and eventful bearing on a later date. Subsequently, when the Nobel Prize jury met to select the winner for the 1902 Prize in Medicine and Physiology, it was Koch, who was himself locked in animus with Grassi and was filled jealousy ever since his Italy visit and Grassi's hurried publication of *Plasmodium* cycles in *Anopheles* mosquito and human in 1890 without any reference or acknowledgement to Koch's own hypothesis, railed against the idea of conferment of the medal jointly to Grassi and Ross. The result: Grassi was dropped!

MOSQUITOES UNDER SUSPICION

Truly speaking mosquitoes were suspected for long to have connection with certain illnesses, particularly intermittent fevers. Malaria was already known widely and adequately characterized clinically amongst the folks in various parts of the world. It indeed occurred as a special malignity of the environment. Europeans did know through expeditions to South America from 17th century onwards that cinchona bark extraction was able to cure intermittent fevers but they had classified all fevers as one phenomenon. Europeans, whose dominance around the world then was phenomenal, associated malarial sickness with wetlands, hence the name 'paludisme' [which in French means from swamp; *palus* (Latin) = swamp]. Rome was particularly a malarial inferno owing to Pontine swamps in vicinity, hence the local name 'Roman fever'. Clearly malarial fevers related to swamps, as unhealthy and notorious fever breeders, along with disturbed lands, forests under denudation and lands under ploughing, particularly the last one, hence the emergence and acceptance of the "telluric" and 'miasmic' theories for centuries.

Yet, though traditional epidemiologists stuck to their unhealthy soil (telluric) and malicious air (miasmas) theories, some noted 18th century Italian physicians, like Lancisi, tried to find out etiology of the disease by explaining the fevers were caused due to inhalation of a living contagium or animalcules, called 'serafici' invisible to naked eyes, which multiplied in human body under commensurable conditions. A hypothesis subsequently advanced by Rasori noting that different kinds of fevers were caused by different types of species – an extraordinary thought so close to the true interpretation of malaria which would still take several decades until discovery of advanced microscopes good enough to spot micro-organisms. The relationship between swamp-emanated inhaled bad air, comprising whatever toxin of Lancisi and others, with the fever was so obvious that the Italians, possibly in the 18th century, naturally saw the cause of fever as *mal' aria* (bad air), which was later refined as *malaria* in Great Britain. Subsequently in late 19th century when Laveran, and others, discovered causative parasites-*Plasmodium* species, the word 'malaria' was readily accepted unchanged for the disease itself.

Both the telluric and miasmic explanations circulated for a long period, added by another explanation of mixing of the toxin or poison from the soil into water which resulted in fever after being drunk. This hypothesis changed the course of thinking about malaria which was now classified along with other important gut diseases as cholera and typhoid. This meant that, as typhoid and cholera - which were associated with unhealthy aquatic, land and living environments - had to be taken care of through improved sanitary measures, malaria too – considered as a disease of rural decay, spawn of an ill-kempt, savage and uncultivated land, yielding to good husbandry and civilization – fit the pattern and invited similar countermeasures. This thinking of control of disease through good sanitation is still largely practiced.

So, water was understood beyond doubt the key factor to malaria prevalence. Interestingly, it was discovered inadvertently that malaria distribution virtually overlapped the distribution of anopheline mosquitoes (for details on *Anopheles* mosquitoes see Chapter 4). Why the miasmas were associated with these mosquitoes is due to the fact that most of the malaria-carrying species bite following the night is set in, a pointer to the fears surrounding night miasmas. Another interpretation is the feeble-bodied adult females, in particular, face the danger of desiccation if exposed to the heat of the day, hence they venture out and about more often when the air is damp than when it is dry which is very much likely during the daytime. While a school of thought of the day prudently tried to build the connection between intermittent fevers and the mosquito prevalence, another school silently advocated the presence of the pathogenic micro-organisms. Unfortunately, the fewer doctors who ventured out to bridge observed associations between mosquitoes and intermittent fevers were actually scientifically greatly impoverished due to lack of demonstrable facts. Some approached the solution in a different way; Dr. John Crawford, an American physician, while working on 'yellow fever mosquito theory' opined in 1807 that these are the eggs 'insinuated ... into our bodies' which caused the malarial fever. He was faced with all round criticism, but little did the world know that he was more nearly right than anyone else would be for ninety years in future. In 1854, Louis Beauperthuy, a French physician, better known as the prototypical author of the mosquito theory, boldly connected mosquitoes with intermittent fever – in warm weather but not in the cold. He was however firm in narrating that mosquitoes *injected* toxins, picked up from the putrefied vegetation in swamps, into humans. Rasori, an Italian physician, thought mosquitoes might carry the contagium or animalcules. All these early thinkers had no idea or a scrap of evidence of what exactly that imaginative contagium was or how and where it came from, albeit the basic notion that parasites might shuttle between different animal species having been already proposed in 1790 by Peter Christian Abilgaard!

MOSQUITOES UNDER SCRUTINY

Mosquito has been conceived to cast evils on man for centuries and by 1870-80s some concrete evidences also began to emerge manifesting mosquito's role in transmitting certain diseases. The first important work on the role of insects was done by Dr. Patrick Manson in China during the 1870s when he showed that mosquitoes might be vectors of the disease colloquially known as elephantiasis from the gross swellings, particularly of the legs and scrotum, that it often produces (Harrison, 1978). The hypothesis underlying mosquito's role in transmitting filariae precisely dawned on Manson in 1875 when on a furlough to England. Reasons which he forwarded were absolutely irrefutable: he found out the young microfilariae, which unlike their procreator moved freely in blood stream and were always enclosed in a delicate membrane, could neither feed nor move of their own volition, and therefore must complete their life cycles somewhere else – the mosquito?! But, Manson whose inquisition into the ingenious ways of nature was ebullient and effervescent conjectured by working meticulously and arduously for seven long years on his 'filariasis mosquito theory', all along with his fulltime and highly successful medical practice, and despite excruciating illness due to general gout which he contracted at the age of twenty six and suffered from it throughout his life – often totally crippled for days or weeks. Yet, finally in early 1880s, Manson succeeded in demonstrating that mosquito, *Culex pipens quinquefasciatus*, was responsible to effect man to man transfer of the not so deadly but fiercely dreaded and debilitating lymphatic filariasis worm. This discovery was by all standards a great victory in furthering human knowledge about mosquito's placement in the man-mosquito-disease triangle. However, despite the partly true nature of his discovery (Manson, a surgeon-physician, had made two major mistakes while proclaiming mosquito's intermediate role in disease transmission: (i) First, he thought that these worms completed their development in mosquitoes by five days after feeding – which we today know is more than 10 days - which implied that worms had still not matured, and (ii) Secondly, believing that the mosquito fed only once in her lifetime, he postulated that the parasites infected people only when they consumed drinking water having disintegrating body of an infected mosquito), it bore no adverse significance, considering Manson's phenomenal recognition as the tropical medicine genius and his firm demonstration that mosquitoes performed an essential role in generating human diseases, particularly his guidance to Dr. Ronal Ross about malaria and mosquito – a subject that would soon reveal the mosquito's role in producing explosive outbreaks of the disease, both in terms of human maladies of morbidity and mortality exacted by the *Plasmodium-Anopheles* nexus, as well as the unprecedented show of scientific rivalry by geniuses of the day. The filariasis-mosquito discovery by Manson went largely unopposed.

Taking cue from his 'filariasis-mosquito connection' theory, Manson began to speculate on malaria. Like Laveran, he too was baffled with the encounter of 'exflagellation' phenomenon of the parasite. Following Laveran's assertion that these flagellae were the mature stage of parasite, as well as the speculation that mosquitoes picked them up in that stage for transfer to a new host, Manson, in 1894, began to navigate his thinking the filarae way, though quite faultily. Already, for the benefit of Manson, the emergence of flagella in only a half minute, or more, old drawn blood was highlighted by the Russian Sakharov in 1889 and confirmed by Grassi in 1892. By this time, too, Manson – a warm, patient and avuncular doctor of 50 years of mature yet youthful age -

had emerged without question the outstanding expert in tropical diseases in England, and it had become widely known in medical circles that he, following his remarkable 'mosquito filariasis theory', was seriously embarking to explain the malaria tangle by way of mosquito's role in transmission. However, Manson's newly found filaria-related mosquito theory weighed heavily on him and he drew the conclusion that the red blood cells which enclosed the malaria parasite were comparable to the filarial sheaths. Consequently, he landed with a totally misplaced explanation. Even so, Manson's mosquito theory (first presented primordially in 1894 and later in the fully developed form in lectures of 1896) was a brilliant imaginative construction to pave the way for future research. Ross picked up the evidence where Manson had left!

Almost around the same period, precisely in 1880 in Havana, Dr. Carlos Finlay aired his finding of *Aedes aegypti* (then known as *Stegomyia fasciatia*) transmitting the yellow fever disease. He arrived this adrenaline-charged conclusion by pursuing a scientific methodology, otherwise next to impossible at present due ethical reasons. He made yellow fever patients to be bitten by mosquitoes, only to be let loose to interact in course of time with the uninfected healthy persons. He did succeed partially in effecting yellow fever in some healthy subjects, but failed to realize the minimum time period the infected mosquitoes needed to be fully infective (actually, Finlay only conjectured about the disease based on its jaundice-implicated nature impaling the patient, without any knowledge about the pathogen, a virus, which was not yet known to the science). But, unlike Manson in case of filariasis mosquito, Finlay rightly discovered that mosquitoes bite more than once – a trait that truly pointed out towards mosquito's role as a biological transmitter of the disease. However, as it would normally go with every exciting new discovery, Finlay was not only widely ignored, he was ferociously repudiated for this observation of his – a Cuban, particularly by the then dominating English and German workers, but also by his contemporaries in Latin America. Finlay's yellow fever mosquito theory meandered a long journey through all thick and thin to be finally accepted, almost in the name of another person, a truly famous mosquito-man himself though, the great Dr. Walter Reed, when in 1990 a universal acceptance was accorded to this great piece of scientific inquiry. Such was the deep impact of world-wide recognition of the scientific genius of Dr. Reed, a U.S. Army officer, that the 'discovery of yellow fever mosquito theory' was without any further innuendoes accepted globally. However, unlike malaria where the pathogen, a protozoan parasite belonging to genus Plasmodium, was discovered much earlier than its mosquito vector, the world had had to wait for some thirty years before the yellow fever virus could be identified, along with another *au fait* discovery regarding feral primate populations in the American and African forests harbouring a reservoir of the disease. Luckily to Finlay, his notion on mosquito theory for the yellow fever was propped up by the ultimate discovery of mosquito's prowess as a vector of malaria by Ross and the Italians, amongst which Grassi was most prominent, in 1897.

THE INTERNATIONAL COMPETITION

The three major mosquito-related researches in connection with demystifying transmission cycles of some of the most deadly and debilitating tropical diseases included well-placed and rather globally important 'mosquito yellow fever theory', 'mosquito filariasis theory' and 'mosquito malaria theory' – all during the fourth quarter of 18th century. Of these three milestones, the yellow fever and filariasis connections to mosquito went rather peaceful or unchallenged, though low-voiced disagreements or oppositions could always be audible. Both these mosquito-borne revelations were made much earlier than the one for malaria or, more importantly than anything else, the event of establishment of the highest prize money in the name of Nobel Prize in 1896, to be commenced for awarding from the year 2001. Malaria thus asserted some special claim, as already was paralleled - with still greater attention due to its prowess to stay on year after year - with the population-decimating deadly plague, a reputation no other tropical disease was bestowed with! Enemies not in clear vision then included *Plasmodium* parasites and the *Anopheles* vector mosquito species, and, true to the demand of Nature, they were to be found out by none other than the two military men, Doctors Alphonse Laveran of French Army and Ronald Ross of British Indian Medical Services, who made basic discoveries that unlocked the secrets of malaria. Both of these scientists had characteristic opposite basic human qualities (Harrison, 1978): Dr. Ronald Ross, a complicated and chronically maladjusted man, was nevertheless filled with ceaseless energy, endurance, positive thinking and, above all, the desire to do something extraordinary – something that nobody had earlier done and which would single him out amongst the world's greatest men of honour. He took life as a struggle and a challenge, and himself as a soldier eternally pitched against the people and forces who sought, from stupidity or maliciousness, to frustrate him and hold back the course of human progress. Ross was exaggeratingly ambitious for progress and would not accept anyone as hindrance in his way. Dr. Alphonse Laveran was by nature a sober and noble human having a special penchant for scientific research and professional progress.

Initially it all started as a pure international competition, with the noted German pathologists Dr. Heinrich Meckel, in 1847 and Dr. T. Frerichs, in 1858, along with Dr. Rudolf Virchow, having described and characterized the black particles in the blood smears of malaria patients of being formed of degraded haemoglobin from the affected red blood corpuscles, respectively, and many a French doctor like F.C. Mallot, Leon Colin, P.J. Keiner and Acheille Kelsch, particularly the last one, who had enormous experience in malaria pathology, showing presence of innumerable tiny black particles, especially conglomerated in spleen, liver and, sometimes, brain to give the affected organ a tinge of slate grey. Many explanations were floored as to the origin of these malaria specific black particles, and the science of understanding moved ahead without much hassles. Dr. Kelsch, who worked in Algeria in early 1870s, apparently missed the discovery of the first malarial parasite viewed in the human, which was some six years later done by Laveran in Algeria only. Professor Kelsch, true to the nature of his profession of a pathologist, looked at everything malarial with a classical pathologist's eye only, even having probably routinely killed the contents of the freshly drawn blood from the malaria patients – in which there existed every likelihood of finding what Laveran had subsequently discovered, the 'crescents'. Though, Kelsch had passed by the brink of discovery, he nevertheless had demonstrated that the black pigments emerged with the onset of fever. He obviously did not follow the question of uniqueness of black pigment in malaria, and focused on other important lines of malaria research pertaining to the disease's impacts on red and white blood cells. Thus, the central theme was left unattended and unraveled, for the man who had a keen eye for observation and a determination to find answers to most baffling queries – Charles Louis Alphonse Laveran.

Laveran, sent to Algeria in 1878 as an official necessity and certainly not on any mission to research on malaria, was nonetheless forced to go into studying malaria as military bosses were desperately keen to cease the outbreaks (of malaria) that constantly wrecked their soldiers there and exacted astronomical costs in replacing dead and the sick as well as in boosting up their soldiers' morale. He was familiar with the work of Kelsch and others and soon understood the value of studying alive the contents of the freshly drawn blood to which he had access in plenty. He met with success rather early on 5th November, 1880 when he clearly saw 'spheres' (the exflagellating male gametes). For the next four years he worked to know more about the malaria parasite which he described *in extenso* in a book titled '*La Nature parasitaire des accidents de l'impaludisme*', published in 1881, followed by two more elaborative books in 1884 and 1898, respectively. When he reported his findings to the Academy of Medicine in Paris he did not forget to underline his claim to have been the first to discover the malaria parasite. It was indeed the first glimpse of one form of the malarial parasite, *Plasmodium malariae*, a protozoan – still not known by that name. Ironically, even such a spectacular discovery had to go unattended in appreciation. There were contemporary scientists who stooped low to belittle the significance of Laveran's grand discovery by giving one or another bizarre explanation. Others, more vocal, including the celebrated haematologist Dr. William Osler, Baltimore (USA), totally refused to believe his observations and an event such as the 'exflagellation' in blood.

Disagreement with Laveran's single-celled protozoan (which he referred to as the germ) led to divert the attention of several scientists of the day to consider an alternative possibility – the bacteria. The bacillus theory of *Bacillus malariae* prevalent in Italy's infamous malarious swamps, raised by two Italians, was as soon sidelined as it arose in spite of undercurrent of support from a newly emerging highly promising discipline of bacteriology directly under the guidance of as immensely famous bacteriologists as Louis Pasteur and Robert Koch. However, this dethronement of malarial bacillus hypothesis did not quite help ready acceptance of Laveran's germ theory but, on the contrary, Laveran had to take upon himself the responsibility to convince in person by practical demonstration of 'exflagellation' to the fellow scientists both in Algeria (Dr. Richard) and in Italy (Drs. Ettore Marchiafava and Angelo Celli), but largely in vain, since both the Italian experts had their opinion reserved in favour of *Bacillus malariae*, thought as the most likely agent of malaria. Not only this, Marchiafava and Celli mischievously hid from him their own research outcome in next two years conforming to Laveran's findings, which they published without any reference to Laveran's remarkable original work with a view point of making their claim as first for the discovery. Laveran was immensely exasperated by this fraudulent act of Italians who, in turn, believed what they experienced in fresh blood preparations was not seen by Laveran and was certainly different. In an attempt to put their stake for original claim of the discovery they named their form as *Plasmodium malariae.* Dr. Camillo Golgi, whose primary interest was the nervous system, confirmed their findings, with some startling advancement in knowledge. With this the initial coolness of trust in comradeship and scientific fraternity seemed to have volatilized, and the first seeds of an embittered future world war of priorities and primacy had been sown sans healthy international competition even when new human malaria parasite species continued to be discovered with a large number of excellent scientific inputs by Golgi and others.

Laveran, as said earlier, was a never-say-die type of scientist and his consistently good relation with scientists worldwide paid dividends before long. Starting in 1885, Laveran's discovery soon spread to America, Great Britain and India, and by the end of 1890 it was said that acceptance had been universally granted. The technique with a lot of malaria-related knowledge would in time pass on to Dr. Patrick Manson in Great Britain who, in return, descended the wealth of facts to his protégé, Dr. Ronald Ross. Both Grassi and Ross would later take their own time to accept or reject these wonderful discoveries, but never to sit quite even a bit moment when it came to their own research findings. The stage for gruesome "First 'Mosquito' World War" had just been set in.

MOSQUITO-MALARIA WORLD WAR OF PRIORITY AND PRIMACY

Dr. Ronald Ross, who had been ushered into the medical profession against his intention to become a writer or a musician, did not even imagine ever to pursue research in his early life in British Indian Medical Service, leave aside the exemplary Nobel Prize winning mosquito-malaria research which of course he did beating everybody else with his unmatched discipline and focused attention. The transformation was not subtle. In fact, Ross seemed to have never come to terms with the extreme tropical climate of India, rife with blood-sucking mosquitoes everywhere almost round the year. It was after all his father's plan, not his, to serve in the British Medical Services in India where the senior Ross was lucky to have risen himself to the rank of a full General! Ronald Ross, a good son, obeyed his father to this extent, but felt eternally distant from medicine and research, and quite closer to matters deeply imprinted on his mind – the music, poem and novel writing (he did succeed in publishing a romance novel, *The Spirit in the Storm*, in London in 1896) and, of course, mathematics. In his initial days Ross had felt tormented by both the dusty heat of Indian plains (he once even got a brief posting to Nasirabad close to the eastern edge of the Thar Desert in the erstwhile Rajaputana Estate) and the callous behaviour of his seniors in sending him quite frequently on postings to less cozy and recreational, and monetarily least attractive places, still mostly as a forced bachelor having had to leave the family behind either in the native Great Britain or the previous posting. Consequently, he confronted his military bosses on one or another pretext, and, out of agony and exasperation, even threatened to quit the military job and return to England. Frustration enveloped Ross – hook, line and sinker! He had plenty of time to golf away and compose music, but the impact of continued humiliation in service kept constantly blasting off his ever alert mind to do something extraordinary that nobody in the world had earlier done and doing what he could rise to the top honours of the world. He was eternally restless but did not know where to go, or navigate his life to reach his objective. Greatly lost in the labyrinthine medical disciplines and researches, he decided to coolly think meticulously within the precincts of his medical assignments and carefully narrowed down to malaria research which indeed was the hottest topic of the day! Ross immediately jumped with a burning desire in his eyes. He thought he had got the objective, but still did not know how to strike the bull's eye!

The problem was, however, that Dr. Ross, with almost total ignorance of the literature, knew very little about malaria, had never seen a parasite, did not know the necessary techniques to examine malaria parasites (despite the fact that Laveran and, subsequently, the Italians and Americans have demonstrated microscopically the presence of various stages of *Plasmodium* species throughout the second half of the 1880s) and could not at all distinguish one mosquito from another. Overcame he did yet most of these one by one as the inexorable laws of time and destiny had in store for him something the whole humanity will be later proud of. As if providentially designed to reach his objective, he first met the great Dr. Patrick Manson under whose careful tutelage Ross rose to enormous heights of success, winning finally the race to pinpoint *Anopheles* mosquito's role in malaria transmission in 1897 and, more rewardingly for Ross, the Nobel Prize in 2002 (actually awarded in 2004). His discovery of the role of *Anopheles* mosquito (which he, ignorant of any scientific nomenclature, colloquially referred to as 'dapple wing') in malaria transmission in an avian model is a classical saga of exemplary determination, perseverance, courage, patience and ambitiously focused application.

At a time when Ronald Ross was groping in dark and ceaselessly endeavouring to grapple with the necessary microbiological and microscopic techniques, the Italians, particularly the genius of Giovanni Batistta Grassi and his team of as celebrated scientists as Drs. Marchiafava, Celli and Golgi, were already at the problem of solving the malaria mystery through implication of mosquito. It appears that both Ross and Grassi knew about each other's moves in demystifying the hidden secrets of malaria cycle, but none really showed any signs of this awareness. Ross, true to his military training was meticulous in keeping his day-to-day's account of mosquito-malaria experiments as well as bringing the findings to the notice of all through Manson in England and quick dispatches to the scientific journals of international fame. Thus, not only his phenomenal express reporting of the pigmented cells in two mosquitoes fed on malaria patient (Ross, 1897), even publication of his proof of mosquito's involvement in an avian model did not take long to get public within a few months since revelation of pigmented bodies in *Anopheles* mosquito (Ross, 1898a,b). He carefully planned his bracing the tape first by every possible, logical means in the international race for staking the claim for the unique discovery that coincided knowingly or unknowingly with the news of institution of the Nobel Prizes, including one in Medicine and Physiology, following the death of their willer, the dynamite king Alfred Nobel, in 1896, which, if won, meant unimaginably colossal prospects ahead both monetarily and from honours point-of-view.

Around the same time, and possibly a bit earlier to Ross's revelation of August 20, 1897, Professor Grassi, who had been working relentlessly on the mosquito-malaria connection for several months, then immediately unaware of Ross's discovery, true to his reputation and genius patiently waited for the last iota of truth to emerge to foolproof his own 'first' discovery of the same matter as Ross's. Grassi's first publication from his observations taken since mid-July, 1898 through summer 1899 came to see the Sun in October, 1899 and then in December, 1899, wherein he announced, referring to his associate colleagues (Celli, Bignami, Dionisi and Bastianelli), having arrived at a definitive solution of the question of mosquito-malaria riddle and that the victory was completely theirs. The rush to publish the results and air the news was unusual and inclined to beat Robert Koch who too had been advocating his own mosquito-malaria theory since as early as 1883. Ironically, it was Ross, and not Koch, who was hit. Ross felt treacherously ignored by the Italians for his 'first and original' discovery.

Italians' superior techniques and work infrastructure were always highly revered in scientific circles. Even Ross was full of praise for both Grassi's and Bignami's ingenuity in medical research and in identifying the vector, and confirming the mosquito cycle of the *Plasmodium*. In return, Grassi returned courtesies full of applause to Ross for his efforts.

These pleasantries between Grassi and Ross were only ephemeral and all the sense of mutual respectability effervesced with the initial news aired about Ross's having brought first his claim for the primacy of the discovery of *Anopheles* mosquito being invariably responsible to transmit malaria. The competition readily reached its lowest and ugliest point with the war for credit of priority and primacy of the discovery. Ross charged Grassi's work of several years as childish which he could have accomplished in a few weeks. Grassi, who was rightly considered by many to be the preeminent medical zoologist of his time, shot back calling Ross as an amateur and an interloper. Ross was stung to an instant wrath. He was a person once launched in anger was never to pull back.

Dr. Ronald Ross and Professor Giovanni Batistta Grassi envied each other for the rest of their lives, broaching over priority in various different mosquito-malaria related researches. Ross, feeling eternally insecure for his place on various discoveries, was more vociferous and hostile, having spared no opportunity to attack Grassi and inflating his own achievements. Ross engaged in a lifelong polemic to quash all Italian pretensions and reassert with ever-diminishing qualifications that he and only he had demonstrated the transmission of malaria by mosquitoes. Grassi, who in course of the continuing controversy on primacy, lost association of two of his best colleagues Bignami and Batianelli – the chief architects of Grassi's mosquito-malaria work, due mostly to his growing arrogant and egoistical contentions, battled Ross ceaselessly with matching intensity. Ross's obsession with garnering every possible laurel was so intense that he did not even hesitate to hurl insinuations on his mentor – Dr. Patrick Manson, his science and scientific integrity, just because Manson, truly a generous, impartial and honest grand man of tropical disease sciences, indicated to credit Grassi for his outstanding contributions in malaria research.

Ross, in spite of having lost association of many friends in course of the never-ending battle with Grassi, did have many others to support his cause and value. One such well wisher was Dr. Robert Koch. Haunted by Grassi's ignominious treatment to his Italian visit in late 1890s and later the surreptitious correspondence, Koch had turned into an eternally implacable foe of Grassi. Koch's persistently vehement attacks on Grassi's scientific integrity, climaxing into his violent opposition for the Italian genius against a consideration for awarding the 1902 Nobel Prize jointly to both Grassi and Ross, in all likelihood cost the former the fortune. Ross won the Nobel Prize, besides the prestigious knighthood a few years later.

For whatever happened between these two geniuses the humanity was greatly benefited from their tireless inquiries in the mystique mosquito-malaria relationship and what benefits we reap today in saving precious lives to the savage attacks of the insidious malarial disease is due entirely to their, and other similarly devoted scientists tireless and focussed hard work.

CHAPTER - 11

WARS WAGED AGAINST MOSQUITOES:
VICTORIES, DEFEATS AND TRUCES

Mosquitoes, despised for their blood-sucking habits and, in the process, also transmitting several dreaded and/or debilitating infections, are a relatively much greater source of nuisance and disturbance amongst arthropods all over the world, but particularly in the tropical and subtropical countries (Table 11.1). Many significant attempts have been executed in Americas, Africa and Asia to wipe out the mosquito which, barring a fewer examples, remained invincible (Ross, 1903). Actually, any attempt to exterminate forms of life, such as the mosquito, which compete too successfully with us may be as mistaken as it has generally proved in practice. While successes were restrictive in glory, failures, so universal and apparently ineluctable in their exacerbation, were instructive to remind that we, having basically misconstrued the mosquito problem *per se*, have proved incompetent warriors.

India during the British regime was for obvious and comprehensible reasons of widespread colonial prowess and unquestionable administrative governance in most parts of Asia, in particular, on one hand, and the ever increasing impoverished state of health and economy, on the other, glorified as a perfect experimental model ground for a variety of research investigations and demonstrations of control programmes especially about malaria - the greatest scourge of humans at that time, killing more than a million people every year across the country. This philosophy of "Indianness of all vector-borne diseases experimentations" providentially also brought in light to witness some of the most startling and bold researches ever conducted on mosquitoes, such as Ronald Ross's epoch-making discovery of an *Anopheles* mosquito transmitting malaria, in the late 1890s. It is highly interesting and stimulating to know about these pioneers and their brave moves against the mosquito, which at times appear to take the form of a real battle-field between two enemy forces hellbent to try against each other every possible arsenal of knowledge, energy and experience through all thick and thin. From man's point of view 'he' overpowered 'her' in some encounters, but not without tasting a pinch of defeat or frustation in several others. Notwithstanding, victory or defeat, each of these ventures was a saga of man's enormous courage and determination to wipe off the villainous mosquito to inherit a life without any bite or infection!

Table 11.1. Facts and figures about mosquitoes as well as some other arthropods and diseases they transmit

Human disease	Pathogen(s) involved	Vector(s) involved	Date of of confir-ming vec-tor's role	Country	Authority
Filariasis	*Wuchereria* and *Brugia* spp.	Mosquitoes	1886	China	Patrick Manson
Malaria	*Plasmodium* spp.	Anopheline mosquitoes	1897	India	Ronald Ross
Yellow fever	Togavirus	Mosquitoes	1890	Cuba	Finlay
Dengue	Flavivirus	Mosquitoes	1906	Africa	?
American Sleeping sickness	*Trypanosoma* spp.	Tritomine bugs	1895 1908	S. America S. America	David Bruce Carlos Chagas
African sleeping sickness	*Trypanosoma* spp.	Tsetse flies	1903	Africa	David Bruce and David Nabarrow
Onchocerciasis	*Onchocerca* spp.	Blackflies	1926	Sierra Leone	D.B. Blacklock
Encephalitides - WEE, EEE, SLE etc. - Lyme borreliosis and Ehrlichiosis	Virus/Bacteria	Mosquitoes Ticks	1930-1990	Various countries	Various scientists

Mosquitoes' prowess to transmit disease pathogens is matchless in the whole animal kingdom, as they are involved either solely (e.g., malaria, yellow fever, dengue) or mainly (e.g., lymphatic filariasis and several types of encephalitides) in effecting these infections in humans. At the same time, of no less significance are the discoveries of other non-mosquito arthropod vectors involved in transmission of the diseases prevailing outside India, in Africa and the Americas, in particular, although the arthropod/insect group sans the concerned transmitting vector species along with the pathogen involved do exist in the Indian environment; in 1895, David Bruce, while working in South America, has incriminated triatomid bugs (*Triatoma* spp.) for transmitting trypanoso-miasis; Bruce and David Nabarrow, while working in Africa, determined the role of tsetse flies (*Glossina* spp.) in transmitting African sleeping sickness in 1903, while around the same time Carlos Chagas found that triatomine bugs carried the *Trypanosoma cruzi* infection; in

1926, D.B. Blacklock, while working in Sierra Leone, confirmed the role of blackflies (Simuliidae) in transmitting river blindness or onchocerciasis, and, between 1930 and 1990, various virus-borne encephalitides transmitted by mosquitoes (by *Culex tarsalis*), but also equally deadly infections, effected by ticks, were correctly understood about their transmission. Arthropods, particularly mosquitoes, have developed several novel schemes to tide over difficult situations and the vertical mode of virus transmission is able to help the vector species maintain the virus in nature even if the same is eliminated from the human population. Various mosquitoes exercise this mode of transmission, but the first instance of vertical transmission by any arthropod was exemplified by *Boophilus annulatus*, a cattle tick, in 1891 in USA by Theobald Smith and F.L. Kilbourne.

THE VICTORIES

1. Yellow fever at Panama canal

This was the daunting task that the Yellow Fever Commission scientists (Reed, Caroll, Agramonte and Lazear) faced in Cuba more than 100 years ago. How could these four men possibly stop this dreaded killer? They had spent their first full month, July 1900, in unsuccessful attempts to prove a relationship between yellow fever and the bacteria, *bacillus icteriodes*. Because there was little hope of identifying the etiologic agent, the commission members decided to focus their efforts determining how yellow fever was spread. They had found mounting evidence against the commonly held fomite theory of transmission. Among the several other theories, Dr. Jesse Lazear was most intrigued by the possibility of a mosquito vector. The mosquito theory of transmission of yellow fever had been around for many years. Possibly the first to discuss it was Dr. Josiah Nott of Mobile, Ala., in 1848. It had been proposed again by Dr. Carlos Finlay of Havana in 1881. But try as he might, Finlay had been unable to prove his theory in almost 100 experiments on human subjects. Many in the medical field belittled him because of his passion for his mosquito theory, but the discovery in 1897 by Dr. Ronald Ross of the British Indian Medical Service that malaria was transmitted by the mosquito gave added weight to this possibility.

The tale of the Panama Canal, or "the big ditch", as it was then coloquilly called, is a passionate one, embodying within its innumerable folds dramas of every kind – the joy, the sorrow, the disappointment, rewards and honours, and even comeuppons and death (Fig. 11.1). It is the story of an impassioned and persistent struggle against the forces of nature, marked with the blood, sweat, and tears of tens of thousands of labourers. Engineering challenges, political turmoil, and bureaucratic headaches each added its own seasoning to an already spicy and exciting chapter in history exemplifying man's indefatigable ventures to command victory over the seemingly looking invincible land-mass, and, of course, the real villain – the yellow fever disease and its vampish carrier, *Aedes aegypti*.

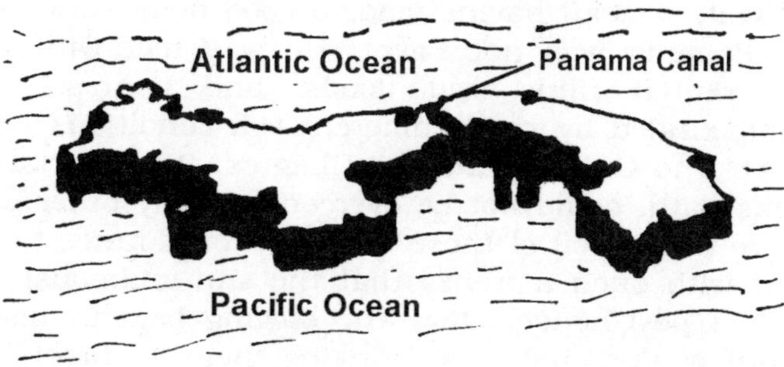

Fig. 11.1. Panama Canal joining two great Oceans.

Cutting a 50-mile swath through mangrove swamp and mountains, the Panama Canal was considered the most daring and ambitious engineering feat of its time. But before completion of "the big ditch" in 1914, tens of thousands would die, international tempers would flare, and the biggest villain had to be held at bay. There were many starts and stops before scientists, sanitary engineers, and canal builders would come to agree that the biggest enemy was actually so small as to pass unnoticed, yet it was so powerful that it very nearly destroyed the centuries-old dream of a passage connecting the world's two greatest oceans – the Atlantic and the Pacific. Before the canal was finally built, more than 30,000 people would succumb to this tiny twin-enemy, *Anopheles* spp. and its distant ally *Aedes aegypti*, the mosquitos that carry malaria and yellow fever. The dream of opening a waterway across the isthmus of Panama dates back to 1524, when Emperor Charles V commissioned a study for constructing a canal to link the Chagres River, which flowed into the Atlantic Ocean, with the Pacific Ocean. Charles' successor, Philip II, quashed the project, justifying that the act would be against God's wish. Interest in the idea waned until the mid-nineteenth century, when gold was discovered in California, igniting man's rapacious desire to rush to the newfound golden land. This panchant for gold inevitably also engendered a rejuvenated interest in finding a quicker east-to-west route than the tedious trip around Cape Horn at the tip of South America.

France, Great Britain, and the United States - the three commerce giants at the time - drew up different plans for cutting the distance. One resulted in a 50-mile railway built across the Panamanian isthmus in the 1850s. Here, railroad workers had their first run-in with what canal laborers would later face: malaria and yellow fever waged war on all comers to their tropical soil. The non-immune labourer population was particularly sickened. The French were the first to tackle canal-building in Panama. Ferdinand de Lesseps, a great engineer who had just successfully completed the Suez Canal, linking the Mediterranean and the Red Sea, was basking in international glory of bringing the Far East nearer to the West. He was naturally the first choice to undertake the job and, looking at the enormous financial prospects, he was only too eager to repeat the feat, without having made a good homework. But, he soon realized that Panama was not Egypt. He was met with thick jungle vegetation, torrential rains and floods, and oppressive heat and humidity, compounded by vastly different soil conditions than those of the arid desert. The Culebra and the Chagres, the mountain and river that barred his path, could not be overcome in any other way. Disease and death were taking toll of those brought from France to work at the Panama canal with such a frenzy that the site infamously earned the ghost-name of a pest-house. Once the digging began, the mosquitoes were driven out of their habitats, bringing them in direct contact with canal workers. Yellow fever and malaria killed an estimated 3,500 people a year during the 1881-1889, forcing de Lesseps to terminate the attempt precociously, at huge costs of character assassination, charges of financial bungling and amassing public wealth, and of course international repudiation about his engineering skills. The malaria and dengue mosquitoes were obviously the main villains in driving out the genius of de Lessips, who, following phelonious charges of fund-swindling and failure to complete the task of building the canal, was abominably tried out, with comeuppons settled – heavy fines and imprisonment - by the French Government. De Lesseps died on December 7, 1894, in his ninetieth year. Mosquito-engineered malaria and yellow fever had beaten man in their first encounter!

At the time, widespread disagreement existed about what caused these epidemics. One popular belief was the "marsh miasma" theory, which held that the invisible mists and vapours given off by swamps and decaying organic matter produced yellow fever and malaria. Even men of science became locked in fierce disagreement. Debate raged between those who believed that these pestilences were communicable diseases caused by some infectious agent and spread by materials contaminated by patients, and those who clung passionately to the idea that they were air- or waterborne. These differences in opinion gave rise to wildly different approaches to prevention and treatment. For instance, during a yellow fever epidemic in Philadelphia in 1793, Dr. Benjamin Rush, a well-known professor and physician, attempted to cure his patients by bleeding them to get rid of the environmental contaminants that he believed had invaded their bodies. Around the same time, in Washington, D.C., Dr. Albert King hypothesized that mosquitos were the real culprits, but he became the object of much ridicule when he proposed that a gigantic mesh screen be erected around the U.S. capital to keep the insects out. Another hundred years would pass before scientists realized that King was on the right track.

In 1880, as de Lesseps was en route to Panama to begin work on the canal, Dr. Charles Alphonse Laveran, a French physician working in the then French territory of Algeria, discovered *Plasmodium*, the organism that causes malaria, in the blood of a patient. The news came too late for de Lesseps. On the other hand, it might not have mattered because de Lesseps did not believe in the mosquito theory. The next year, a Cuban physician, Dr. Carlos Finlay, made a discovery that would revolutionize disease control and ultimately make possible the construction of the Panama Canal. He identified the *Aedes aegypti* mosquito as the carrier of yellow fever. Finlay's breakthrough had glitches, however. He correctly had surmised that the mosquito was spreading the disease by biting people previously infected with the disease and then transmitting the disease agent to other people through its bite. But when he attempted to prove his theory by having infected mosquitos inoculate healthy volunteers with the disease, he failed. What Finlay didn't know was that the virus needed to incubate for approximately 12 days inside the insect in order to develop. He was attempting to inoculate his volunteers with the virus using mosquitos that had only been infected a few hours before. More years would pass before these missing pieces were added to decypher the enigma. His discovery got a pat when Sir Ronald Ross, a British physician in India, identified a malaria parasite in the stomach of an *Anopheles* mosquito. Finlay formulated a detailed plan for combating both yellow fever and malaria.

In 1898, the United States declared a war with Spain colonized in Cuba. During the hostilities, it soon became apparent to the U.S. military that yellow fever was claiming more lives than the actual war itself. True to its name, the diseased person turned yellow and, in most instances, readied to follow a nearly certain death. It had taken its toll on thousands of U.S. Army recruits during the brief Spanish-American War, and it was thought that the death due to yellow fever among American troops was much more common than that from bullets. Yellow fever had been the scourge of the Caribbean and the coastal and southern United States for centuries. Despite years of efforts by the best medical experts, no one knew how it was spread, why some were stricken and other family, friends or neighbors simply passed over. It appeared that no defense, no barrier, no precaution could stop its deadly progression. In response, the U.S. Government dispatched a medical commission, whose members comprised Walter Reed, James Carroll, A. Agramonte and Lazear, to find a way of halting the epidemic. Under the leadership of Dr. William C. Gorgas, the commission mounted a vigorous campaign to rout the disease, but since neither Gorgas nor Reed put much stock in Finlay's mosquito theory, efforts focused initially on ridding the Cuban capital of the filth and contamination presumed to cause the disease. But when the yellow fever death toll rose despite a massive cleanup effort, Reed decided to test Finlay's theory.

Reed constructed two small buildings through which air could circulate freely (to first show that the disease was not airborne). In one building he had volunteers sleep for 20 nights in the foul bedsheets and garments of yellow fever patients. In the other building, volunteers slept next to infected mosquitoes, but were separated by screens. In this experiment, Reed did not use recently infected mosquitoes like Finlay had, but rather those mosquitoes that had been infected for some time. Then one volunteer was exposed to infected mosquitoes for three successive nights. This volunteer was the only one to develop yellow fever. Proof at last was conclusive: *Aedes aegypti* was the culprit. Reed had yielded the knowledge that would free Havana from yellow fever. Despite this proof, many in the medical community, including Gorgas, remained unconvinced of Finlay's theory. But he set about eliminating the mosquito and its breeding places in Cuba where he succeded in virtually eradicating yellow fever within one year, forcing him to agree with Finlay's mosquito theory. Gorgas then proceeded to destroy *Anopheles* species' habitats and scored similarly dramatic success against malaria. The Havana anti-mosquito campaigns, the first large-scale systematic public health effort of its kind anywhere in the world by that time, not only provided a formula for controlling epidemics elsewhere, they served as an early model for effective health promotion and community involvement. This victory over yellow fever laid the foundation of yet another significant and mammoth task of cleaning up the area of the future Panama Canal.

However, there were many serious political constraints. At the time, Panama still belonged to Columbia. A bitter civil war ensued from 1900 until 1903 when, with the U.S. support, it finally won its independence. Meanwhile, a second French company had bought the canal franchise in 1894, largely to keep the project alive. But, they sold their holdings to the U.S. for $40 million in 1904, whereafter plans to resume construction began. Taking stock of the huge numbers of workers the French had lost to malaria, yellow fever, and other ills related to Panama's swampy environment, the newly formed Panama Canal Commission decided that the first step was to tackle the sanitary situation in the Canal Zone. President Theodore Roosevelt sent a group of experts that included Gorgas, Reed, and others who had taken part in the Havana triumph. Since they knew now who the enemy was, they made plans to wage an all-out mosquito war, not without being shown a contretemp"s by the bureaucrats who dismissed the mosquito theory – hook, line and sinker!

In the meantime, a veritable health crisis was brewing. As hundreds of non-immune workers reached Panama, the number of yellow fever cases and deaths began to grow steadily, raising the specter of an uncontrolled epidemic. By June 1905, fear of the disease had grown so intense that three-quarters of the canal work force, including the chief engineer, had fled Panama. It appeared that, once again, *Aedes aegypti* would foil attempts to build a canal. Faced with the prospect of failure, President Roosevelt fired all the commissioners, appointed new ones, and named John F. Stevens the new chief engineer. Stevens believed in Gorgas and gave him free reign. Without the bureaucracy, Gorgas was able to get the isthmus rid of the mosquitos and yellow fever. His group began by eliminating *Aedes aegypti* breeding places. At the same time, they carried out a sweeping fumigation campaign. In Havana, which had 200,000 inhabitants, only the houses of yellow fever patients and neighboring houses had been fumigated. Panama City, which had only

20,000 inhabitants, and Colón, which had a much smaller population, were fumigated house by house. For the first time in the history of the isthmus, yellow fever was stamped out and the victory was truly fantastic. Yet, it was eclipsed by an even bigger threat that remained - Panama's most serious problem was malaria, which most believed to have killed even more people than yellow fever.

The two diseases differed in several ways. Yellow fever confers immunity, but malaria does not. Repeated attacks are frequent, and people who have been infected remain capable for years of transmitting the disease to others through the bite of an anopheline mosquito. Gorgas estimated that most of the 12,000 people living along the Canal Zone carried the infection. The mosquito carriers also had very different habits. Whereas *Aedes aegypti* laid its eggs in fresh, still water, *Anopheles* mosquitoes bred in swamps which were prodigious in Panama. So Gorgas and his men researched and, noting that *Anopheles* mosquitoes could fly only short distances, ordered the drainage or destruction of all potential breeding places located close to human dwellings which resulted in their elimination. Obligatorily doors and windows were screened to keep the adult mosquitos out, and underbrush that might shelter them was cut. In addition, the antimalarial drug quinine was dispensed to all inhabitants as a preventive measure. The results were simply spectacular: the number of malaria deaths fell to zero and the number of cases among canal workers also dropped dramatically. The isthmus had been freed from the grip of yellow fever and malaria. At last, the canal could be completed after a long peregrination of time spanning over 34 years, on August 15, 1914!

Thus, making of the Panama canal is also a saga of the fierce battle between man and mosquito in which the former finally edged out the later by freeing the isthmus from both malaria and yellow fever, largely due to the invaluable public health contributions, directly or indirectly, of indefatigable men like Laveran, Ross, Finlay, Gorgas, and Reed, but never to forget in the least men of great courage like Carroll, Agramonte and Lazear, the last of whom even sacrificed his life for a noble cause of deciphering a centuries-old puzzle – the cause of transmission of yellow fever. The malaria and yellow fever campaigns carried out in Panama provided a working blueprint for the control of these diseases that was subsequently applied around the world.

2. Mosquito/Malaria control in Kheda district (Gujarat State), India

In the first half of the 20th century there were an estimated 100,000,000 cases of malaria each year in India – roughly 25% of the population. The insecticide DDT, introduced in 1950s, was dramatically successful in controlling malaria vectors and incidence plummeted to 100,000 cases per year in the mid-1960s. The momentum of the then

National Malaria Eradication Programme [NMEP, which was first re-christened as National Anti-Malaria Programme, in late 1990s and more recently subsumed alongwith other similar national programmes into a newly erected integrated scheme called, `National Vector Borne Disease Control Programme (NVBDCP)' by a government decree dated 2nd December, 2003] could not, however, be sustained due to multiple reasons, such as the vector resistance to insecticides, decreasing community acceptance of house-hold spraying and high costs of alternative insecticides, compounded by various financial and administrative constraints. At the same time malaria incidence rose rapidly to get around 2 million mark per year for only a couple of years, with *Plasmodium falciparum* proportion rising steadily, notwithstanding. Recognizing plausible drawbacks in controlling malria vectors with insecticide spraying, the Indian Government through the Indian Council of Medical Research (ICMR)/Malaria Research Centre (MRC) launched, under the able stewardship of Dr. V.P. Sharma, Director of Malaria Research Centre in Delhi, a malaria control programme with bio-environmental approaches to vector control (including engineering modifications, environmental manipulation involving local communities and the use of larvivorous fish and other predators to reduce mosquito abundance) at a dozen sites throughout India in a variety of urban and rural locations, of which Kheda district was most significant.

Kheda is basically a rural site, typical of much of the high malaria risk areas, located close to the industrial capital of Gujarat State in western India. The project was started in mid-1983 and soon various details were gathered about mapping, surveys, malaria stratification, site selection, fish production, requirement and procurement of equipment. The project carried out work in seven villages with a population of 25,000 but expanded yearly to cover a population of 700,000 by 1987. The project undertook several field level activities, like (a) environmental management to reduce mosquito breeding sites by minor engineering works, drainage, soak pits and application of polystyrene beads to cover water sources, (b) biological control of mosquito larvae through the introduction of larvivorous fish, (c) education of the community concerning malaria, its transmission and control, (d) surveillance and chemotherapy to reduce transmission of malaria, and (e) other environmental improvements and income generating activities (e.g. social forestry, fish breeding, the introduction of smokeless stoves. In addition, the project also undertook considerable amount of research, management, supervision and training activities.

As can be seen from the table below, a substantial case reduction was achieved in environmental control area villages than those which were conventionally insecticide-sprayed by the NMEP, although the annual parasite incidence (API) remained more or less same in both types of localities (Table 11.2).

Table 11.2 . Malaria incidence in sprayed and environmental control project area in Khera district

Year	Sprayed area (NMEP)		Environmental control area (MRC)					
			Village Complex A		Village Complex B		Village Complex C	
	Malaria cases[a]	(API)[b]	Malaria cases[a]	(API)[b]	Malaria cases[a]	(API)[b]	Malaria cases[a]	(API)[b]
1981	75.71	(32)	4.04	(155)	2.03	(60)	11.63	(51)
1982	58.70	(25)	2.36	(91)	1.97	(58)	11.24	(44)
1983	26.42	(17)	0.41	(16)	0.47	(14)	3.25	(12)
1984	13.52	(6)	0.14[d]	(5)	0.14	(4)	1.37	(5)
1985	4.72	(2)	0.07	(3)	0.09[d]	(3)	0.91	(3)
1986	5.25[c]	(2)[c]	0.06	(2)	0.07	(2)	0.48[d]	(2)

[a] In 000s.

[b] Annual Parasite Incidence (cases detected per 1000 population).

[c] No spraying took place in this year.

[d] Indicates change of control strategy from insecticides to environmental methods.

It was effectively demonstrated that environmental control measures have an edge over the insecticide-spraying methodology. The main vector of malaria in Kheda district is *Anopheles culicifacies* which has developed resistance against DDT, dieldrin/HCH and malathion. Therefore, entire emphasis was laid on reducing populations of this vector through various different environmental measures.

There are obviously a variety of associated benefits from the environmental intervention. A few examples are worth quoting here; (i) the introduction of fish to reduce larvae and mosquito densities has provided a lucrative additional benefit in the form of fish for consumption. The gross income from the ponds during 1986-1987 was Rs. 78,000 and was retained within the villages for their development fund, (ii) tree plantations in wasteland and marshy areas to reduce malaria breeding sites have encouraged the development of village nurseries which are expected to generate income either during their lifetime or on felling (in one village alone the tree yield is estimated to be worth Rs. 200,000). The net cost per head for environmental control is thus clearly less than Rs. 4.5.

This project was obviously a great success, igniting a fresh hope to abate both mosquito population and the malarial disease they transmit, in a true 'Gandhian' way – by practicing safe environmental measures. Dr. V.P. Sharma, the dynamic crusader of this unique experiment, was a clear victor!

3. Mosquito/Bancroftian filariasis control in Pondicherry, India

Most part of the Coramandel coastal region along the south-east peninsular India is endemic for lymphatic filariasis caused by *Wuchereria bancrofti*, a nematode belonging to family Filariidae, and transmitted to man by the infective bites of haematophagous mosquitoes. *Culex quinquefasciatus* is the major vector responsible for transmission of bancroftian filariasis in India. As such, the areas in and around Pondicherry Union Territory have also been affected badly with both the disease and mosquito infestation, and in order to develop a model for their containment, the Indian Council of Medical Research through its institute of Vector Control Research Centre, Pondicherry has undertaken a filariasis control demonstration project, better known as the Pondicherry Project, through an integrated disease vector control approach, under the pioneering leadership of the Centre's Director, Dr. P.K. Rajagopalan, ably assisted by Dr. P.K. Das. Project's Assistant Director (and now the Director, VCRC) and a team of dedicated scientists and technicians. I was associated with this project between 1984 and 1986.

Pondicherry, a former French dominion with a total area of 59.38 sq. km and population of 272,250 (at 1981 census level), lies between 11.45° to 12.15° latitude north and 79.35° to 80.00 longitude east on the Coramandel coast, about 165 km south of Chennai (old name Madras). Its climate is dissymmetric type, with high humidity, rainy days and a relatively cool season. Despite general hygiene in urban agglomeration in Pondicherry being of relatively high order, it nevertheless bore enormous mosquitogenic conditions largely due to unplanned urban growth, choking or obliterating in the process the existing drainage storm-water canal system of the township. The major contributory breeding sites, of course were cesspits, cesspools and choked drains etc.

Over the years Pondicherry presented a grave situation of the disease prevalence. In 1957, of about 12% people surveyed, out of a total population of 59,835 resident in the town, for various epidemiologic parameters the *mf* rate was found as high as 10.3%, with the disease rate, infection rate and infectivity rate as 2.7%, 3.1% and 1.4%, respectively. In 1980, according to the National Filariasis Control Project, the *mf* rate, disease rate, infection rate and infectivity rate obviously declined to 4.89%, 0.62%, 1.20% and 0.24%, with both *mf* (aver. 8.59%) and disease rate (45.81%) being highest in the core urban area. In this core urban area the vector mosquito, *Cx. quinquefasciatus*, also was found in the highest per-man hour density (32.24). The control of disease was brought about mainly by reducing mosquito population through integrated vector management strategy comprising (i) inter-sectoral cooperation, (ii) community participation, and (iii) integrated vector control by source reduction and allied methods. A detailed plan of activities was drawn which was subjected to evaluation and re-evaluation periodically to assess the vector preponderance in each classified area. Amidst a control programme utilizing extensively only environmental and biological methods, insecticides were also deployed where inevitably required.

At the conclusion of the project, in 1985, the *mf* rate dropped in Pondicherry from 8.41% in 1981 to 6.34% in 1985, and to 6.35 in 1986 (showing a reduction by 24.51%) from 8.93 in 1981 in the VCRC control demonstration area, which was however comparable to the NFCP zone figures of 6.33 in 1986 and 8.93 in 1981. Of great interest was the observation that *mf* rate in the 0-5 yrs age group wherein it was 2.22% in 1981 and 1.76% in 1986 in the NFCP area, but got dropped to 0.35% in 1986 in the VCRC control demonstration area. The disease rate in the VCRC area was found to be 2.26% in females and 13.67% in males, with older age group people manifesting more of disease symptoms. Both the diseases and mf rate had clearly been shown to decline.

The ingenius leadership of Dr. P.K. Rajagopalan, ably assisted by his team especially Dr. P.K. Das, emphatically demonstrated that the filariasis vector mosquito, *Cx quinquefasciatus*, could be substantially controlled through known, simple and yet strictly implemented environmental measures and with minimum use of insecticides. I, being a part of the project team, witnessed the major part of development of this project between 1984 and 1987. Victory in the end was surely theirs - Dr. Rajagopalan and his team.

4. The Klang and Port Swettenham Project

Almost contemporary to the Mian Mir experiment of mosquito and malaria control in India in the beginning of the 20th century, malaria control activities were also initiated in Malaysia in 1901, aided with the new information on the role of mosquitoes in the transmission of malaria (*cf.* Konradsen *et al.*, 2002).

The town of Klang, harbouring 3,500 people, was located in hilly terrain on the Klang river in one of the most important districts of rubber and coffee cultivation. The town experienced enormous malaria problem afflicting both civilian and service personnels as well as labourers who were largely incapacitated and prevented to attend to work (Watson, 1905a). Anophelines were found breeding copiously in the swamps in and around the town as well as large number of open well and pools along the road. The foothill lands being rich in high water table often created flooded conditions following rains (Watson, 1924).

Control measures aiming at bringing down the mosquito population and reducing malaria burden included:

(i) Dense forest vegetation and undergrowth from within and without the town was removed.

(ii) As many as possible swamps present in the township were filled or drained.

(iii) Large-scale drainage systems were put into use, based on construction of concrete turf-lined inverts. This type of drain was found to be relatively inexpensive and the turf sides facilitated water inflow from subsoil.

(iv) Special constructions were designed to prevent tidal water from entering the drains.

(v) Open drains that followed the contour of the hill at its lowest point were constructed along the foothills. The idea behind this was to capture the water from the springs and surface runoff before it could accumulate on the plains outside the town.

Within a few years the entire town perimeter was drained with the cost primarily covered by the State and, to a lesser extent, by private landowners and enterprises. After some years experiments with subsoil drainage systems were introduced with a good result in terms of improved drainage and reduced mosquito populations.

Mosquito menace and malaria were also rampant further down the Klang river at Port Swettenham surrounded by mangrove swamps. Though not very old, the port immediately began to report malaria cases, with virtually 100% labour force falling sick. The infection later spread even to crews from the labourers returning to work. To abate vector mosquito breeding, embankments were raised at the edge of port township to prevent inflow of the sea water, aided with a network of surface drains to channel any tidal water. An iron pipe was attached with each drain in such a way that the intruded water would be fluxed out during low tide but the pipe was closed during high tide. However, the mosquito menace could be partially checked because of a railway line on a raised bund which prevented clearing of deposited water across the bund albeit frequent culvert constructions. Mosquito breeding was prevented, therefore, by both chemical and environmental methods and the cases were treated with quinine. Basically it was a joint effort of several organizations working in coordination with a meticulously drawn plan. The result was spectacular both on Klang township and the Port Swettenham as both malaria cases and mosquito menace were reduced substantially. This experiment proved to be highly successful and replicated in many parts of Malaysia, Singapore and elsewhere in Asia, as well as in United States of America.

THE DEFEATS

1. Freetown, Accara and Lagos in West Africa

Freetown, Sierra Leone, the was first of Britain's West African colonies. The Freetown malaria control programme had failed because Dr. Ronald Ross, who was posted there to fight malaria, underestimated the number of breeding pools and the sheer number of vectors that he was trying to control. Ross had very limited funding and the best available technology was to pour oil on the numerous breeding sites around Freetown.

The problem was not only that Ross severely underestimated the number of breeding sites, but he didn't appreciate that as soon as the oil treatments stopped, breeding would begin again. Ross redoubled his efforts with increased funding from private sources and ensured the removal of all potential breeding sites, including rubbish, broken bottles and other potential water containers. Despite these concerted efforts, the programme was remembered more for its impact on the Freetown's rubbish than with malaria control. Despite the fact that Stephens and

Christophers had seen the programme fail in Freetown, they both believed that malaria could be dramatically reduced at Mian Mir.

2. Mosquito control in Mian Mir

One of the earliest and highly challenging mosquito control campaigns was undertaken in India, or for that matter in the entire Asia, at the beginning of the 20th century. Then Mian Mir was a part of undivided Punjab in the pre-independent India, and is now a suburb in Lahore in Pakistan. Malaria was particularly severe in Mian Mir and was therefore taken up as an experimental site and model for intervention programmes elsewhere with equally high malaria endemicity. Mian Mir comprised a huge and sprawling army garrison, which continued to be severely hit by malaria despite extensive sanitation activities, and traditional farming areas punctuated with scattered residential localities. The physiography of Mian Mir, though alluding towards an arid environment, included very flat terrain allowing limited natural drainage (cf. Konradsen et al., 2002). Soil of the region was impervious to water after 1-2 hr of rains which would result into formation of indiscriminate puddles inveigling mosquito breeding if persisted for longer time. Since the gradient in land was very poor, even the storm-water drains did not keep constant flow, leading to form a perennial breeding ground for certain mosquitoes. Another significant feature of Mian Mir was its extensive irrigation network of shallow channels from which peasants drew water by means of Persian wheels to irrigate field crops and home gardens. Christophers (1904) has described in detail the baseline investigations carried out before initiating vector control activities, which included assessment of malaria endemicity in the area, seasonal transmission dynamics of malaria and stratification of malariogenic conditions particularly the vector breeding sites.

Besides malaria being rampant amongst the civilians, the infection also came down heavily on the army personnel. Thus, a special focus of the investigation was the burden of disease among the services men and the vector breeding in the vicinity of army barracks. Soon, it was revealed that *Anopheles culicifacies*, which bred profusely in the smaller irrigation channels growing some kind of vegetation along the edges, was the vector of endemic malaria in Mian Mir, signifying importance of special attention on types of irrigation network in the region (James, 1903a). Sometimes, these water courses would overflow, or even get breached, creating breeding opportunities along the channels. The poorly designed and maintained small earthen drainage canals were other important breeding sites. The practice of providing excess irrigation water on daily basis to small gardens immediately surrounding homesteads created breeding opportunities, too. The army quarters being very close to the irrigation canals, numbering ten, were the worst affected (James, 1903b).

The Mian Mir mosquito-malaria menace was conceived as a problem of environment degradation, therefore, most of the methods adopted to reduce vector breeding comprised (i) repairing of the banks of irrigation channels, (ii) clearing silt from the irrigation channels to maintain good flow without allowing overflow, (iii) reclamation of low-lying areas otherwise tending to collect water with the smallest of rains, (iv) removal of detritus and/or debris from the irrigation canals near to the army barracks so as to reduce chances of pool formation, (v) drying up of seasonal water pools, as much as possible, (vi) emptying the domestic water containers undesired for any good domestic purpose and which unnecessarily collected rain water or otherwise for unduly long periods, (vii) creating superficial small drainage lines from large temporary ground pools in order to dry them up by thinning under the sun, and (viii) during the most troublesome rainy period of June and July, sustained efforts were made to reduce the breeding potential of large pools formed in drainage canal either by using buckets to empty the pools or by applying kerosene oil. Simultaneously, malaria cases were treated with through mass distribution of quinine (Christophers, 1904).

The Mian Mir case was one of the first documented cases resulting in heated debate over the success achieved in reducing vector population and the malaria incidence through mainly the environment-based methods. Since, however, very little reduction in adult vector population was achieved, along with nearly indiscernible reduction in malaria cases particularly in the army residential areas, the prestigious Mian Mir campain was branded a failure. Ross (1904 a,b), and others, were critical from the outset itself about the improper design of the operation. Ross (1910) claimed that not much efforts were actually put in to environmentally sanitate the region, besides insignificant investment into the campaign. If only the costs on sick soldiers at Mian Mir as made by the government was to be considered then the overall expenditure for malaria prevention made up only a $1/400^{th}$ proportion of this cost. Watson (1924) charged that the area was never sufficiently drained and proper outlets should have been constructed from the drainage canals, possibly with pumping devices, to assist the drying of canals.

The Mian Mir case was also one of the first cases resulting in a debate over the role of irrigation in the spread of malaria and how to combine the necessity of food production and livelihood with public health (Giles, 1904; Anonymous, 1910). It also gave rise to some of the fundamental questions for effective malaria vector control, i.e., how large an area should be covered to have an effect on the protection of human settlements and how much reduction in adult population will have to be achieved to measure an impact of malaria on human beings? The subject of cost effectiveness in environment-based control methodology became the central point of subsequent discussions and enquiry (Watson, 1905b).

CHAPTER - 12

MOSQUITOES SEIZED:
DDT AND BEYOND IN INDIA

Mosquitoes are responsible for transmitting some of the most dreaded communicable diseases in humans, such as malaria, filariasis, dengue (including dengue haemorrhagic fever), Japanese encephalitis and several other types of viral infections. Of all these diseases occurring in India malaria is of greatest importance, though dengue too has recently emerged as a major health problem. In the pre-DDT era, malaria alone was responsible for about 0.1 million deaths and over 100 million cases in India each year, resulting into unlimited human misery involving economy as well as physical and intellectual standards (Sinton, 1935; Sharma and Mehrotra, 1986). With the discovery of insecticidal properties of DDT in 1939, and its availability for health programmes in late fourties, it became the mainstay for controlling vector mosquitoes and curbing the disease transmission. However, soon it was discovered that the target malaria mosquito *Anopheles culicifacies*, which was responsible for nearly two-third of the total malaria cases in the country, had developed resistance against DDT, rendering it ineffective even at a maximum dose of 2 gm (*ai*) per sq.m. This situation immediately led to the employment of other insecticides, although some of these were even lethal to mammals and birds (*e.g.*, dieldrin) while others especially pyrethroids, proved relatively less residual, on one hand, and far more expensive, on the other. The control of vector mosquitoes appeared a near possibility when less than 0.1 million malaria cases without a single death in 1965 were reported under the aegis of National Malaria Eradication Programme (NMEP). The euphoria was, however, short-lived and malaria resurged with a vengeance within a few years. In the ensuing years it had become extremely difficult to control or even suppress effectively vector population with insecticides like DDT, BHC, malathion (as adulticides) and Baytex and Abate (as larviicdes) due to various technical constraints, although these all still remain to be major chemical weapons against vectors. Their injudicious usage in the past have largely polluted the environment. Consequently, insecticides are becoming increasingly unacceptable to the community.

Need for Alternative Strategies

In spite of the well recognised fact that the bare usage of insecticides can no longer suppress the vector population, the national disease vector control agencies are even today practically solely dependent on these chemicals for instant results. Three main reasons can be cited here to discourage the employment of conventional insecticides in future. First, many vector species have developed resistance against most of the insecticides in vogue. All over the world, 392 species under Arthropoda have developed resistance, of which 136 (35%) species are vectors of public health importance. These include 47 anophelines resistant to dieldrin (and BHC), 34 species to DDT, 10 species to organophosphate compounds, 4 species to carbamates and 7 species to synthetic pyrethroids. There are also 42 species of culicine mosquitoes that have developed resistance against these insecticides. In India, the most important malaria vector, *Anopheles culicifacies*, has become resistant to DDT in most parts of the country and to BHC in some areas. In gujarat and Maharashtra, it has become resistant to DDT, BHC and malathion.

Secondly, costs involved in the insecticidal operations are very high. It has been estimated in 1990s that to produce one effective insecticide nearly Rs. 4000 million (US$ 20 million) were needed, whreas the chance for discovering a successful insecticide was only 1/1500 compounds screened (or 0.06%). It can thus be understood that the exhorbitant costs involved in insecticides' production and their subsequent employment in vector control operations are highly discouraging for any government particularly in the developing countries. Yet, between 1960 and 1980 only, a total of 77.75 mt DDT, 123 mt BHC and 68.7 mt malathion had been applied under various health programmes in India. The Government of India is currently spending approximately Rs. 2000-3000 million annually on control of malaria alone, and there are persistent demands for higher allocation each subsequent year.

Thirdly, the persistent and injudicious use of insecticides has brought about drastic changes in our environment. DDT, which has particularly been used excessively due to its low mammalion toxicity and high residuality on sprayed surfaces, has been often detected in the human blood, milk and tissue. According to one estimate in the mid-1980s, DDT occurred 20 times the permissible limit in the vegetables in Delhi. In Maharashtra, the bottled milk was found to contain 4.8-6.3 ppm of DDT which is 7.2 - 9.5 times the permissible limit. The daily intake of DDT in an average Indian diet was estimated to be 0.27 mg or around twice the dose accepted by WHO, while the DDT residues in human milk was estimated around 0.51 ppm which is quite alarming. As the DDT continues to be the cheapest insecticide available its injudicious use in health programmes can be judged from the fact that 110 mt DDT had been employed for vector control against only 38.23 mt

DDT in agriculture in India till 1978. All insecticides have close interaction with the environment and their use in any way is bound to affect man and his associates severely.

In view of these facts intensive researches have been carried out to seek alternative measures for vector control which are cost effective, non-hazardous, acceptable to the community and do not pose the danger of resistance development in the target vector species. Hereunder are few examples which have been tried not only in the field but have also proved promising for the vector control in India and/or elsewhere.

CHEMICAL-BASED MEASURES

Following different kinds of measures have been found effective.

(i) Insect Growth Regulators (IGRs)

These are also called third generation insecticides. The IGRs interfere with the hormonal mechanisms of target organism and result into various kinds of morphological, anatomical and physiological, abnormalities so that the target species does not reach the final, viable adult stage of development. There is no likelihood of resistance development against these IGRs. However, in case the phenomenon somehow occurs, it will first of all take a long time to develop and, secondly, it can be easily overcome by the addition of synergists like piperonyl butoxide and S' S' S' tributyl phosphorothioate. A large number of IGRs, both juvenoids and chitin synthesis inhibitors, have been evaluated for the vector control but only very few of these are found effective and commercially feasible *e.g.*, diflubenzuron, methoprene, fenoxycarb (Tyagi, 1987; Amalraj et al., 1988). The IGRs have an added advantage of being used at a relatively very low dose compared to the conventional insecticides.

(ii) Surface active agents

These are nonionic, biologically degradable chemicals which on application to the aquatic media (mosquito breeding habitats) spread spontaneously to form a monomolecular film on the water surface, *e.g.*, Arosurf ISA 20E [polyoxyethylene (2) isostearyl ether (2) mole ethoxylate of isostearyl alcohol] (Tyagi et al., 1985). These films bring about mortality in the immatures by a "physical" rather than a chemical factor as observed with the conventional larvicides. The monomolecular films form a barrier on the treated surface and the larvae and pupae of the target species die off due to anoxia. The films also lower the surface tension of the treated water body (25-29 cm/dynes) causing suffiocation in the tracheal system on one hand, and disallowing inflation of the wings of the emerging adults, on the other. At the constant rate of 11.21/h Arosurf has been effectively used to control *Culex quinquefasciatus*, *Anopheles stephensi* and *Aedes aegypti* in different types of breeding habitats such as cesspits, cesspools, drains and wells. Due to their self-spreading properties, it has been constantly advocated that Arosurf and other surface active films can be employed in combination with a fast acting and residual larvicide like fenthion for a better coverage of the breeding habitats and long effective life of the film in such inaccessible habitats like marshes and lagoons.

(iii) Insecticide-impregnated paint:

This technology has perhaps never been realized on large scale in India, but these paints are in use for decades in some developed countries. Das et al. (1986) have demonstrated the usefulness of one such an insecticide-impregnated paint, Vernacide, against both the mosquito *Cx. quinquefasciatus* and the cockroach, *Periplanata americana*, in urban conditions. In both water- and oil-based formulations, this paint was found 100% effective for over one year. Its effectiveness was regained after a cold water wash to the painted surface. This could be possible because the insecticide is embodied in an acrylic medium with small openings which help to prevent its fall-out from the surface. Consequently, these paints are safer than the conventional water dispersible powder (WDP) formulation of the adulticides, and can be conveniently employed in public places such as hospitals, airports, hotels, cinema theatres, ships etc. where pestfree conditions are desirable for considerably longer time.

Apart from these paints, insecticide-impregnated fabrics have also been developed and used effectively primarily to protect from the mosquito bite. While Jambulingam *et al.*, (1989) have demonstrated the usefulness of permethrin-impregnated bednets (at the rate of 0.5 g (ai) / sq.m.) in reducing the vector density of *An. culicifacies* in the tribal villages of Orissa, which were highly endemic for malaria, for more than 3 months, Sharma *et al.*, (1989) have demonstrated the practical significance of impregnated (deltamethrin) ropes in the human dwellings.

(iv) Controlled slow - release formulations

These formulations now commercially available in different forms have invariably a biodegradable base carrier which is imbibed with a relatively high concentration of some larvicide so that when placed in the mosquito breeding habitat the base carrier spontaneously biodegrades releasing the larvicide slowly in desired low, yet toxic, concentration. Such a method is particularly very desirably in highly polluted media where repeated spraying of the conventional larvicide (*e.g.*, fenthion) is highly cost prohibitive and time consuming. *Culex quinquefasciatus* and *Cx. tritaeniorhynchus* can be significantly brought under check by this methodology. These formulations are required to be sprayed on weekly basis to bring about good reduction in the larval density.

NON - CHEMICAL MEASURES

Expanded polystyrene (EPS) beads are a synthetic product of very light thermocol-like material widely available commercially at a price of

Rs. 80 per Kg. These beads are spherical in shape and measure about 0.2 - 0.8 mm in diameter. These are non-biodegradable and can remain effective for months on the treated surface such as the wells and laterine pits. The EPS has a property to keep afloat irrespective of the level of the breeding habitat's water at any time. Curtis and Minjas (1985) and Sharma *et al.*, 1985) have shown tremendous potential of the EPS in abating breeding of *C. quinquefasciatus* and *A. stephensi*, in laterine pits in Tanzania and in overhead tanks in India, respectively. Since these beads are non-polluting in nature, they can be easily accepted in the community. To efectively control breeding of the vectors, atleast 3-6 layers of EPS should be formed on the treated surface for complete prevention of breeding or emergence at the treatment site. About 1 kg EPS is required for a surface area of 4 sq.m. Considering its long-life, non-hazardous nature and one-time application, EPS is considered cost-effective and feasible.

BIOLOGICAL CONTROL AGENTS

There is a wide range of biological agents which are now used in controlling vector larval populations in different breeding habitats.

(i) Bacterial agents:

Bacillus thuringiensis H-14 and *B. sphaericus*, commonly called as *B.t.* and *B.s.*, respectively, have shown great promise to control a variety of vectors, particularly those breeding in highly polluted water (Poopathi et al., 2003). While *B.t.* is now commercially available as Deltor, Bactimos and Tekner in different formulations (*e.g.*, liquid concentrate and water dispersible powder), *B.s.* is still undergoing the suitable marketing process and is expected to be commercially available before long. Both *B.t.* and *B.s.* are highly effective against all larval stages of *Anophelines*, *Culex* and *Aedes* mosquitoes. Deltox and Bactimos at the rate of 1 gm and Tekner at the rate of 3 ml per sq.m. of surface area were found to reduce the larval density drastically. Also, these bacterial species, albeit being pathogenic to the target vecter species, are non-hazardous to other beneficial and non-target organisms.

(ii) Fungal agents

Because of their ability to destroy larval populations under natural conditions, several fungi are considered to be potentially suitable for development as biological agents against vector mosquitoes. Candidate fungi currently being evaluated include *Culicinomyces clavosporum* and *Tolypocladium cylindrosporum* (fungi imperfecti), *lagenidium gigenteum* (*Oomycetes*) and several species of *Coelomomyces (Chytridiomycetes)*. In India, extensive work has been done on for *Culicinomyces*, *Lagenidium* and *Coelomomyces* species isolated from natural resouces. Conidia of *Culicinomyces* spp. can be suspended in water and applied with manually-operated pressure sprayers or motor-driven knapsack sprayers at the application rate of 10^{13} conidia/ha. *Lagenidium* spp. is introduced (seeded) into breeding habitats at a rate of 10^{11} zoospores/ha or 10^{10} oospore/ha during early phase of the breeding season, although repeated applications are desired subsequently. *Coelomomyces* is introduced by placing in medium resistant sponrangia or moribund larvae that contain resistant sponrangia. While in the breeding season, single introduction of 1000 infected moribund larvae/ha is good enough to check the larval breeding effectively in the early stage, yet repeated applications will be desired subsequently for a long-term control.

(iii) Nematode against mosquitoes

There are certain mermethids that attack and kill vector larvae. *Romanomermis culicivorax* and *R. iyengari* have been consistently mass-cultured in India. No commercial formulations are, howver, available at present. It is the infective preparasite stage (nema) which penetrates the mosquito larvae. After 6 - 8 days when the parasite emerges out as a postparasite, this invariably results into the death of the immature mosquito form. Since, however, both these mermethids tolerate very low levels of salinity, alkalinity and pollution, these can be safely released only in fairly freshwater bodies, such as the rice-fields..

(iv) Larvivorous fishes and insect predators

Larvivorous fishes are considered an important vector control tool because they are effective predators and are prevalent over a wide range of ecological habitats, as well as because the operational costs involved in the integrated control programmes are low. Fish like *Cyprinus carpio, Ctenopharyngodon idella, Catla catla* and *Gambusia affinis* have been found to have special liking for *Anopheles* spp. and, to some extent, *Culex, Aedes* and *Armigeres* mosquitoes.

Amongst insects, *Anisops bouvieri* (Notonectidae) and *Toxorhynchites splendens* (a non-biting mosquito) have been found to control vector larval population under specific conditions. Both these insect species can be mass-cultured and employed annually. Single fourth instar larva of *T. splendens* can consume 114 larvae of the vector species. Recently, in some south Asian countries, some dragonfly species, too, have been successfully used to control the yellow fever vector, *Aedes aegypti*. Sebastian *et al.*, (1980) employed larvae of the dragonfly, *Crocothemis servillia* in suppressing *Ae. aegypti* breeding in vases and other ornamental pots, and demostrated the advantage dragonflies had over other insect predators. Corbet (1986) has also advocated the application of dragonflies in eliminating vector larval populations in certain specific conditions, such as cement tanks, pools and ornamental pots *etc.*, as they have effective preying mouth parts and relatively longer larval life.

INTEGRATED VECTOR CONTROL STRATEGY

This is defined as "the utilization of all appropriate technological and management techniques to bring about an effective degree of vector population's suppression in a cost-effective manner". The significance of such a strategy has already been successfully demonstrated in rural areas of Kheda district for controlling malaria and in the urban agglomeration of Pondicherry for controlling filariasis (see Chapter 11). Integrated vector mangement is based on sound ecological principles and integrates multi-disciplinary methods in developing ecological principles and integrates multi-disciplinary methods in developing ecosystem management startegies that are practical, effective, economical and protective to both public health and environment. However, it is important to know that this approach is specific to local needs and problems and the operational strategy, therefore, changes from area to area, accordingly. This method offers a desired degree of control at a lower cost or with a greater long-term benefit.

Health education, participation of community and supportive legislation are three major factors that make the integrated vector control management successful. In India, for control of vectors in town, there are effective legislative provisions under the Epidemic Diseases Act dealing with elimination of potential mosquito breeding places in a specified time and manner and also securing entry into houses for the purpose of insecticidal spray and case treatment. Successful control of *An. culicifacies* in Kheda district (Sharma *et al.*, 1986), *Culex* quinquefasciatus in Pondicherry (Rajagopalan and Das, 1988) and *Aedes* mosquitoes and *A. stephensi* in Bombay are classical examples.

CHAPTER - 13

LIVE AND LET LIVE:
IS THE CO-EXISTENCE OF MAN WITH MOSQUITO INEXORABLE?

Mosquitoes have been known to the inhabitants of India from ancient times. Even though the medical practitioners of yore did not pinpoint on a particular mosquito for being responsible to transmit the disease pathogens, nevertheless there exist several important scriptures implicating mosquitoes for both the nuisance and unspecific human maladies and scourges, particularly in context with malarial fever, through their painful bites. Some of the earliest scholarly works, such as the Atharva Veda, around 1500 B.C., had described and classified malarial fever. Even there is an evidence about the usage of mosquito nets during, and possibly much before, the 13th century in some parts of India (Raina, 1991). They have been known by different names in various of the Indian States, often depicting their relationship with some disease or particular social situation (Rao, 1981). Few of these vernacular names are presented in Table 13.1. Susruta Samhita, a Sanskrit work of ancient India, which identified five types of mosquitoes (Mashakah), has referred to mosquitoes as: "Their bite is as painful as that of serpents, and causes diseases resulting from the three humours joined together; the bite as if burnt with caustic or fire, is red, yellow, white or pink colour accompanied by fever, pain in limbs, hair standing on end, pain, vomiting, diarrhoea, thirst, heat, giddiness, yawning, shivering, hiccup, burning sensation, intense cold, vesicles or pustules increasing, swellings, knots under the skin, circles etc." Mosquitoes referred to in this scripture of yore were: 'Samudrah' (marine or coastal mosquito), 'Parimandalah' (very small mosquito), 'Hastimashakah' (elephant mosquito), 'Krishnah' (Black or dark mosquito) and 'Parwatiyah' (hill mosquito). Their bite were identified to have caused 'swelling as well as itching' on the part bitten.

Table 13.1. Names of mosquitoes in different Indian languages

Sl. No.	Lnaguages	Names
1	Sanskrit	Masha, Mashakh
2	Assamese	Masha, Mashak
3	Bengali	Masha, Mashak
4	Gujarati	Das
5	Hindi	Machchar
6	Kannad	Solle
7	Kashmiri	Moh
8	Malayalam	Kothuku
9	Marathi	Das
10	Oriya	Masha, Mashak
11	Punjabi	Machchad
12	Tamil	Koshu, Kochu
13	Telugu	Domia
14	Urdu	Machchad

Mosquitoes and the malarial disease were both prolifically rampant and devastating during the 19th century; the mystery surrounding *Anopheles* mosquito's role in transmitting malaria was, however, still to take several years to be solved when Ronald Ross would enter the scene! Accounts of Ross's encounters with these blood-sucking creatures during his painfully cumbersome and never ceasing extensive service-bound peregrinations find repeated references in his *Memoirs* (Ross, 1923). He understood the significance of mosquito nets since those who defied them fell sick almost instantaneously. The British government in India maneuvered sporadic, infrequent and half-hearted mosquito control operations in selected areas on small scale, but these were largely evanescent without any appreciable output. Indian folklore during the British regime was always suffering with mosquito menace. Mostly locally available means such as the neem-leaves (*Azadirachta indica*), lemon-grass oil (*Cymbopogon citratus*) and smouldering dung-cakes were there to resort to for prevention from mosquito bites. Mosquito had become an inseparable part of the human life, particularly in rural areas, and the use of locally available means to repel them was ritually practiced. Now, we know that mosquitoes are the most dangerous animals in the world, killing an estimated 2-3 million people every year by way of various deadly infections. Even though the mosquito-borne infections had in remotely distant past, however, a history of worldwide distribution, most of them presently restrict in their epidemic proportions in the tropical countries, including India where over 300 species and subspecies of mosquitoes occur.

When Ross struck the 'Mosquito Malaria Theory' in 1897, mosquitoes instantly became the target of control with more enthusiasm and zest (see Chapter 11). In years that followed, beginning from the end of 19th century to nearly throughout 20th century, some very impressive attempts were made to abate mosquito vectors in different parts of the country under the redoubtable stewardship of such great malariologists as Christophers, Barraud, Covell, Viswanathan, Ramachandra Rao, to name a few (see Chapter 2). Still, by and large, Indian peoples could not realy isolate themselves from their foremost foe, the mosquito, despite countrywide national malaria and filariasis eradication schemes running during past nearly five decades. Mosquito has in the process become a subject of day to day tete-a-tete, so much so that when sun would be lowering to setdown nothing but mosquito is the first thing to be remembered, to be cautioned about!

Over the last one century, since the epoch-making discovery by Ross in 1897, man has made several important advances in researches about the mosquito with which our familiarity too has improved a great deal. We must, however, be absolutely cautious to make our *au fait* claims, since even after a century of intensive investigations many a mystery remains an enigma and continues to haunt the genius of human mind. Why does a mosquito look for human blood at all when other more 'bare skinned' alternatives persist? Is it only maturing of eggs in females for which the mosquito risks to engorge on human blood? How does a mosquito scan and assort a blood vessel deep under the skin to draw blood from it? What kinds of cues does a mosquito use to identify one host from another before landing for feeding, particularly in pitched dark nights? What type of sucking force does a mosquito develop to pull blood through a feeding tube with a diameter that is so infinitesimal that the resulting friction should make it impossible? How does a salivary pump work through even finer salivary tubules? Why do certain types of mosquitoes support the development of one kind of pathogen and not other? Whether the mosquito is aware of the impending parasitism or pathogenecity to her body following engorgement on human or other animal blood? These and many other baffling and arcane inquiries make a mosquito the most sought after organism in the animal kingdom.

In their enormous variety, and their adaptive genius, the ubiquitous mosquito offers us with a microscopic view of nature's logic. For those mosquitoes which share human's living environment and subsist on his blood, their dependency appears to be their strength. Like weeds, vector mosquitoes, too, tend to thrive best where their environment is disturbed, which had so often been done to bring a large number of otherwise innocuous species in contact with man, and to feed on his blood eventually. Having thus been driven away from her original sylvan homes, the mosquito has been forced to earn her livelihood against an enormously dangerous environment. Yet, this seemingly simple and feeble-bodied organism, bestowed with gigentic powers, continued to thrive. Everything about its body, life and behaviour is a perfect example of great ingeniusness, as she commences all her jobs from feeding to egg-laying with exquisite precision and economy, on one hand, and keep meeting high challenges such as the man's continuous use of insecticides as a weapon against her, deprivation from her original habitats, and, on the top of all, a cadence of inhospitable climate, on the other. It is indeed surprising to see the speed with which the mosquito, which was subjected to all kinds of adversaries by being pushed into the alien human environment, has adapted diversely and successfully to live with man.

MOSQUITOPHOBIA

The beauty of the mosquito is that, notwithstanding her being infamously known as an ugly, irritating and dangerous animal, it is also the one most remembered by the humans! The sheer feeling of fall of the evening reminds man about this notorious creature who stalks from close quarters of man's own living conditions to allow him neither the peaceful outdoor recreation during the dusk or the sound sleep anywhere at anytime. Although malaria, dengue, yellow fever and encephalitis – all apread by mosquitoes – cause a great concern to human health and economy, but perhaps nothing matches the hair-raising, fearsome feeling of the sight of a person suffering with those grotesque and disfigured lower limbs due to elephantiasis, a type of extreme manifestation of lymphatic filariasis caused by *Wuchereria* and *Brugia* worms transmitted in Asia by *Culex quinquefasciatus* and certain *Mansonia* species (see Chapter 6). No one who confronts a patient suffering from this disease ever forgets it. Not only this, the imprint is so deep, and almost indelible, that mosquito, who is responsible for its transmission, takes on the form of a devil and frightens all, rich and poor. Elephantiasis is one of the most stigmatizing diseases in the world (Spielman and D'Antonio, 2001). It occurs when the flow of lymphatic fluid is blocked and parts of the body – frequently the legs and genitals – swell enormously. Under the constant pressure of swelling, the skin stretches until it becomes thiner to the hair's breadth. Fluid collects, often forming thin sheaths that flow from a person's hips down his legs. The swollen testicles in males can match the size of a shot-put ball or even larger, dangling in between the legs to the utter embarrassment and shock of the patient. In females, the vaginal labia can swell to nearly the same volume. If legs are affected (often with dissimilar impacts), feet and ankles are the most vulnerable sites, though the swelling can envelope more of the leg upwards, giving patient a semblance to an elephant's leg. The mosquito, *Cx. quinquefasciatus*, notwithstanding a ubiquitous presence in all environments of India, is more rampant in urban set ups.

The situation of 'mosquitophobia' is almost the same in both rural and urban areas, as well as plain, (low altitudenal) mountainous and forested regions. Cities and township are presently worst affercted simply because of evolution of enormous number and types of breeding habitats supporting different variety of mosquitoes. Bigger cities and towns today mean bigger mosquitogenic problems, despite bigger shares of financial inputs in sanitizing them. Therefore, whether it is Delhi, Kolkata, Mumbai, hyderabad, Bangalore or Chennai – all great cities and the capitals of respective states, in addition to Delhi being also the national capital, the 'mosquitophobia' thrives unchecked.

MOSQUITOGENIC TALE OF TWO INDIAN CITIES

Bigger cities and towns, or metropolises, in India as well as many other countries both in New and Old worlds, are regarded parallelly with the mosquito menace. In India, *Wuchereria bancrofti* related lymphatic filariasis is principally a urban disease transmitted by the urban black-brown domestic mosquito, *Cx. quinquefasciatus*. Malaria in Indian urban set ups, spread mainly by *An. stephensi*, is presently sharing as high as 40% of the total cases in the country from a virtually negligible in the early 20th century. Dengue infection, too, though lately erupting in rural and sylvan environments, is characteristically associated with urban style of living. A mosquitogenic cross-section of two major metropolitan cities in the country, the Chennai and Bangalore metropolises, where living is far better, organized and systematic than many others, will still reflect their face punctuated heavily with mosquito-donned havens.

(1) Chennai Metropolis

Chennai, (old-timed Madras) is a metropolitan city with a much checqered history, but is also sadly known for a high incidence of malaria and, in recent times, dengue – both transmitted by mosquitoes, the former infection by *Anopheles* mosquitoes while the latter by *Aedes* mosquitoes. Over the years, unorganized and uncontrolled urbanization, population boom, poor quality of housing and a myriad kinds of unhygienic situations rife with different kinds mosquito breeding sites have all contributed to eruption of heavy mosquito density and infections.

The city of Madras stretches over nearly 172 sq. km. along the west, from Thiruvanmiyur in the South to Thiruvottiyur in the North. Madras is one of the fastest growing metropolitan cities in the country both in terms of population and employment growth. Important buildings and structures such as the Fort, the Marina - the road parallel to the sea - and the fronting public buildings, the High Court, the Moore Market, the Connemara Public Library and other public buildings are the land marks of the city.

The Madras Metropolitan Area consisting of Madras city and its environment extends over 1,166.76 sq. km. The estimated population of Madras urban agglomeration in 1991 using the exponential growth model was 5.56 million and by the component method 5.43 million. The population estimates for MMA and city area estimated by MMDA will be 10.0 million in 2011. It also indicates that between 1991 and 2011, 3.5 million persons will be added to MMA of which the share of the city is 2.0 million.

Three distinct areas are identified within the metropolitan area: (i) First, there is the city itself within the limits of Madras corporation, (ii)

Secondly, there are the urban centres and extensions contiguous to the city area, and (iii) Thirdly, there is the rural area with a number of rural settlements. *Madras city* means the area under the jurisdiction of the city corporation, *Madras urban agglomeration* means the area consisting of Madras city and the urban centers, and *Madras Metropolitan area* means the area delineated as Madras Metropolitan area for planning purposes which is also the Madras standard urban area defined by the census.

Till a few years now, malaria counts in Chennai city surpassed all estimates from rest of the State. Chennai also had the highest annual parasite incidence per 1000 in India, besides contributing 10 per cent of total cerebral malaria cases caused by the deadly parasite *P. falciparum.* Mosquitoes, which breed in the city's numerous open overhead tanks, wells, drains and stretches of stagnant water, function to courier the disease pathogens. According to an estimate the city has 95,023 wells (82,287 uncovered), 89,523 overhead tanks (47,367 uncovered), 23,368 sumps, 72,834 cisterns, 12,300 cess pools, 15,000 cess pits, 13,000 septic tanks, 700 km of storm water drains, 35 km of minor waterways and 56 km of major waterways, besides open water-filled tar drums left at construction sites, sewage treatment plants of Metro Water and trenches dug by government water, power, and communications departments and others - each a potential breeding spot for mosquitoes.

With the advent of independence, industrialization of the state zoomed with many important industries both in the public and private sectors getting established in and around Madras. The Industrial areas within the city are mainly in the Northern and Western parts where they are closely mixed up with residential developments. Only in much recent times, new industrial establishments were located outside the city limits; in fact, many important industries are bracing the outskirts. People who came to work in these areas could not afford or rent any house and thus began the growth of slums in Madras City.

Slums are important places for mosquito proliferation. The vast majority of the people who migrated to the city were attracted by opportunity and comforts offered by modernization. They belonged to the working class and found it difficult to secure accommodation within their means. So, they squatted on every open space available, as near their workplaces as possible and put up huts with cheap building materials. In this way slums grew in number and population in Madras City. The Govt. of India slum Areas (Improvement and Clearance) Act of 1954 defines a slum *'as any predominantly residential areas, in which light or sanitary facilities or any combination of these factors are detrimental to the safety, health or morals*". Slums are predominantly residential areas of blight. They are chaotically occupied, unsystematically developed and generally neglected areas, overpopulated and ill repaired. The other horrors of slum life are insufficient communications, inadequate amenities for physical and social health, the absence of minimum needs and comforts. A slum is taken to mean "*hutting areas with huts erected in a haphazard manner without proper access, without protected water supply and drainage arrangements and congested as to allow little free air to get in*". Slums generally present the most unhygienic ugliest, nauseating scene. During rainy season, the whole area gets flooded, the pathways become swampy and the entire colony become a fertile breeding place for mosquitoes, exposing the slum dwellers living in the area to all sorts of diseases and mosquito-bite related maladies.

Some of the fierciest mosquitogenic situations and or factors complementary to copious preponderance of mosquito populations in the metropolis are briefly cited below:

Slums in Madras City

The slums are scattered throughout the city. As South Madras has larger tracts of open space than North Madras, which is predominantly an industrial area, the number of slums are more in South Madras. Due to scarcity of land in North Madras, huts are built mostly on corporation dumping grounds and even over the drainage and sewage pipe lines. It is seen that 56.89% of the slum families are put up on public lands. The other 43.11% of the slums families are on private land and on lands of religious and other institutions.

Cattle population

Another menacing problem of environment deterioration is caused by cattle. There are about 70,000 cattle in the city maintained in 2,000 cattlesheds. More than 2,500 families of which 1,900 depend solely on the cattle involving over 1,300 people. About 350 families maintain cows for their own use.

Garbage and Waste Disposal in CMDA

The city generates about 2,400 tonnes of garbage of which about 1,900 to 2,000 tonnes are cleared every day. The civic body has 302 heavy vehicles, 96 light vehicles and 150 autorickshaws. Often in the event of heavy storm, the city is left behind in garbage and refuse, turning the metropolis of about 5.5 million people into a vast rotting garbage dump.

Water Ways of Madras City

The rivers that can be treated as water ways of transportation are Coovum, Adyar and Buckingham Canal. These rivers are crisscrossing the city meeting at several places. But they are used as no more than open drains to carry the city's sewerage sullage, garbage and industrial wastes, leaving them hazardous.

Coovum

It is uncontaminated at its entry point to the city, collects the surpluses of 75 small tanks from Kancheepuram where it originates and runs 16 km from the upper reaches of Madras till it reaches the Bay of

Bengal. The Coovum river runs along middle of the city, bifurcates on the southern side of the Central Jail and forms an Island and the two arms join again slightly upstream of the Napier Bridge. There are three types of waste discharges that pollute the river.

The first flush of storm water is a severe pollutant and as bad as raw sewage and in addition nutrients, sediments and heavy metals are carried in the storm water. House owners throw garbage and other hazardous and non-hazardous wastes into drains leading to Coovum. Further along the Coovum there are 27 points through which waste raw sewage enters into Coovum contributing more than 25 million litres per day. All these contamination lead to the increased pollution of Coovum. It is said that the velocity of flow of the river falls far below the self-cleaning velocity as the terrain is even and flat and accordingly the silt and sedimentation are heavy. The problem is much more with the sandbar at the mouth of the Bay of Bengal stagnation in the estuarine area. The sandbars prevent free water flow into the sea for most part of the year. The sandbars form a pool of water between the Chetpet bridge and the river mouths of Coovum. Nearly 10 lakhs to 20 lakhs tonnes of sand deposits collects at the mouth of Coovum per annum. However during incessant and nonstop rains the floods carry a substantial quality of Coovum water into the sea and the increase of population at the estuarian area every year complicates the situation.

(2) Bangalore Metrolpolis

Bangalore, the garden city of India with the reputation of a 'salubrious environment' and 'peaceful living', offers a rather glaring example of how the once finest city could virtually turn out to be a haven for these blood sucking creatures – all for the reasons associated with man's needs and activities for prosperity! Incidentally, this is also India's first metropolitan city which drafted a `Master Plan for Mosquito Control'.

Bangalore cty, known for long as the garden city of India and appreciated for its salubrious environment, has also been lately plagued by mosquito nuisance, notwithstanding a good infrastructure for mosquito control and reasonably sufficient funds (>one crore, at early 1980s level). The solution to the metropolis's mosquito nuisance was offered by Indian Council of Medical Research's Vector Control Research Centre, based at Pondicherry, by practically working out all the reasons underlying emergence of the mosquito menace in the city and subsequently tabling it in the form of a unique "Mosquito control master plan for Bangalore city" (Rajagopalan et al., 1987). Bangalore city, with all its impeccable glory of being one of India's cleanest, well-organized and hospitable cities, provide the perfect model for understanding how a nearly mosquito-free environment could be transformed into a varitable paradise for mosquitoes, only to vex its inhabitants and rob them of their hard earned sleep! More than anything, man himself is to bear the blame for this mess. Bangalore, at one time – during the 1980s, in particular, was rated the fastst growing metropolitan city in the country when the population boomed from mere 25 lakhs in 1981 to close to 45 lakhs by 1987. Huge influx of humans, in search of vocation or in connection with enterpreneureship, coupled with unorganized growth of living conditions like large slums and shanties, added much to convert the area, particularly the peripheral areas, into mosquitogenic situations. The black domestic or house mosquito, *Culex quinquefasciatus*, is the wicked culprit, although *Aedes aegypti*, the dangerous vector for dengue, is also often present.

REASONS ACCOUNTING FOR MOSQUITOGENIC CONDITIONS AND ERUPTION OF MOSQUITO-BORNE DISEASES

These are aplenty. Almost everything on earth that collects water is liable to become a viable breeding place for some or the other mosquito species, many of which can be serious vectors of human diseases. Some mosquitoes, particularly those belonging to the genus *Aedes*, require but a little moistened soil, in close vicinity to an underlying watersource though, to breed and thrive, obfuscating the 'mosquito phenomenon' a further great deal. Even so, any of these, or all of these naturally existing breeding conditions, hardly ever trigger onset of a devastating epidemic, unless some catastrophic and colossal vicissitudes or interferences with the nature are committed by man. Often wars at borders cause driving of people across the border (from hyperendemic areas to non- or hypoendemic areas) and concentration of refugees in relief camps, adding further to the already long list of mosquito breeding conditions and factors effecting spread and exacerbation of a parasitic infection like malaria.

Sometimes drastic climatic variabilities, with a human hand, are a culprit; (i) Heavy rains in otherwise xeric environment, such as the 1990 heavy rains in the Thar Desert, caused onset of an unprecedented spate of malaria epidemic in the State of Rajasthan (Mathur et al., 1991), or (ii) Prolonged deficient rains or droughts, again in the Thar Desert between 1997-2000, forcing migration of local-folks along with their cattle over to other States (with hypoendemic status), became the cause of malaria epidemic on return of these people who also brought back along with them the parasite to a non-endemic or hypoendemic home population for a sudden conflagration, or (iii) simply the irreversible changes in physiography, once again in the Thar Desert, with innumerable groundwater-based breeding conditions developed in an otherwise a *terra incognito* for a mosquito species with an infamously high potential for transmitting malaria, like *An. culicifacies*, converting the virgin land into a hotbed for disease epidemics, on one hand, and a virtual paradise for mosquitoes, on the other. The countrywide dengue epidemics in as recent times as mid-2003 have rocked the nation once again as it looked very much like bearing a human hand in inviting the problem that infected over 3500 people and killed 66 others, many of which were children. A non-stop annually cyclic malaria epidemics in the Thar Desert since 1992, the worst of which occurred in 1994 when 452 deaths were reported on the soil of Thar Desert alone, mostly in the command area covered under Indira Gandhi Nahar Pariyojana (IGNP), is an eye-opener to public health managers and disease control planners in the country. Drastic changes the IGNP brought about in the Thar Desert's physiography, besides formation of innumerable breeding grounds due to canal's seepage to several mosquito species, including the dangerous *An. culicifacies*, earlier unknown to the interior of the desert region, has turned the land as one of the 'hotbeds' for malaria outbreaks (Tyagi, 2002, 2003). These outbreaks and epidemics are a testimony to our constant struggle with mosquito-borne-diseases. The great conundrum, however, in our relationship with the mosquito is that much of the trouble is caused in part by our own deeds; air traffic, sea voyages and different kinds of commerce *modus operandi* intercontenentally.

CO-EVOLUTION OF SPECIES JUSTIFIES CO-EXISTENCE

Mosquitoes occur worldwide, in most climatic regions and situations, and in many parts persist in enormous numbers. They have been collected on mountains and below ground-level; with reference to India at altitudes of 14,000 feet or more in Kashmir, and, as exemplified by *Culex quinquefasciatus*, 903 feet below sea-level in the Kolar Gold Mines. As it appears more outbreaks of mosquito-borne diseases are expected in the future. The issue of mosquitoes is now extremely important with dengue fever on the rise, in addition to prevailing grim situation of malaria, filariasis and JE in most parts of India. These insects have long been the bane of human beings. This fascinating little creature has drawn now an attention which was never seen before.

Despite the weight of previous failures and variety of threats that human endures, man can still learn to live with mosquitoes. To accomplish this, there can not be a single formula since there are vast differences in economic status between developed and those who are still not developed, putting the 'co-existence theory' at a different footing for each group of countries. Those with means and methods, that is the rich countries, can design their needs while others with lesser resources, that is the developing and/or the underdeveloped nations, will have to look for what suit them best.

One thing that is more than sure is that, in the event of a do-or-die conflict between man and mosquitoes, the latter will likely survive us. Mosquitoes were here on earth much before humans and certainly have more tricks up their arms (in fact, legs or, may be, proboscis) than the entire human race can conceive to device with all the modern facilities available. The earlier thought of annihilating vector mosquitoes from the face of earth, therefore, was derived more by passion and overconfidence than wisdom and logic. Both species can live, without one causing harm to other. Man being the thinker has to learn and define, how?

All the while, however, reserchers may go on finding effective and novel ways to bring the mosquitoes under control. One such research is currently embarking on creating transgenic mosquitoes of a given vector mosquito species, and scientists are hopeful to suceed early. Meanwhile, following experiences obtained through past encounters with man's deadliest foe mosquito, a few time tested anti-mosquito measures can surely help mitigate the mosquito-borne dangers!

1. General living hygiene upgradation

Environment sanitation is a time-tested tool to bring about a long lasting reduction in mosquito breeding and preponderance. Most of the North American and European countries in early 20th century as well as some of Asian countries as late as 1980s, have raised their living standard to such an extent where mosquito breeding in and around houses is not only difficult to find but allowing intra-domestic mosquito breeding is regarded a punishable offence. Peoples' general awareness there is very high and they work both by themselves as anti-mosquito brigades and in close coordination with government agencies meant to control mosquitoes and/or mosquito-borne diseases in their countries. Man in such developed nations have made their life style to discourage mosquito blood-feeding on them, restricting within houses in late evening hours, wearing full sleeves shirts and trousers, and using repellants in evening recreations outdoors. House construction styles play an important role to check on mosquito entry in houses during night. In India, too, regulations governing anti-mosquito breeding are in force in some metropolises, at least, but these are generally insufficiently followed and are overtly ignored in the outgrowth of haphazard and unplanned expansion of cities and towns. A lot of efforts, both at government and people's levels, will have to be meticulously conceived, planned and executed to do what is necessary to check mosquito breeding in and nearby houses.

2. Improved housing conditions

As referred above, the improved housing style which make it unimpregnable to mosquito without sacrificing the essential necessities of *fang*, *sui* and *chi* (meaning air, light and water – in Chinese). To effect this not only the design of house, but also the material to be used in making the house in different ecological situations is also important. Adequate databases are presently available to develop the infrastructure indigenously in any country and it only requires definite regulations and implementing bodies to execute the act.

3. Obligatory mosquito-netting of all entry/exit points in homes

Houses unimpregnable to mosquito means a properly meshed one on all doors, windows and exhaust-fan points like ones in kitchens and toilets. In some countries houses are to be put up invariably with a mosquito netting as per law requirements. This is also found as a foolproof method to check mosquito's entry in the house even in the event of some negligence on the part of inhabitants. In India, houses are properly screened only either under statutory provisions applicable to government quarters in certain services or with the individual awareness of mosquito menace. In south India, particularly in Kerala, houses are excellently designed and constructed but still mostly without screening, albeit a nearly round the year mosquito menace. Several modes and designs are currently available to place it on doors, windows and exhaust-fan points, but this exercise requires a strong conviction, with a legislative provision to back up, to the people about their long term benefits.

4. Compulsory use of bed nets where affordable

Mosquito bed nets, though in use in different forms in several countries including India, have never really got as strong a sense of deployment in personal prophylaxis as in last three to four decades, especially with the invention of the insecticide-impregnated mode. Use of bed net is simple, cost effective and highly effective against mosquito biting. In fact, in Indian military services it is mandatory for everyone to use a bed net while sleeping. At the same time, any one found to flaunt the instruction has to face punishment. Yet, in public life the usage is generally restricted to either those who sense its significance or by the people who are simply overwhelmed by the mosquito nuisance day in and day out. The deployment of bed net, however, requires a strict discipline since otherwise, in the wake of improper deployment – which generally is the case, the bed net would be rendered completely useless for the purpose. As a solution to this frequent folly, insecticide-impregnated bed net provide a good recourse since not only mosquitoes are prevented from attacking humans but are kept at bay even when the bed net is not properly folded under the mattress or, in extreme cases, it is worn out here and there due to excessive and/or faulty use. This method has arisen as a possible alternate to the long-timed indoors application of residual insecticides.

5. Use of local, indigenous repellents

While outside home in the evening hours, for recreation or work, certain repellents are very handy to keep the mosquito away (e.g., DEET, DMP, DEPA). In recent times, in view of certain physiological disturbances by some of these synthetic repellents, due emphasis has been laid to rely on plant-based repellents. Use of repellents bars mosquitoes to make a contact with human and, thereby, prevents not only from the painful bite but also the risks of disease transmission.

CHAPTER - 14

MOSQUITO IN INDIAN PSYCHE:

PUBLIC HEALTH, DISEASE ECONOMICS AND SEMANTICS OF CONTROL OPERATIONS

There are lots of reasons why people should be interested in mosquitoes. Not only mosquitoes trouble people, young and old, male and female, by their irritating and painful bites, but, in the process, they also mediate transmission of many life-threatening and debilitating diseases (Tyagi, 2001). Through their bites, however, mosquitoes levy unbearable uneasiness, ill temperament and loss of sleep, thereby affecting human psychology in attending to important jobs at hand. A mosquito bite *per se* may not bring instant havoc but it is capable to cause disasters if it carried disease pathogens. Now, since there is both fear and awareness about them, it is only reasonable to put the whole subject in its right perspective to help comprehend mosquito-related fears (mosquitophobia) especially in context with the Indian people's psyche. As it would be seen in Chapter 13, mosquitoes were referred in the Indian scriptures of old times, the Vedas, in 1500-1600 BC, were in fact classified on the basis of size, colour and behaviour, and even related with intermittent fevers. The fact that mosquito has been known by as amny as 14 different names is a grand testimony to its being known to folklores in India in a much diversified manner.

MALARIA, ROSS AND INDIA

Destiny even planned Ross's discovery of *Anopheles* mosquito transmitting malaria in India only, as if to re-emphasize a grand tradition of knowledge seeking in this great country. Mysteries surrounding malaria, prolifically rampant and devastating in the warmer countries, were being attempted to be solved in many countries during the 19th century, both in the tropic and temperate regions, much before the entry of Ronald Ross in the arena. Many European nations, especially Britain, France, Spain and Holland, had under their rule several countries in the fragmented Africa, Americas, Asia and Africa, but the mosquito-borne diseases like malaria and filariasis always dreaded them to make their effective governance and living. Dr. Patrick manson had already determined the role of mosquito in transferring filarial worm from man to man in China in early 1880s, a team of great scientists like Prof. Grassi, Dr. Golgi and Dr. Bignami were already working on malaria (and possible role of mosquitoes) for several years in Italy around the same period – the seed for which were sown about one hundred years ago by Giovanni Lancisi, and even the most plausible interpretation on possible mosquito-association with malaria dissemination was already sounded by Dr. Daniel Beauperthuy in Venezuela and Dr. Josiah Nott in Alabama more than two decades ago, still the 'mosquito malaria theory' was destined to be written in India, by Ronald Ross, who despite his total ignorance about mosquito systematics and biology, and but surfacial information on malaria parasite (which too he gained courtesy Dr. Patrick Manson in England), proved through dint of his perseverance and application the real cradle of great malaria invention in this country. Indian scientidfic community take upon themselves this moment of great discovery as a matter of pride and dignity, as well as a grand research legacy. No surprise, therefore, when the Indian scientists, under the banner of Indian Society of Parasitology, logically organized an international symposium, with malaria as the central theme, in 1997 at Hyderabad, to commemorate hundred years of Ross's discovery of malaria mosquito in 1897. Not only this, very recently a malaria seminar was organized by the Godrej Sara Lee Pvt. Ltd., a great commercial name in manufacturing mosquito-repelling products, on 20th August, 2003 in Mumbai (old-named Bombay), to coincide with the "Mosquito Day" coined by this great malriologist – Dr. Ross. Ever since Ross's famous 1897 discovery, India has been in forefront of malaria related researches ably contributed by internationally famed malriologists or culicidologists like Drs. Christophers, Barraud, Sinton, Covell, McDonald, Viswanathan, Puri, Ramachandra Rao, Rajagopalan, (V.P.) Sharma etc. to name a few (see Chapter 3). Thus, the mosquito continues to be incredibly closer to Indian human's psyche, by virtue of its place both in their life and the research programmes aiming at her. Interestingly, in Indian life, one is aware of quite a few sayings referable to mosquito for its bearing on man: for example, when a human is extremely emaciated, he is refered to resemble a 'mosquito'; when somebody is annoyingly exacerbated with a hateful opponent, the former reminds the latter of crushing like a 'mosquito' is done to; when a person is uncontrollably, but secretively, talking close to another person's ear the former is urged not to whine like a 'mosquito'; and, when an already big human population continues to grow unchecked they are often referred to increasing like flies and 'mosquitoes' do.

THE "MOSQUITO FACTOR" IN INDIVIDUAL, AND NATIONAL HEALTH AND ECONOMY

Mosquitoes can be an annoying and serious problem in man's domain. In fact, the animal responsible for the most human deaths world-wide is the mosquito. They interfere with work and spoil hours of leisure time. Their attacks on farm animals can cause loss of weight and decreased milk production. Some mosquitoes are capable of transmitting diseases, such as malaria, yellow fever, dengue, filariasis and various types of encephalitides {*e.g.*, St. Louis encephalitis (SLE), Western Equine encephalitis (WEE), Eastern Equine encephalitis (EEE), LaCrosse encephalitis (LAC), Japanese encephalitis (JE), West Nile virus} to humans and animals (see Chapters 5-7). Malaria, the severest of all, is also humanity's greatest bane triggered by mosquito. Throughout the world more than 40% of population is exposed to varying levels of risk for malaria in 91 countries. Of a total world population of approximately six billion, 1.7 billion (35%) live in areas where endemic malaria has been considerably reduced or even eradicated, but where transmission has begun again and in which the situation is now unstable or deteriorating. In some areas, severe malaria problems developed following major ecological or social changes, particularly the agricultural or the exploitation of forests or socio-political interests. Each year 300-500 million cases of malaria, 90% of them in Africa, and between 1.5 and 2.7 million deaths are reported worldwide. Severe malaria morbidity and mortality are linked with areas endemic for *Plasmodium falciparum*, the predominant species in Africa. According to a report malaria kills a child somewhere in the world every 30 seconds (Alnwick, 2000), whereas it has been estimated that 800,000 children die each year of malaria in Africa alone. Mortality is lower but nevertheless important in the Americas and Southeast Asia, especially where *P. falciparum* drug resistance is widespread. The Southeast Asia region embodying India is of particular interest to us as there occur approximately 3.54 million cases every year.

Once seen as little more than a small sacrifice for the lush vistas of tropical climes, mosquitoes, and the diseases they transmit, have international agencies and the medical community up in arms. Mosquito -borne diseases exert unfathomable economic losses to humans. These losses are not merely monetary but relate to health as well. Diseases like malaria, filariasis, dengue and Japanese encephalitis incur annual losses worth several trillions of US dollars all over the world, if only expressed in terms of Disability Adjusted Life Years (DALYs) (Table 14.1). In Sub-Saharan Africa, a major malaria infested region, only malaria-related annual economic burden in 1987 was to the tune of US$ 0.8 billion (0.6% of gross domestic product) which was expected to further hike to US$ 1.7 billion (1.0% of GDP) by 1995.

Economic losses have a tendency to rise along with the national income and improved socioeconomic conditions. Also, there are a number of direct and indirect factors which are considerable when to calculate economics of malaria. Often, the related documentations are faulty and grossly incomplete. In spite of this, even if only the recorded facts are taken into consideration, the economic loss due to malaria in India comes somewhere between Rs. 15,777.85 and 29,934.85 million at the 1991 level. This estimate obviously did not consider major areas like enormous losses in agriculture, industry, mining, railways and other important productive sectors in the country susceptible to malaria-related perils, in addition to indiscernible yet apparent losses to the physical and intellectual growth of the population living in endemic areas. Keeping in mind some of the factors like estimates of morbidity and mortality due to malaria expressed as DALYs, an overall loss is postulated to be falling in the range of Rs. 16,192 – Rs. 20,192 million, provided prevalence figures of only 2 million malaria cases are considered (which, according to the WHO SEARO, is at least 6-7 times lower than probably the actual incidence in the country!). The above information is a clear indication about the enormity of seriousness malaria disease needs to be dealt with in India. It needs be understood emphatically that the management of mosquito-borne diseases in India takes away more than half of the financial layout earmarked for the entire health sector!

Table 14.1. Global public health importance of insect-borne diseases with their vectors and distribution

Disease	Vector species	Burden of disease[1]	Distribution
Malaria filariasis	*Anopheles* spp.	40,213,000	Tropics, main burden Sub-Saharan Africa
Lymphatic filariasis	*Culex* spp.	5,549,000	Tropical urban areas
Onchocersiasis	*Simulium* spp.	951,000	W.- Africa, C.- America
Leishmaniasis	Sandflies	1,810,000	Patchy, Old World
Old world	(*Phlebotomus* spp.)		(semi) arid zones,
New world	(*Lutzomyia* spp.)		New World humid forests
Trypanosomiasis		1,585,000	Patch, in W.- and southern Africa
African trypanosomiasis (Sleeping sickness)	Tsetse flies (*Glossina* spp.)		
American trypanosomiasis (Chagas' disease)	Triatomid bugs	680,000	S.- and C.- America linked to poor housing
Arboviral diseases			
Dengue	*Aedes* spp.	433,000	Urban tropics
Japanese encephalitis	*Culex* spp.	426,000	S. and SE Asia, linked to irrigated rice/pigs
Yellow fever	*Aedes aegypti*	n.a.	Africa, South America

[1]Global estimate of Disability Adjusted Life Years (DALYs) in 2000.

MALARIA - *NUMERO UNO* AMONGST ALL MOSQUITO-BORNE DISEASES

Malaria offers perhaps the best example for determining the economic burden levied by any arthropod-borne disease in a country, since (i) among all the vector-borne diseases malaria continues to be one of the biggest contributors to disease burdens in terms of death and suffering, and from the public health point of view it is undoubtedly the most devastating parasitic disease in the tropics, and (ii) it is definitely the most extensively studied subject. By undermining the health and capacity to work of hundreds of millions of people, it is closely linked to poverty and contributes significantly to stunting social and economic development. Nearly four decades ago there were hopes that malaria could be eradicated in a fairly short time, mainly through the intensive use of insecticides. However, malaria still remains a serious threat in endemic countries. Those at greater risk of dying from the disease are children under age 5 in malaria-endemic areas, pregnant women, people moving from non-malarious to malarious zones for reasons of work, migration, refuge, war or tourism, and travelers who visit endemic countries and return home with the disease. In many parts of the country, malaria is becoming even a greater problem than it has been before. Epidemics are recurring in areas where transmission has been interrupted, and are generally associated with deteriorating social and economic conditions.

ECONOMICS OF MALARIA IN INDIA

In India, before malaria control programme was launched in 1953, nearly 100 million people suffered from the disease annually and close to 0.1 million people died every year (Sharma & Mehrotra, 1986; Tyagi, 1994). The latter figure clearly appears to be a gross underestimate since during frequent epidemics so common in the country those days the overall number would have been much higher (Sinton, 1935), as in 1931, the number of deaths indirectly or directly resulted from malaria had exceeded 2 millions. Besides, malaria affected the national as well as individual development as economic losses of the order of Rs. 2936 million per year were forced on India.

Soon after the country-wide National Malaria Eradication Programme was launched in 1958 (in fact, the formal approval for the initiative was accorded by the Government of India as early as the late 1957), malaria cases were reduced to less than 0.1 million in 1965-66, without single death, and it appeared that malaria would be wiped off from India before very long! It is noteworthy here that decline in cases was in case of *P. vivax* and not in *P. falciparum*. Within the next ten years malaria epidemics were widespread and the disease, recording the highest ever number of malaria cases (6.47 million) by 1976, invaded even those areas which were earlier not known for malaria transmission or were once freed from its scourges. Subsequently, during 1980s and early 1990s, following a series of critical reviews of the malaria eradication programmes, the malaria incidence was brought down to less than 2 million, but soon increased to nearly 3 million in the mid-1990s, as reported by the National Malaria Eradication Programme (NMEP). Ironically, the malaria incidence and deaths reported by NMEP were based on the parasitologically proved cases, thus excluded a large number of asymptomatic carriers and/or those who approached other medicinal systems for cure of disease or even resorted to self-medication. As a consequence, both morbidity and mortality were apparently under-estimated. While the NMEP reported 2.5 to 3 million malaria cases and about 1000 malarial deaths annually, the WHO South East Asia Regional Office put the estimates to 15 million malaria cases (nearly 6 times) and 19,500 deaths due to malaria (Sharma, 1999). Therefore, it appears safer to assume that the malaria incidence reported offers at most the disease trend and not its true incidence, knowledge of which, for logical reasons, is inevitably indispensable for effective implementation of malaria control in the country.

Control of malaria in India

Malaria control in India is presently hampered mainly by two reasons: (i) drug resistance development in malarial parasites, and (ii) insecticide resistance development in malaria vectors. *Plasmodium vivax* and *P. falciparum* are dominant species in India, with the former responsible for 60-65% cases and the latter for 35-40% cases (Sharma, 1999). Among the six major vector species prevalent in India, *An. culicifacies*, the main transmitter of rural and peri-urban malaria in the peninsular India, is alone responsible for 65% malaria and 55% *P. falciparum* annually. This mosquito is resistant to DDT and HCH all over the country, and to malathion in Maharashtra, Gujarat and certain other States. *Anopheles stephensi* transmits nealy 12% malaria cases in the country, mostly in the urban and industrial areas. This species, too, has developed multiple resistance against various insecticides, many of which are however used as larvicides. Yet, another vector species, *An. fluviatilis,* which like above two species is common in Rajasthan, transmits 15% of total malaria cases and 30% *P. falciparum* in the country.

The strategy of malaria control in India has been to cause interruption in the disease transmission by spraying residual insecticides in the rural areas (351.8 million population) and by source reduction and larviciding in urban areas (62.1 million population). Because most of the primary malaria vectors in India are endophilic and/or endophagic in behaviour, indoor residual spraying of insecticides in the rural areas and anti-larval operations in the urban set-ups have been the major means to control malaria in the country. Led by DDT's widespread usage in controlling vector population, now largely jeopardized owing to development of resistance in vector mosquitoes, other insecticides like HCH (later banned in 1997), malathion, and certain synthetic pyrethroids such as cyfluthrin, deltamethrin and lambdacyhalothrin have also been selectively employed in the control of malaria, although spray targets were rarely achieved due to logistic and operational constraints.

Major anti-malaria activities launched in India since the independence can be summed up briefly as follows:

(1) ***National Malaria Control Programme*** (NMCP): It was launched in 1953 with comprehensive indoor residual spray with DDT. By 1957, 165.7 million population was effectively protected.

(2) ***National Malaria Eradication Programme*** (NMEP): The success achieved under the NMCP was so spectacular and encouraging that the Government of India launched National Malaria Eradication Programme, in 1958, with the objective of eradicating the disease from the country. This brought down malaria cases to 0.1 million and no deaths due to malaria were reported in the country in 1964. The euphoria of victory however proved to be ephemeral as malaria incidence began to escalate in the late sixties owing to multiple factors comprising financial, logistic, administrative and technical constraints, on one hand, and a feeling of complacency due to a near victory over the disease in the country, on the other. Consequently, malaria resurged with a vengeance and reached its peak in 1976, when about 6.47 million cases of malaria were recorded in the country.

The NMEP originally began as a completely centrally sponsored health programme and, till 1979, the entire expenditure on malaria control throughout the country was borne by the Centre. Thereafter, the programme was implemented on a 50:50 cost sharing basis with the States, with the exception of the seven north-eastern States which, from December 1994 onwards, were brought under 100% Central assistance as a special case. All other States in the country, while meeting out the entire operational costs for malaria control as their part of the cost sharing, continued to receive 50% of the costs from the Centre in the form of required equipment and material. The Indian Government spend nearly Rs. 350 crores on malaria control annually. Additionally, the World Bank also recently provided a sum of Rs. 1075 crores for five years for the Enhanced Malaria Control Programme.

(3) ***Modified Plan of Operation* (MPO):** The successes and failures under the NMEP were reviewed in depth and a Modified Plan of Operation was introduced in the year 1977 with three main objectives (i) prevention of deaths due to malaria, (ii) reduction of morbidity due to malaria, and (iii) maintenance of industrial and green revolution due to freedom from malaria as well as retention of achievement gained so far. Under the MPO a new strategy was adopted from eradication of malaria to the containment of the

disease in the country, *i.e.*, prevention of death and reduction in morbidity due to malaria. As a strategy under this Plan areas with 2 annual parasite incidence (API) or more were earmarked for regular rounds of spray. Surveillance activities in the whole country were decentralized to Primary Health Centres (PHC) along with laboratory services. People's participation was solicited by involving volunteers in the distribution of anti-malarials through Drug Distribution Centres (DDC) and Fever Treatment Depots (FTD). Malaria control in urban areas (<40,000 population) is carried out under the Urban Malaria Scheme (UMS) of the NMEP, launched in 1971-72. This led to significant reduction in malaria incidence in the country to a level of around 2 million cases by 1987 for one decade. Since 1994, however, focal outbreaks of malaria have been continuously reported from different parts of country, particularly the north-eastern States and the Thar Desert region in Rajasthan State in the north-west India. This unwarranted situation resulted in resurgence of malaria, generally pronounced by *P. falciparum* dominance, as well as deaths due to malaria in the country.

(4) ***Plasmodium falciparum* Containment Programme (PfCP):** The *P. falciparum* problem attained new dimension when a focus of chloroquine-resistant strain was discovered in Assam in 1973. Subsequently, new foci of chloroquine resistance were traced in different north-eastern States, Orissa, Madhya Pradesh, Maharashtra, Rajasthan *etc.* This situation led to the creation of PfCP in 1977 to tackle the rising threat of *P. falciparum*.

(5) ***Malaria Action Programme Based on Expert Committee Report – 1995*:** An expert committee that took cognizance of the fact of availability of appropriate technologies for control of malaria in different epidemiological situations or paradigms in the country attributed the various setbacks in malaria eradication programme, particularly recurrence of periodic epidemics and high mortality due to various different imminent factors like the administrative indifference, the organizational weakness, the low prioritization to malaria under the health services and apathy of middle level and peripheral workers in the States. Accordingly, the Expert Committee recommended intensification of malaria control activities throughout the country with focus on high risk areas through an integrated malaria control strategy, with following mentioned components: (i) Early case detection and prompt treatment, (ii) Selective vector control, (iii) Promotion of personal protection methods, (iv) Early detection and containment of epidemics, (v) Information, education and communication towards personal prevention and community participation, and (vi) Institutional and management capacity building, trained

manpower development and efficient management information system.

(6) ***Enhanced Malaria Control Project* (EMCP):** It was launched in September 1997 with the help of World Bank assistance and covered, in addition to 19 towns in 10 States, 62.2 million high risk tribal population living in 100 districts and 1045 PHCs of the seven peninsular States, *viz.*, Andhra Pradesh, Bihar, Gujarat, Madhya Pradesh, Maharastra, Orissa and Rajasthan. These PHCs, which a tribal population of more than 25%, had contributed 65.5% malaria cases and 78.60% *P. falciparum* cases in 1997. The benefits of the programme are conceived to reach other malaria endemic areas as well since the strengthening of the components of the Information, Education and Communication (IEC), trained manpower development and Management Information System (MIS) covers the whole country.

(7) ***Roll Back Malaria* (RBM):** In January 1998 the World Health Organization proposed Roll Back Malaria initiative taking into consideration large scale morbidity and mortality due to malaria in developing countries and also observing the changing epidemiological scenario, and the experience of the past control efforts. The basic concepts of the RBM initiative are not different from the existing control strategy in India, but it lays more emphasis on social mobilization, inter-sectoral collaboration, effective partnerships, strengthening of Primary Health Care, investment in development of effective methods of control, shaping of existing tools by operational research and it is expected that this will result in sustained reduction in disease burden by the year 2010.

Early Malaria Epidemics

In the 19th century, tropical diseases, particularly vector-borne infections, among which malaria ranked first, was a major concern for the colonial British government, as its empire reached around the globe. In several African and Asian countries, including India, colonial officers and their families were terrified by a host of illnesses, which claimed lives, slowed the march of conquest and exacted an enormous financial price (Spielman and D'Antonio, 2001). Unfortunately, no systematic country-wide studies were performed prior to 1961, and details are available through scattered surveys in different parts of India. Records show just how devastating malaria was in the early days of the British occupation of India. Table 14.2 details recorded deaths from malaria in the Punjab between 1869 and 1908. It is not difficult to comprehend as to how much the State of Punjab, one of the greenest and watered in the country, had had to suffer in maintaining health, particularly the children and women, two most vulnerable people among the society, and maintain economic and social activities. Mortality due to malaria were not enumerated but it could be safely gauged to be simply stupendous.

Table 14.2. Malaria Epidemics in the Punjab by the beginning of 20th century

Year	Area	Period	Number of Deaths	Mortality rate
1869	Punjab	Oct, Nov	116,540	Unknown
1876	Punjab	Oct, Nov	174,238	Unknown
1878	Punjab (Ludhiana, Jullinder & Hoshiarpur)	Oct, Nov	180,356	Unknown
1879	Punjab	Oct, Nov	141,996	
1884	Punjab	—		Records show mortality rate to be 30 higher than normal
1892	Punjab	—	283,223	Unknown
1894	Punjab	Oct, Nov	132,767	Unknown
1900	Punjab (Ludhiana, Ambala, Karnal, Gurdaspur & Raya)	—	254,580	Unknown
1908	Punjab	—	307,316	Unknown

Malaria epidemics occurred throughout India with varied cadance (Covell, 1927). Sir Gordon Covell, the one-time head of the Malaria Institute of India, records epidemics of "great intensity" in Murshidabad (Bengal) in 1821. In 1852, one malaria epidemic wiped out the entire village of Ula and then spread across the Bhagirathi River to Hooghly and continued to devastate populations for many years in Burdwan. The development of the Indian railways under the British administration contributed to the spread of malaria. The rather haphazard construction of railway embankments provided a number of breeding sites for the malaria vectors. It is also likely that the labourers that worked on the construction of the railways introduced different strains of the parasite to the areas in which they worked, possibly increasing the incidence of the disease. In year 1908 saw a countrywide pandemic during which huge mortality and morbidity occurred. In the early 1920s, Bengal suffered a severe malaria epidemic which resulted in over 730,000 deaths in 1921 alone. Thereafter, the number of deaths from malaria slowly decreased to within 300,000 to 400,000 per annum. During the Second World War however malaria deaths rose again, when Bengal recorded over 680,000 deaths in 1943, and 763,220 deaths in 1944. In Bombay Presidency, there were significant outbreaks of malaria during the construction of the Colaba causeway between 1821 and 1841 and during the construction of Alexander Dock and Hughes Dry Dock. The number of malaria cases in Bombay increased relatively sharply after 1901, during which time the construction of the railway was completed, ensuring an increase in traffic and a greater movement of people.

Among British troops in 1921, over 30% of all hospital admissions were for malaria and in 1922, 10,552 cases of the disease were reported, with 26 deaths out of a total military force of 60,166. This represents an incidence rate of 175 per 1000 which increased slightly to 180 per 1000 in 1923. The number of cases among the Indian troops is broadly similar to that of the British troops, with 173 cases per 1000 in 1922. However, the mortality rate among Indian soldiers was around 3 times higher than for British soldiers who possibly did not have access to the same level of medical care and treatment offered to the British soldiers.

Early Malaria Control

Early malaria control efforts tended to concentrate on the removal of breeding sites and later used chemicals such as the Paris green and kerosene for larviciding mosquito immatures. One of the first formal operations to control the disease took place at Mian Mir, near the city of Lahore. Mian Mir later became a British military post and the focus of a concerted effort to eradicate malaria. Mian Mir had an intricate system of irrigation canals which provided excellent breeding grounds for the malaria vectors. The incidence of malaria was astonishing, with an average incidence of slightly under 100%, rising to over 300% during epidemic years. The malariologists Dr. J.W.W. Stephens and Dr. S.R. Christophers, arrived at Mian Mir in 1901 with ambitious plans to remove all the breeding sites, evacuate the infected people and administer quinine as both a curative and preventative measure. Their programme began in 1901 and eventually developed into a massive effort, with between four and five hundred soldiers set to work full time at filling in the irrigation canals. The programme of constantly filling in ditches and removing puddles and any other potential breeding site continued until 1909, when to their dismay a serious malaria epidemic broke out, forcing to abandon the programme mainly on accounts of high costs and low output. The lesson learnt from the Mian Mir debacle was that mosquitoes, well entrenched, tenacious, adaptable, are difficult to control. The enemy (mosquito) in Mian Mir proved more formidable than had previously been admitted, especially in its adaptability to attack, its capacity to cover considerable distances and make use of secondary breeding places, when those of first choice were closed. It proved above all that in malaria control every local situation needed to be thoroughly studied and a specific method employed (Harrison, 1978).

Similar failures were conceivable in Bombay, Jhansi, Poona, Meerut, Secunderabad and at all other military posts where similar activities concurrent to Mian Mir were initiated. In 1917, the Bengal Nagpur Railway (BNR) and the East India Railways formed a separate malaria control organization, specifically to control the disease in and around stations. There every person suffered 3 bouts of malaria during a single year. Similar larviciding and breeding pool removal programmes were undertaken during the 1920s in the tea plantations of Assam and in Mysore by the Rockefeller Foundation.

The Rockefeller Foundation began using pyrethrum sprays experimentally in India to great success and this method of malaria control was recognised as enormously valuable. Pyrethrum was first used in malaria control in Delhi in 1938, following its phenomenal successful usage by William Gorgas in Cuba about two decades ago. The use of pyrethrum was then expanded to Assam in 1942. All of the above interventions would have had some impact of the incidence of malaria, yet they were unable to sustain the control of the disease.

With availability of DDT, the wonder insecticide, in late 1940s, however, changed for a moment the whole concept of malaria control through vector control.

DDT – That wonder insecticide

DDT was first synthesized by Othmar Zeidler in 1874 but the chemical remained in obscurity until 1939, when Dr. Paul Müller discovered its insecticidal properties. It's best use was obtained during the Second World War where the ectoparasites like lice felled many troops by the deadly relapsing typhus fever. DDT was doused over people and it instantly was a success in controlling diseases. In India, DDT was first used by the armed forces in 1944 for the control of malaria and other vector-borne diseases. In 1945, DDT was made available for civilian use in Bombay to control malaria and shortly afterwards in 1946, pilot schemes using DDT were set up in several areas, including Karnataka, Maharashtra, West Bengal and Assam. Between 1948 and 1952 the WHO set up DDT demonstration teams in Uttar Pradesh, Rayagada, Wynad and Malnad. The distribution of DDT in Bombay produced some remarkable results within a very short period. On 1st July 1945, the first civilian home was sprayed in India with a 5% solution of DDT mixed in kerosene. This initial spraying ensured that not a single adult *An. fluviatilis* from any house was found either during daytime or night time catches. The use of DDT was not only popular among the locals for control of the deadly *Anopheles* mosquito, but the insecticide also ensured that lizards, cockroaches, scorpions, bedbugs and ticks all disappeared after a single spraying.

After the spraying in the Kannara district, it was reported that the population began to grow precisely because of a decrease in the death rate as a result of the DDT spraying. Prior to DDT being used, the district reported an average of 50,000 malaria cases every year, which was reduced by around 97% to only 1,500 cases after DDT was introduced. The spray programme in Bombay took place during a period of great upheaval in India and the withdrawal of the British Raj from the Sub-continent. During 1949, it is estimated that over 6 million people in Bombay were protected from malaria through the use of DDT and that at least half a million cases of malaria were prevented. Along with the control of malaria came the control of a number of other diseases. The incidence of human plague was completely eliminated and there were reports of reductions in the incidences of diarrhoea and dysentery. It is likely that this was because the numbers of insects that contributed to the spread of these diseases were all dramatically reduced. The benefits were not restricted to people; cattle also enjoyed improved health as the ticks that caused festering sores and infections were all but destroyed.

In the early 1950s India's population was estimated to be around 360 million and every year around 75 million people suffered from malaria and approximately 800,000 died from the disease. The disease put almost the entire population at risk and the country had, and still has, over fifty different anopheline species. Of these, *An. culicifacies, An stephensi, An. minimus and An. fluviatilis* were defined as the major vectors of malaria. The apparent health and economic benefits that came from using DDT prompted the formation of the National Malaria Control Programme (NMCP) in 1953. The control programme first set out to control the disease in the endemic and hyperendemic areas with 125 control units. Each of these control units consisted of between 130 and 275 men and their plan was to protect approximately 1 million people each. By 1958, the malaria control programme had been increased to protect at least 165 million people from the disease with 160 control units. Controlling the disease became an issue of economic importance. A study conducted in the early 1950s estimated that there would be a net gain of Rs. 200 million if the government spent Rs. 40 million to control malaria.

After consultation with the WHO and with the assistance of UNICEF and the US Agency for International Development (USAID) the decision was made to attempt the eradication of malaria. The programme was expanded to cover the entire country with an additional 230 units and the National Malaria Eradication Programme (NMEP) was thus born in 1958.

The control units were divided up into sub-units, sectors and sections where the sections each had the task of protecting around 10,000 people from malaria. This meant that on average each section would be responsible for spraying 2,000 houses once during the high transmission season (if the area was in a seasonal malaria area) or twice every year in those areas where transmission occurred year round.

The entire eradication plan was based on the spraying of DDT and the provision of chemotherapy to those in need. Based on the experience from the NMCP, it seemed reasonable to expect that eradication could be achieved in just a few years and originally the officers would not be forcing the DDT onto the insect, rather the insect was choosing to enter the house, take a blood meal and then kill itself by landing on the walls. The prescribed manner of spraying DDT was to mix 5% suspension of DDT wettable powder in water into a compression sprayer and to pump it to a pressure of 40 pounds per square inch. The sprayer would then stand about 18 inches away from the wall and spray the mixture onto the wall in a sweeping motion so that the DDT would be left on the walls at a concentration of 2 gm/m^2.

Despite some outstanding results initially, the programme began to encounter setbacks. By the early 1970s it became apparent that controlling one of the most devastating diseases known to man was not going to be achieved solely with the use of DDT.

REFERENCES CITED

Alnwick, D., 2000. Roll Back Malaria – What are the prospects? Bulletin of the World Health Organization 78: 1377.

Amalraj, D. *et al.*, 1988. Evaluation of two new juvenile hormone compounds against mosquito vectors. Indian Journal of Medical Research 87: 19-23.

Anonymous, 1910. The lesson of Mian Mir. Journal of Tropical Medicine 13: 146-151.

Banerjee, K., 1994. Dengue in India. Proceedings of national brain-storming session on dengue and dengue haemorrhagic fever, Pune.

Barraud, P.J., 1934. The Fauna of British India, including Ceylon and Burma. Diptera, Cullicidae. V. Tribes: Megarhinini and Culicini. Today & Tomorrow, Delhi.

Beaty. B.J. and W.C. Margquardt (Eds.), 1996. The biology of disease vectors. Univ. Press of Colorado, Niwot, Colorado, pp. 632.

Bouma, M.J. and H.J. van der Kaay, 1994. Epidemic malaria in India and the El Nino Southern Oscillation . Lancet 344: 1638.

Bouma, M.J. and H.J. van der Kaay, 1995. Epidemic malaria in India's Thar Desert. The Lancet 346: 1232-1233.

Bouma, M.J. and H.J. van der Kaay, 1996. The El Nino Southern Oscillation and the historic malaria epidemics on the Indian subcontinent and Sri Lanka: an early warning system for future epidemics? Tropical Medicine & International Health 1: 36-42.

Bouma, M.J. *et al..*, 1997. Tropical Medicine and International Health 2: 1122-1127.

Brandner, G., et al., 1992. Preservation of HIV infectivity during uptake and regurgitation by the stable fly, Stomyx calcitrans, L. AIDS-Forschung 5: 253.

Carde R., 1996. Odour plumes and odour-mediated flight in insects. In: Olfaction in mosquito-host interactions. Wiley, Chichester (Ciba Found Symp 200) pp. 54-70.

Champagne, D.E., 1994. The role of salivary vasodilators in blood-feeding and parasite transmission. Parasitology Today. 10: 430-433.

Charlab, Rosane *et al.* 2000. The Salivary Adenosine Deaminase from the Sand Fly Lutzomyia longipalpis. 95: 45-53.

Chouhan, G.S. *et al.*, 1991. Clinical and virological study of dengue fever in Jalore city, Rajasthan, 1985. Indian Journal of Medical Research 91: 414-419.

Christophers, S.R., 1904. Second report of the anti-malarial operations at Mian Mir, 1901-1903. Scientific Memoirs by the Office of the Medical and sanitary Departments of the Government of India. 1-37, Calcutta.

Christophers, S.R., 1933. The Fauna of British India, including Ceylon and Burma. Diptera, Culicidae. IV. Tribe: Anophelini. Today & Tomorrow, Delhi.

Corbet, P.S., 1986. Using dragonflies to suppress mosquitoes in domestic water-storage containers. Waterlines 4:10-11.

Cork, A., 1996. Olfactory basis of host location by mosquitoes and other haematophagous Diptera. In: Olfaction in mosquito-host interactions. Wiley, Chichester (Ciba Found Symp 200), pp. 71-88.

Covell, G., 1927. A critical review of the data recorded regarding transmission of malaria by different species of Anopheles, with notes on distribution, habits and breeding places. Indian Medical Research Memoirs 7: 87-99.

Cross et al. 1993 a. Antibody responses of BALB/c mice to salivary antigens of hematophagous black flies. Journal of Medical Entomology 30: 725-734.

Cross et al. 1993 b. Modulation of murine immunological responses by salivary gland extract of Simulium vittatum. Journal of Medical Entomology 30: 928-935.

Cross et al. 1994. Modulation of murine cellular immune responses and cytokines by salivary gland extract of the black fly Simulium vittatum. Tropical Medicine and Parasitology 45:119-124.

Cupp, M.S. et al. 2000. Horn fly saliva Targets Thrombin Action In Hemostasis. Entomological Society of America. 37(3): 416-421.

Curtis, C. F. and J. Minjas, 1985. Expanded polystyrene for mosquito control. Parasitology Today1: 36.

Darsie, R.F. Jr. et al., 2002. New Addition to the Mosquito Fauna of United States, *Anopheles grabhamii* (Diptera: Culicidae). Journal of Medical Entomology 39: 430–431.

Das, Bina Pani, et al., 1990. Pictorial key to the species of Indian anopheline mosquitoes. Zoology (Journal of Pure and Applied Zoology) 2(3): 131-162.

Das, P. K. et al., 1986. Vernacide, a new insecticide impregnated paint for controlling mosquito vector, Culex quinquefasciatus, and cockroach, Periplanata Americana. Indian Journal of Medical Research 83: 268-270.

De Jong R, and B.G.J. Knols, 1995. Olfactory responses of host-seeking Anopheles gambiae ss Giles (Diptera: Culicidae). Acta Tropica 59: 333-335.

De Jong R and B.G.J. Knols, 1996. Selection of biting sites by mosquitoes. In: Olfaction in mosquito-host interactions. Wiley, Chichester (Ciba Found Symp 200) p. 89-103.

Do Nascimento, T.F.S. and R.L-de-Oliviera, 2002. Anopheles halophylus, a new species of the subgenus Nyssorhynchus (Diptera:Culicidae) from Brazil. Mem. Inst. Oswaldo Cruz 97: 801-811.

Edman, J., J. Day and Walker, E., 1984. Vector-host interplay-factors affecting disease transmission. In: Ecology of Mosquitoes: Proceedings of a Workshop. Florida Medical Entomology Laboratory, Vero Beach p. 273-285

Edman, J.D. and T.W. Scott, 1987. Host defensive behavior and the feeding success of mosquitoes. Science Applications. 8: 617-622.

Fontenille, D. et al., 1997. High annual and seasonal variations in malaria transmission by anophelines and vector species composition in Dielmo, a holoendemic area in Senegal. American Journal of Tropical Medicine & Hygiene 56: 247-253.

Giles, H.M., 1904. The anti-malaria operations in Mian Mir. Journal of the Royal Army Medical Corps 3: 131-133.

Goddard, J., 2000. Physicians guide to arthropods of medical importance. CRC Press, London. pp. 422.

Gubler, D.J., 1998. Dengue and dengue haemorrhagic fever. Clinical Microbiology Review 11, 480-496.

Gubler, D.J., et al., 2001. Climatic variability and change in the United State : Potential imparts on vector- and rodent borne diseases. Environment Health Perspectives 109: 223-233.

Gunn, H.M., 1993. Role of fomites and flies in the transmission of bovine viral diarrhoea virus. Veterinary Records 132:584-585.

Guzman, M. G. and G. Kouri, 2002. Dengue: an update. Lancet Infectious Diseases 2: 33-42.

Hales, S. et al., 2002. Health in a warm country. http://image.thelancet.com//extros/01arp11175 wef.pdf.

Harrison, G. 1978. Mosquitoes, Malaria and Man : A History of Hostilities Since 1880. J. Murray, London.

Harrison, B.A. and J.E. Scanlon, 1975. Medical entomology studies - II. The subgenus Anopheles in Thailand (Diptera: Culicidae). Contributions of American Entomological Institute, Philadelphia, No. 12, pp. 307.

Humphrey-Smith, I., G. Donker, A. Turzo, C. Chastel and H. Smidt-Mayerova, 1993. Evaluation of mechanical transmission of HIV by the African soft tick, Ornithodorus moubata. AIDS 7: 341.

Jambulingam, P.K. et al., 1989. Effect of permethrin-impregnated bednets in reducing population of malaria vector, Anopheles culicifacies, in a tribal village of Orissa State (India). Indian Journal of Medical Research 89: 48-51.

James M.T. and Harwood, R.F. 1969. Medical Entomology. The Macmillian Company. Library of Congress.

James, S.P., 1903 a. A report of the anti-malarial operations at Mian Mir (1901-1902). Reports to the Malaria Committee of the Society, London, pp. 27-77.

James, S.P., 1903 b. First report of the anti-malarial operations at Mian Mir, 1901-1903. Scientific Memoirs by the Office of the Medical and sanitary Departments of the Government of India, Calcutta, pp. 1-53.

Jetten, T.H. and D.A. Focks, 1997. Potential changes in the distribution of dengue transmission of HIV by the African soft tick, Ornithdorus moubata. AIDS 7: 341.

Joshi, V. et al., 1996. Transovarial transmission of dengue 3 virus by Aedes aegypti. Transactions of the Royal Society of Tropical Medicine & Hygiene 90: 643-644,.

Jouzel et al., 1993. Extending the Vostok Ice-core record of palaeoclimate to the penultimate glacial period. Nature 364: 407-412.

Kennedy, J. S. 1939. The visual responses of flying mosquitoes. Proceedings of Zoological Society, London 109: 221-242.

Kiauta, B., 1975 a. Introduction to insect cytotaxonomy. Part I. Lectures delivered at the Tribhuvan University, Kathmandu Nepal.

Kiauta, B., 1975 b. Cytotaxonomy of dragonflies, with special reference to Nepalese fauna. Pt. II. Lectures delivered at the Tribhuvan University, Kathmandu Nepal.

Knaus, R.M. *et al.*, 1993. Insect blood meal studies using radiosodium ^{24}Na. Journal of American Mosquito Control Association 9: 264-268.

Knight, K.L. and A. Stone, 1977. A catalogue of the mosquitoes of the World (Diptera: Culicidae). The Thomas Say Foundation, Entomological Society of America, pp. 611.

Kogan, P.H., 1990. Substitute blood meal for investigating and maintaining Ae. aegypti (Diptera: Culicidae). Journal of Medical Entomology 27: 709-712

Konradsen, F. et al., 2002. Engineering and malaria control: learning from the past 100 years.[In: Malaria in irrigated agriculture.(Eds) E. Boelee, F. Konradsen and W. van der Hoek]. Papers and abstracts for the SIMA Special Seminar at the ICID 18th International Congress on Irrigation and Drainage, Montreal, 23 July 2002. IWMI Working Paper 47. SIMA Document 2, Colombo, Sri Lanka: International Water Management Institute.

Leon P.D. and W.J. Tabachnick. 1996. Apyrase activity and adenosine diphosphate induced platelet aggregation inhibition by the salivary gland proteins of *Culicoides variipennis*, the North American vector of bluetongue viruses. Veterinary Parasitology. 61: 327-338.

Martynov, A. V., 1938. Etudes sur l'histoire geologique et de phylogenie des orders des insects (Pterygota). Part I. Paleoptera et Neoptera-Polyneoptera. Ttrav. Inst. Paleont. Acad. Sci. U.R.S.S. 7(4), pp. 155.

Mathur, K.K., G. Harpalani, N.L. Kalra, G.G.K Murthy and M.V.V.L. Narsimham, 1992. Epidemic of malaria in Barmer (Thar desert) Rajasthan during 1990. *Indian Journal of malariology* 29: 1-10

McMichael, A.J. and S. Kovats, 1999. Weather, climate and health. Bulletin WHO 77: 14-18.

Miike, L., 1987. Do insects transmit AIDS? Staff Paper 1. Office of Technological Assessment, U.S. Congress, Washington, D.C., pp. 43.

Mumcuoglu, K.Y. *et al.*, 1996. Antihemostatic activity in salivary glands of the human body Pediculus humanus. Journal of Insect Physiology 42, 1083-1087.

Nagpal, B. N. and V.P. Sharma, 1995. Indian anophelines. Oxford, New Delhi. 416 pp.

Napoli, V.M. and J.E. McGowan, 1987. How much blood is in a needlestick? Journal of Infectious Diseases 155: 128.

Ostfeld, R.S. and F. Keesing, 2000. The function of biodiversity in the ecology of vector-borne zoonotic diseases. Canadian Journal of Zoology 78: 2061-2078)

Pasen, G.C. et al., 2000. Tick histamine-binding proteins: lipocalins with a second binding cavity. Biochimica et Biophysica Acta 1482: 92-101.

Pittaway, A.R., 1991. Arthropods of medical and veterinary importance. CAB International, Weltingford, Oxon, UK, pp. 178.

Peyton, E.L. and J.E. Scanlon, 1966. Illustrated key to the female Anopheles mosquitoes of Thailand. Department of Medical Entomology, US Army Medical Component, SEATO, Bangkok, Thailand.

Poopathi, S. et al., 2003. Control of Culex quinquefasciatus (Diptera: Culicidae) by Bacillus sphaericus and B. thuringiensis *subsp.* israelensis, produced on a new potato extract culture medium. Biocontrol Science and Technology 13: 743-748.

Puri, I.M., 1949. Anophelines of the Oriental region. Boyd's Malariology. W.B. Saunders & Co., New York, pp. 483-505.

Puri, I.M., 1955. The distribution of anopheline mosquitoes in India. Health Bulletin No. 17: Malaria Bureau No. 7, Government of India. pp. 104.

Puri, I.M., 1960. Synoptic keys for the identification of the full-grown larvae of the Indian anopheline mosquitoes. Health Bulletin No. 16. Malaria Bureau No. 7. Government of India Press, Calcutta. pp. 104.

Raina, B.L., 1991. Introduction to malaria problem in India: Prevedic times to early 1950. Commonwealth Publishers, New Delhi, pp. 187.

Rajagopalan, P.K. et al., 1987. Bangalore mosquito control project : Master plan. Vector control Research Centre, Pondicherry, pp. 309.

Rajagopalan, P.K. and P.K. Das, 1988. Fileriasis control Demonstration Project. Vector Control Research Centre, Pondicherry.

Rao, T.R., 1981. The anophelines of India. Indian Council of Medical Research, New Delhi.

Reid, J.A. 1968. A note on Anopheles subpictus Grassi and Anopheles indefinities Ludlow (Diptera: Culicidae). Journal of Medical Entomology 3: 327-331.

Reisen, W.K. et al., 1993. Effect of temperature on the transmission of western equine encephalomyelitis and St. Louis encephalitis viruses by Cx. tarsalis (Diptera: Culicidae). Journal of Medical Entomology 30: 151-160.

Reuber, P., 1997. Global warming. Sierra 82: 34

Ribeiro, J.M.C. et al., 1985. Anti-hemostatic, anti-inflammatory and immunosuppressive properties of the saliva of the tick Ixodes dammini. Journal of Experimental Medicine 161: 332-344.

Ribeiro, J.M.C., 1992. Characterization of a Vasodilator from the Salivary glands of the Yellow-Fever Mosquito Aedes-aegypti. Journal of Experimental Biology. 165: 61-71.

Ribeiro, J.M.C., 1995. Blood-feeding arthropods: live syringes or invertebrate pharmacologists? Infectious agents and disease - reviews issues and commentary 4:143-152.

Ross, E.S. and H.R. Roberts, 1943. Mosquito atlas. Pts. 1-3. American Entomological Society, Washington.

Ross, R., 1897. On some peculiar pigmented cells found in two mosquitoes fed on malarial blood. British Medical Journal 2: 1786 – 1788.

Ross, R., 1898 a. Pigmented cells in mosquitoes. *British Medical Journal* 1: 550-551.

Ross, R., 1898 b. Preliminary report on the infection of birds with *Proteosoma* by the bites of mosquitoes. Government Press, Calcutta.

Ross, R., 1903. Report on malaria at Ismailia and Suez. *Memoir* (Liverpool School of Tropical Medicine).

Ross, R., 1904 a. The anti-malarial campaign at Mian Mir. *British Medical Journal* 2: 701.

Ross, R., 1904 b. The anti-malarial failure at Mian Mir. *Journal of Royal Army Medical Corp.* 3(2): 210-211.

Ross, R., 1910. Report on the measures taken against malaria in the Lahore (Mian Mir) Cantonment. Lancet 2: 1371.

Ross, R., 1923. Memoirs; with a full account of the great malaria problem and its solution. John Murray, London, pp. 547.

Roy, D.N. and A.W.A. Brown 1970. Entomology, 3rd Edition, Bangalore Press, Bangalore, pp. 855.

Russel, P.F. et al., 1943. Keys to the anopheline mosquitoes of the world. American Entomological Society. The Academy of Natural Sciences, Philadelphia.

Sebastian, A. *et al.*, 1980. The use of dragonfly nymphs in the control of Aedes aegypti. South-east Asian Journal of Tropical Medicine and Public Health 11: 104-107.

Service, M.W., 1996. Medical entomology for students; 2nd ed. Cambridge: Cambridge University Press, pp. 283.

Sharma, R.C. et al., 1985. Field trials on the application of expanded polystyrene (EPS) beads in mosquito control. Indian Journal of Malariology 22: 107-109.

Sharma, V.P and K.N. Mehrotra, 1986. Malaria resurgence in India: a critical study. Social science and Medicine 22: 835-845.

Sharma, V.P. et al., 1986. Bioenvironmental control of malaria in Nadiad, Kheda district, Gujarat. Indian Journal of Malariology 23: 95-117.

Sharma, V.P., 1999. Current scenario of malaria in India. Parasitologica 41: 349-353.

Sinton, J.A., 1935. What malaria costs India, nationally, socially and economically. Records of malaria Survey of India 5: 224-164, 413-489.

Spielman, A. and M.D'Antonio, 2001. Mosquito: The story of man's deadliest foe. Faber & Faber Ltd., London.

Stephens, J.W.W. 1904. The anti-malaria operations at Mian Mir (Punjab). Lancet 1: 637.

Stone, A. 1961. A synoptic catalogue of the mosquitoes of the world. Supplement I (Diptera : Culicidae). Proceedings of Entomological Society of Washington 63: 97-152.

Stone, A. 1963. ibid. 64: 117-140.

Stone, A. and M.D. Delfinado, 1973. Family Culicidae: Catalogue of the diptera of Oriental region. Vol. 1. Suborder Nematocera. University Press, Hawaii, Honolulu, pp. 618.

Subbarao, S. et al., 1999. Malariogenic stratification of India using *Anophales culinfacies* silking species prevalence. ICMR Bulletin 29: 75-80.

Swaroop, S., 1949. Forecasting of epidemic of epidemic malaria in the Punjab, India. American Journal of Tropical Medicine & Hygiene 29: 1-17.

Taylor, P. and S.L. Mutambu, 1986. A review of the malaria situation in Zimbabwe with special reference to the period 1971-181. Transactions of Royal Society of Tropical Medicine & Hygiene 80: 12-19.

Titus, R.G. and J.M.C. Ribeiro. 1988. Salivary lysates from the sandfly Lutzomyia longipalpis enhance Leishmania infectivity. Science 239, 1306-1308.

Tyagi, B.K. 1994. Alternative control strategies for mosquito vectors of human diseases in India, pp.359-372 (In: Perspectives in Entomological Research) Scientific Publishers (India), Jodhpur, pp.406.

Tyagi, B.K. and R. C. Chaudhary, 1997. Outbreak of falciparum malaria in the Thar Desert with particular emphasis on physiographic changes brought about by extensive canalization and their impact on vector density and dissemination. Journal of Arid Environments 36:541-555.

Tyagi, B.K. et al., 1985. Evaluation of a new compound (VCRC/INSA/A23) with juvenile hormone activity against the mosquito vectors. Indian Journal of Medical Research 82: 9-13.

Tyagi, B.K. et al., 1987. Evaluation of three formulations of a chitin synthesis inhibitor (fenoxycarb) against mosquito vectors. Indian Journal of Medical Research 85:161-175.

Tyagi, B.K., 1999. Vector-borne diseases of the Thar Desert. Abstr. Pap. Man in Desert – Medical Problems. CME –99, Military Hospital, Jodhpur, p. 3.

Tyagi, B.K., 2001. Major vector borne diseases in the Thar Desert, with special reference to the changing climate, pp. 103-112 (In; Priorities of research for development of arid regions. Eds.: Dr. S.K. Verma, Dr. N.M. Nahar, Dr. R. Raj-Bhansali, Dr. S.K. Jindal and Dr. Satya Vir) Scientific Publishers, (India), Jodhpur, pp. 238.

Tyagi, B.K., 2002. Malaria in the Thar desert: Facts, figures and future. Agrobios (India), pp. 165.

Tyagi, B.K., 2003. Medical entomology. Scientific Publishers (India), Jodhpur. pp. 262

Valenzuela, J.G. and J.M.C. Ribeiro. 1998. Purification and cloning of the salivary nitrophorin form the hemipteran Cimex lectularius. Journal of Experimental Medical Entomology. 201: 2659-2664.

Waage, J.K. 1979. The evolution of insect/vertebrate associations. Biological Journal of the Linnean Society. 12: 187-224.

Watson, M. 1905a. The effect of drainage and other measures on the malaria of Klang, Federal Malay States. Journal of Tropical Medicine 8: 100-104.

Watson, M., 1905b. Observations on malaria control, with special reference to the Assam Tea Garden, and some remarks on Mian Mir, Lahore Cantonment. Transactions of the Royal society of Tropical medicine and Hygiene 18: 147-161.

Watson, M., 1924. Observations of malaria control, with special reference to the Assam Tea Garden, and some remarks on Mian Mir, Lahore Cantonment. Transactions of the Royal Society of Tropical Medicine and Hygiene 28: 147-161.

Wattal, B.L. 1963. Keys to Oriental anophelines. (In: Russel et al., ``Practical Malariology") Second Edition, Oxford University Press, London, pp. 750.

Wattal, B.L. and N.L. Kalra, 1961. Region-wise keys to the females of Indian Anopheles. Bulletin of National Society of India for Malaria & other Mosquito-borne Diseases 10: 55-138.

Watts, D.M. et al., 1987. Effect of temperature on the vector efficiency of Aedes aegypti for the dengue 2 virus. American Journal of Tropical Medicine & Hygiene 36: 143-152.

Webb, P.A. et al., 1989. Potential for insect transmission of HIV: experimental exposure of Cimex hemipterous and Toxorhynchytes amboinensis to human immunodeficiency virus. Journal of Infectious Diseases 60: 970.

Wells, S., 2003. The journey of man: A genetic odyssey. Princeton University Press, pp. 256.

GLOSSARY

ABBRASION - Damaged area of skin.

ABIOTIC FACTORS - The climatic characteristics (temperature, humidity, saturation deficit, rainfall, light, wind) and edaphic characteristics (type of soil, soil texture) of a region or habitat.

AEDEAGUS - A finger-like evagination of the ventral body wall of male mosquito enclosing the terminal section of the ejaculatory duct.

ALLERGY - A state of hypersensitivity to a given material (called an allergen), acquired through exposure; re-exposure to the Allergen shows altered reactivity.

ANTENNAE - Paired appendage of head of insects and other arthropods. Usually with many segments. Antennae carry chemosensory organs.

ANTERIOR - The front end.

ANTERODORSAL - The front of the upper surface.

ANTEROVENTRAL - The front of the under surface.

ANTHROPOPHILIC - Man-biting (literally: liking man).

ANTICOAGULANT - Stops blood from clotting.

APICAL - At the tip or end of (APEX).

ARBOREAL - Living in trees.

ARTHROPOD - An animal belonging to the phylum Arthropoda in the animal kingdom, having a hard jointed exoskeleton and paired, jointed legs. The phylum includes, among other classes, the Arachnida and the Insecta, many of which are important medically as parasites or as vectors of organisms capable of causing disease in man.

ATTACK PHASE (Of a control programme) - the phase in which all houses are treated, regardless of whether or not the house was infested.

BIOECOLOGY - The study of the interrelationships between plants and animals and their abiotic environment.

BIOLOGICAL CONTROL - Control measures by means of living organisms.

BIOLOGY - The science that deals with the phenomena of life and with living organisms in general.

BIONOMICS - The (quantitative) relation of the development, reproduction and survival of an organism or population of organisms to factors in the environment.

BIOTIC FACTORS - Factors connected directly with the organisms living in a given environment, e.g., nature and availability of hosts,water quality, plant environment, etc.

BIOTYPE - A small topographic unit, including the biotic community.

BRIQUETTES - An insecticide mixed with clay, plaster-of-paris (gypsum) or other substance that will allow it to be moulded and hardened into brick-like blocks that will slowly dissolve in water. Thereby releasing the insecticide over an extended period.

CHITIN - Hard skeletal material found in the majority of Arthropoda.

CHROMOSOME - A linear structure within the nucleus of a cell composed of proteins and DNA which holds the genetic material or genes of an organism, the arrangement of which is unique to each organism.

CIBARIAL PUMP - The anterior part of the pharynx, joined to powerful muscles inside the head of insect, which is used to suck up blood in the course of a blood-meal.

CONTAMINATION - The unintentional entrance, into a material (e.g., food) of my foreign substance, matter, or organism.

CONTROL - Of insects (or other undesirable animals), the restriction of the population density of such insects to a level below that at which they can be harmful to the interests of man.

COPULATION - Physical union of male and female during mating.

COXA - The first basal segment of the leg of an insect.

CUTICLE - A non-living layer secreted by, and overlying the epidermis.

CUTICULAR HYDROCARBONS - Hydrocarbon chemicals in the surface layer of the cuticle which can be used to separate sibling species. Examination requires expensive and sophisticated apparatus, but has the advantage of using dried material..

CYTOSPECIES - Species which because of their morphological similarity, have been identified primarily by cytological charcacters, such as the banding patterns of the chromosomes.

CYTOTAXONOMY - The description of species based on a study of the banding patterns of the chromosomes.

DEC - Diethylcarbamazine, a microfilaricidal drug.

DENSITY (of flies) - Number of flies per given area or other unit.

DENSITY-DEPENDENT - A term used to describe factors that vary according to the population density, and provide a form of "negative feedback" to stabilizepopulation numbers. All living populations must be limited by density –dependent factors operating on either their birth-rate or death-rate (or both), otherwise their populations would increase forever. In some cases, however, populations can be limited by density-independent factors, before density-dependent factors come into operation.

DERMATITIS - Inflammation of the skin.

DERMIS - The true skin or elastic vascualr layer lying below the outer protective layer of the skin (the epidermis).

DIPTERA - An order of insects containing the two-winged flies which have only the anterior pair of wings well developed.

DISPERSION - The movements of flies in and around a locality or area in which they emerged or are found at a given time.

DIURNAL - Daytime - usually relating to a cycle of activity. The opposite of nocturnal.

DIVERTICULAM - A long sac-like pocket, closed at the far end.

DOMESTIC - Associated with houses.

DORSAL - Upper surface or back.

ECOLOGY - The study of the Inter-relationship between organisms and their environment (including other organisms).

ECTOPARASITE - A parasite that lives on the external surface of an animal, e.g. fleas, lice, mites, ticks.

EMBRYO - The early or developing stage of any organism, especially the developing product of fertilization of an egg.

EMBRYOGENESIS - The formation or development of eggs or offspring.

EMIGRATION - The movement out from a given locality or area.

EMPODIUM - Thin claw-like structure arising medially between the claws of the tarsi

ENDEMIC DISEASE - A disease that is caused by factors that are constantly present within a given geographical area or population group.

ENDOPHILIC INSECTS - Insects that regularly enter houses and spend at least part of their life indoors.

ENVIRONMENT - The sum total, at a given moment, of all external influences and conditions to which an organism or object is subjected, including all physical, chemical and biological factors.

ENZOOTIC - Pertaining to a disease that occurs in animals especially one that is restricted to a limited geographical area.

ENZYMATICALLY - Separation of isoenzymes to identify some sibling species of mosquito sibling species complexes.

EPIDEMIC - A period of increased prevalence of a given disease in a population (usually, but not necessarily, as a result of factors that are not constantly present in the populations).

EPIDEMIOLOGY - The study of the factors that determine the frequency and distribution of disease and of other health-related states in populations.

EPIDERMIS - The epithelium covering the body.

ERADICATION - Complete removal or destruction (e.g., of an infectious agent in a given geographical area).

EXOPHILIC INSECTS - Insects that normally do not enter buildings.

EXUVIUM - The cast outer skin which is shed when an insect moults to its next developmental stage.

FACETS - The hexagonal shaped divisions on the surface of the eyes.

FECUNDITY - Reproductive potential.

FEMUR The third segment of an insect's leg, distal to the trochanter.

FILAMENTOUS Thin and thread-like.

FUMIGATION - A process that uses gaseous agents to kill animals.

GENETIC - A hereditary factor controlled by genes of the previous generation.

GENITALIA - The external genital organs of the insect; the gonads and their ducts and all associated accessory organs.

GENUS - (plural, genera) a taxonomic category within a family, consisting of one or more generally similar species.

GONADS - Internal sex organs.

GONOTROPHIC CONCORDANCE - The production of an egg-batch immediately following a blood meal, so that blood-feeding and egg-laying follow each other in a regular sequence.

GONOTROPHIC CYCLE - The events between successive egg-layings, or the time interval between them.

GRAVID - Swollen and full of mature or nearly mature eggs.

HABITAT - The physical place (home) where an organism lives.

HAEMATOPHAGOUS - The term applied to an inesct that feeds exclusively on blood; in mosquitoes only females bloodfeed.

HAEMOCOEL - The physiological fluid in which the internal organs of insects are bathed.

HEAD - The anterior region of an insect, which bears the mouthparts, eyes, antennae and houses the brain.

HOST - The animal parasitized by a parasite or disease organism.

HUMIDITY OR RELATIVE HUMIDITY - The content of water vapour in the air measured as a percentage of the content of saturated air (100% RH).

HYPERENDEMIC - Heavily endemic.

HYPOPHARYNX - A very fine stylet joined to the unpaired salivary duct and used for the injection of saliva during biting.

IMMUNITY - Resistance to parasites or other noxious agents or organisms in a given time, e.g. a year.

INCUBATION PERIOD - The interval of time from the moment of entrance of the microorganisms into the mammal and the first appearance of symptoms of the disease.

INFECTION - The invasion of a host by pathogenic micro-organisms (establishing a host-parasite relationship) and the subsequent multiplication of the latter within the body of the host.

INFESTATION - 1. The invasion of the surface of the body of the host by parasites, 2. The harbouring of parasites on the surface of the body.

INSECT DEVELOPMENT INHIBITORS (IDI) Chemicals which inhibit the development of larvae or pupae

INSECTICIDE A chemical substance or a mixture of substances used to kill insects. It may be applied as a liquid, powder, fine spray or as a vapour. The term "larvicide" is usually applied to an insecticide that kills immature stages of insects, and the term "adulticide" to an insecticide that kills mature or adult forms.

INSECTICIDE RESISTANCE - A characteristic that renders a given population of an insect species resistant to a given insecticide although the species is normally susceptible to that being no longer controlled by the insecticide in the area concerned.

INSECTIVORE -Any insect-eating (insectivorous) group.

INSTAR: - The stage of an insect's life between successive molts, for example the first instar is between hatching from the egg and the first molt.

INTEGRATED CONTROL - Applied pest control that contains and integrates chemical, physical and biological control measures (including the use of natural enemies of pests.

IVERMECTIN - A drug which has been used in the veterinary field for many years, and has now been approved for limited use against human onchocerciasis. So far appears to act as a microfilaricide only.

LABIUM - The lower part of the proboscis forming, together with the labrum, a canal through which the blood is sucked up (food canal) and containing the hypopharynx.

LABRUM - The upper part of the proboscis of mosquito, sealing off the food canal.

LARVA - An immature form occurring in some animals after hatching, differing markedly from the adult form.

LARVICIDES - Chemicals which kill larvae directly or by preventing them from developing into adults.

LIFE CYCLE - Stages of development through which an organism passes between fertilization of the egg of one generation and the same stage in the subsequent generation. The term is sometimes used loosely to mean the number of days between egg- deposition and the attainment of sexual maturity.

MAMMAL - An individual belonging to the class Mammalia, a division of vertebrate animals, including all that possess hair and suckle their young.

MANDIBLE - toothed paired stylets in mouth used for sawing the skin.

MESONOTUM - The notum (tergum) of the mesothorax.

METAMORPHOSIS - In insects, the series of changes through which the insect passes in its development from the larval through the pupal to the adult form. If the pupa is inactive and does not feed, metamorphosis is said to be "complete"; if there is no pupa, or if the pupa is active and feeds, it is said to be "incomplete".

MICROFILARICIDE - A drug which kills microfilaria but not the adult worms. The most common drugs are DEC and ivermectin.

MICROORGANISM - Any microscopic or ultramicroscopic animal or plant, especially any of the bacteria, protozoa, or viruses.

MORTALITY - The action of death on populations; more specifically, the number of deaths in a given population in a given period of time.

NATURAL HOST - A host in which a given pathogenic microorganism (or parasite) is commonly found and in which the pathogen can complete its development. The term "natural host" implies that the host is the usual one; it has the same meaning as "typical host".

NOCTURNAL - Night time, usually applied to an activity cycle. Opposite of diurnal.

NOMENCLATURE - The scientific (Latin) names given to species, genera, families, etc. of animals and plants.

NULLIPAROUS - The term applied to a female mosquito that has not yet laid the eggs.

OBLIGATE - Organism that must live as a parasite of its host.

OLFACTORY - Associated with the sense of smell.

OVARIOLE - Each individual egg in its tube, including the ovarian follicle and oocyte, many of which make up an ovary.

OVIDUCT - An unpaired duct through which the eggs pass, opening into the uterus at its anterior extremity, and dividing into a pair of oviducts, each of which ends in one of the two ovaries of the female mosquito.

OVIPAROUS - Producing young by means of eggs expelled from the body before they are hatched.

OVIPOSITION - The act of depositing eggs.

PARASITE - An organism which derives its nutrition by competing with its host, often with deleterious results to the host, or by feeding upon its tissues.

PAROUS - Having matured and laid eggs. Therefore by inference having taken at least one blood meal. UNIPAR - one gonotrophic cycle; MULTIPAR - more than one gonotrophic cycle.

PATHOGEN - An organism which causes disease.

PATHOGENIC - Causing or giving rise to disease.

PATHOGENS - Microorganisms (germs) such as viruses, bacteria, protozoa and eggs and cysts of worms, which can cause disease if man is infected.

PERIDOMESTIC - Associated with outhouses, storage houses, chicken houses or other animal enclosures, etc., near the house.

PEST - Any mosquito species that is harmful in any way, biting or otherwise, or present in sufficient numbers to be considered a nuisance by man.

PESTICIDE - An agent or a substance or a mixture of substances used to kill species regarded as pests by man.

PHYSIOLOGICAL AGE - The stage of oogenesis (development of the ova) in the female, from which the true age of the insect can be determined with a high degree of precision.

PHYTOPHAGOUS - Feeding on plants.

POLYTENE - Chromosome which have divided many times but have not separated, thereby forming thick bundles of up to a thousand parallel strands. After staining, their banding pattern can be seen under moderate magnification, and they are therefore convenient for examination. They are found in the nuclei of the cells of a few specialized tissues only.

POLYTENE CHROMOSOMES - Chromosomes which have divided many times but have not separated, thereby forming thick bundles of up to a thousand parallel strands. after staining, their banding pattern can be seen under moderate magnification, and they are therefore convenient for examination. they are found in the nuclei of the cells of a few specialized tissues only.

POPULATION DENSITY - The number of individuals of a given population per unit area or volume.

POTENTIAL - Existing and ready for action, but not yet active.

PRECIPITIN TESTING - A serological technique whereby the source of a blood meal can be identified by testing it with antibodies to blood from known host species.

PREDATOR - An animal which preys upon insects or other animals.

PREPARATORY PHASE (Of a control programme) - The planning phase, when areas are mapped and surveyed for infestations, prior to the attack phase.

PROBOSCIS - The mouthparts of mosquito, which form a tube containing the stylets which penetrate the host tissues in order to suck up the host fluids.

PRURITUS - Severe itching.

PULVILLI - Fleshy pads at the extremity of the last joint of the tarsus of the legs of insects; when an insect settles on a surface covered with a contact insecticide, the latter enters the body of the insect through the pulvilli.

PUPA - The free living, highly mobile but not feeding stage of development of a mosquito that occurs between the larval and adult forms.

RESIDUAL TREATMENTS - Treatments with insecticide formulations having a long-term effect (weeks or months).

SIBLING - Very closely related. Literally means brothers or sisters. In taxonomy means the species which make up a single species complex and usually cannot be distinguished by morphological features.

SILVATIC - Not associated with domestic or peridomestic environments, but literally associated with the forest.

SPECIES - The smallest unit of classification commonly used, i.e., the group whose members have the greatest mutual resemblance, are able to interbreed (if not separated in space or time), but not to breed with organisms of other groups (species). Species may be divided into subspecies which differ in certain genetic characters but are able to interbreed.

SPERMATHECA - An organ of the female reproductive system in which mature sperms are stored after copulation.

SPERMATOZOON - The male gamete

SPIRACLES - External openings of the tracheae, the air tubes providing oxygen for the tissues of insects.

STERNITES - Hard chitinous plates.

SUBSPECIES - A geographical group of local populations, differing taxonomically from the other groups of the same type forming part of the species under consideration.

SURVEILLANCE - Observation of a thing, an animal, or a person. The term also implies supervision or inspection.

SURVEY - To examine for some specific purpose; to inspect or consider carefully; to review in detail.

SYNCHRONOUS - Simultaneous or taking place at the same time.

TARSUS - The last segment of the leg of insects; it consists of five joints, to the last of which are attached the claws and pulvilli.

TAXON (plural TAXA) - The basic taxonomical units. Often refers to species, but can refer to an element of the classification at any level.

TAXONOMY - Classification of animals and plants divided into species, genera, families, etc.

TERGITE - Hard chitinous plate forming part of the cuticle of a segment (dorsal surface).

TERMINALIA - The last two or three visible segments and their appendages at the tip of the abdomen which make up the externally visible sex organs of either sex.

TERRESTRIAL - Living on the ground.

THORAX - Flies and other insects are divided into three regions: head, thorax and abdomen. The thorax carries the legs and wings.

TIBIA - The fourth segment of an insect's leg, distal to the femur.

TROCHANTER - The second segment of an insect's leg, distal to the coxa.

ULV - Ultra-low-volume concentrate.

VECTOR - An arhtropod or other animal that carries a parasite from one host to another host. The vector may or may not be essential for the completion of the life cycle or the parasite. If it is not essential, it is referred to as a "mechanical vector".

VENTER - Underneath surface of the abdomen.

VENTRAL - Pertaining to the front or lower surface.

VISCERA - Internal organs of the abdomen.

VOLATILITY - Capability to evaporate rapidly.

WETTABLE POWDER - A water dispersable but insoluble, formulation of an insecticide, in which the active ingredient is bound to an inert insoluble carrier material such as talcum powder. Wettable powder formulations are mixed with water to be sprayed, and usually have greater residual activity than liquid formulations of the same insecticide.

ZOONOSIS - A disease of animals that may be transmitted under natural conditions to man.

ZOOPHILIC - Feeding on animals other than man.

ZOOTIC - Pertaining to animals.

APPENDIX - I

A SYSTEMATIC CATALOGUE OF INDIAN MOSQUITOES

Following Christophers' (1933) and Barraud's (1934) *magnum opuses* on Tribes Anophelini, and Culicini and Megarhinini published under the famous Fauna of British India series, though rife with several dark points and requiring an urgent revision, mainly in view of recent renaming of several genera (e.g., *Megarhinus* is now known as *Toxorhynchites*, while *Harpagomyia* is renamed as *Malaya*), change in status of certain Tribes and/or Subfamilies (e.g., Family Culicidae comprises now only three subfamilies Anophelinae, Culicinae and Toxorhynchitinae; the subfamilies Dixinae and Chaoborinae having been afforded family rank as Dixidae and Chaoboridae), discovery of new species, synonimization of several taxa, raising certain subspecies/ varieties to species level or subsuming a few species into others, and more importantly availability and application of more correctly interpretable sophisticated modern diagnostic tools (e.g. application of PCR in taxonomy and removing ambiguities in identifications of certain closely related taxa or even sibling species), very little has indeed been added to the existing knowledge of mosquito taxonomy and systematics in India particularly when compared to that in certain developed nations like United States of America and Australia, where more and more new species have been described with every passing decade in past, even though the faunal wealth is far more greatly richer and diversified in this country. It may be noted here that most of the species (>90%) have been

---

*Rao, T.R., 1984. The Anophelines of India. Malaria research Centre, ICMR, India.

**Rao, T.R., V. Dhanda, H.R. Bhat and S.M. Kulkarni, 1973. A survey of haematophagous arthropods in western Himalayas, Sikkim and hill districts of West Bengal, a general account. Indian J. Med. Res. 61:1421-1458.

***Reuban, R., S.C. Tewari and J. Hiriyan, 1993. Studies of the mosquito fauna of south India (In: Entomologist Extraordinary: a festschrift in honour of Botha De Meillon). South African Institute for Medical Research, Johannesburg, pp. 97-104.

described during the British India period which, however, originated mostly from a few specific areas, such as the far north-east region, the peninsular India along with the Andaman-Nicobar Ils. and, to lesser extent, from the northeast and -western lower Himalayan ranges. Presumably, this occurred due possibly to the comprehensible reasons of postings on transfers or sometimes commissioned mosquito-related peregrinations of the British culicidologists to these areas. Still more interesting is the fact that Sir S.R. Christophers, immortalized by his famous and unparalleled FBI volume on Indian anopheline mosquitoes, was a medical officer by profession. Mosquitoes in the British India period seemed to have been surveyed and examined systematically more as a need of the Colonial administration to effect better governance in the wake of repeated mosquito-borne disease epidemics, particularly those of malaria, than a necessity to inventorize the specific faunal wealth in the country. This clearly alludes that the presently available information on mosquito species from India is but far from being realistic, warranting more long-term mosquito-based surveys from all the nooks and corners of the country to bring on record the extant mosquito wealth for an everlasting benefit to both the scientists and the science.

Among the most notable works on Indian mosquito taxonomy, including development of annotated check-lists, during last five decades after the country sought independence, following works are indispensably referable: Puri (1949, 1960), Wattal and Kalra (1961), Wattal (1963), Roy and Brown (1970), Rao (1984*), Das et al. (1990) and Nagpal and Sharma (1995). Ironically all these references deal with the anophelines, leaving behind a great void in respect to currently acceptable inventory for the culicines, albeit presence of a few important yet isolated and fragmentary attempts on regional basis such as those in the Western Himalaya and Southern India including Andaman & Nicobar Ils. (*cf.* Rao et al., 1973**; Reuban et al., 1993***). International culicidologists like Ross and Roberts (1943), Russel et al. (1943), Peyton and Scanlon (1966), Reid (1968), Harrison and Scanlon (1977) etc. have published exhaustive taxonomic accounts with comprehensible keys for the Oriental anophelines, while some others like Stone (1961), Stone and Delfinado (1973), Knight and Stone (1977) etc. have dealt with mosquitoes of the world, including the Oriental region that comprised Indian fauna as well.

The present inventory enlists a total of 330 species under 19 genera and three families, along with several subspecies and/or varieties (Table 1). Basically, the classificatory system followed in this check-list is essentially that of Christophers (1933) and Barraud (1934), with inputs from Knight and Stone (1977) and Rao (1984). Names of certain genera/species have been replaced by the ones currently in vogue (*vide supra*). Even so, enlisting authentically all good species is not an easy task, particularly with respect to culicines which have certainly been attended scientifically much less frequently and deeply than anophelines. The present inventory has been developed after meticulously considering all aspects of biogeography and taxonomic status of all the taxa extant in India, although few variations may always occur in view of increasing amount of newer taxonomic and evolutionary interpretations being regularly put forward.

Table 1. Details of species and genera under the three mosquito families represented in India

FAMILY Subfailiy	Genera	Species
CULICIDAE		
Toxorhynchitinae	1	6
Culicinae	15	252
Anophelinae	1	58
DIXIDAE	1	12
CHAOBORIDAE	1	2
Total	***19***	***330***

In the present checklist hierarchical level of a taxon is presented up to species only, with subspecies and/or varieties associated to any species being also mentioned. All the species names are alphabetically arranged under a given subgenus and/or genus under a subfamily or family, as the case may be. As stated above, the family Culicidae is having three subfamilies, viz., Anophelinae, Culicinae and Toxorhynchitinae. Two more families of mosquitoes, viz., Dixidae and Chaoboridae, are additionally listed (*cf.* Knight and Stone, 1977).

(1) FAMILY CULICIDAE

I. Subfamily Toxorhynchitinae

1. Genus **TOXORHYNCHITES** (= Megarhinus) Theobald, 1901

1. *T. albipes* Edwards, 1822
2. *T. edwardsi* Barraud, 1924
3. *T. gravelyi* Edwards, 1921
4. *T. kempi* Edwards, 1921
5. *T. minimus* Theobald, 1905
6. *T. splendens* (Wiedemann), 1819

II. Subfamily Culicinae

1. Genus **TRIPTEROIDES** Giles, 1904

7. *T. affinis* (Edwards), 1913
8. *T. aranoides* (Theobald), 1901
9. *T. edwardsi* (Barraud), 1929
10. *T. powelli* (Ludlow), 1909
    var. *indicus* (Barraud), 1929
11. *T. serratus* (Barraud), 1929

12. *T. similis* (Leicester), 1908

2. Genus **TOPOMYIA** Leicester, 1908

13. *T. aureoventer* (Theobald), 1910

3. Genus **MALAYA** Leicester, 1908

14. *M. genurostris* (Leicester), 1908
15. *M. jacobsoni* Edwards, 1930

4. Genus **HODGESIA** Theobald 1904

16. *H. bailyi* Barraud, 1929

5. Genus **URANOTAENIA** Lynch Arribalzaga, 1891

17. *U. alboannulata* (Theobald), 1905
18. *U. annandalei* Barraud, 1926
19. *U. atra* Theobald, 1905
20. *U. bicolor* Leicester, 1908
21. *U. bimaculata* Leicester, 1908
22. *U. campestris* Leicester, 1908
23. *U. christophersi* Barraud, 1926
24. *U. edwardsi* Barraud, 1926
25. *U. hebes* Barraud, 1931
26. *U. longirostris* Leicester, 1908
27. *U. luteola* Barraud, 1934
28. *U. macfarlanei* Edwards, 1914
29. *U. maculipleura* Leicester, 1908
30. *U. maxima* Leicester, 1908
31. *U. micans* Leicester, 1908
32. *U. nivipleura* Leicester, 1908
33. *U. novobscura* Barraud, 1934
34. *U. orientalis* Barraud, 1926
35. *U. recondita* Edwards, 1922
36. *U. stricklandi* Barraud, 1926
37. *U. testacea* Theobald, 1905
38. *U. unguiculata* Edwards, 1913

6. Genus **THEOBALDIA** Neveu-Lemaire, 1902

39. *T. indica* Edwards, 1920
40. *T. longiareolata* (Macquart, 1938)
41. *T. niveitaeniata* (Theobald), 1907

7. Genus **ORTHOPODOMYIA** Theobald, 1904

42. *O. albipes*, Leicester in Theobald, 1904
43. *O. anopheloides* (Giles), 1903
    var. *maculata* Theobald, 1910

var. *maculipes* Theobald, 1901
var. *andamanensis* Barraud, 1934

44. *O. flavicosta* Barraud, 1927
45. *O. flavithorax* Barraud, 1927

8. Genus **FICALBIA** Theobald, 1903

Subgenus **Mimomyia** Theobald, 1903

46. *F. chamberlaini* (Ludlow), 1904
47. *F. hybrida* (Leicester), 1908

Subgenus **Etorleptiomyia** Theobald

48. *F. fusca* (Leicester), 1908
49. *F. luzonensis* (Ludlow), 1905

Subgenus **Ficalbia** s. str.

50. *F. minima* (Theobald), 1901

9. Genus **MANSONIA** Blanch, 1901

Subgenus **Coquillettidia** Dyar 1905

51. *M. crassipes* (van der Wulp), 1892
52. *M. novochracea* (Barraud), 1927
53. *M. ochracea* (Theobald), 1903

Subgenus **Mansonioides** Theobald, 1907

54. *M. annulifera* (Theobald), 1901
55. *M. longipalpis* (van der Wulp), 1892
56. *M. uniformis* (Theobald), 1901
57. *M. indiana* Edwards, 1930

10. Genus **AEDOMYIA** Theobald, 1901

58. *A. venustipes* (Skuse), 1889

11. Genus **AEDES** Meigen, 1818

Subgenus **Mucidus** Theobald, 1901

59. *A. laniger* (Wiedemann) 1821
60. *A. scatophagoides* Theobald, 1901

Subgenus **Ochlerotatus** Lynch Arribalzaga, 1891,

61. *A. caspius* (Pallas) 1771
62. *A.* pulchritarsis (Rondani), 1872
    var. *asiaticus* Edwards 1926
    var. *versicolor* Barraud 1924
63. *A.* pullatus (Coquillett), 1904

Subgenus **Finlaya** Theobald, 1903

64. *A. albocinetus* Barraud, 1924

65. *A. albolateralis* (Theobald), 1908
66. *A. alboniveus* Barraud, 1934
67. *A. albotaeniatus* (Leicester), 1904
    var. *mikiranus* Edwards, 1922
68. *A.* aureostriatus (Doleschall), 1857
69. *A. auronitens* Edwards, 1922
70. *A. assamensis* (Theobald), 1908
71. *A. cacharanus* Barraud, 1923
72. *A. christophersi* Edwards, 1922
73. *A. chrysolineatus* (Theobald), 1907
74. *A. cogilli* Edwards, 1922
75. *A. deccanus* Barrauds 1923
76. *A. dissimilis* Leicester, 1908
    var. *karwari* Barraud, 1924
77. *A. elsiae* Barraud, 1923
78. *A. formosensis* Yamada, 1921
79. *A. gilli* Barraud, 1924
80. *A. greeni* (Theobald), 1903
81. *A. gubernatoris* (Giles), 1901
82. *A. harveyi* Barraud, 1923
83. *A. inquinatus* Edwards, 1922
84. *A. Khazani* Edwards, 1922
85. *A. lophoventralis* (Theobald), 1910
86. *A. macdougalli* Edwards, 1922
87. *A. mohani* Knight, 1969
88. *A.* niveoides Barraud, 1934
89. *A. niveus* (Ludlow), 1903
90. *A.* novoniveus Barraud, 1934
91. *A. oreophilus* (Edwards), 1916
92. *A. pallirostris* Edwards, 1922
93. *A. poecilus* Theobald, 1903
94. *A. prominens* Barraud, 1923
95. *A. pseudotaeniatus* (Giles), 1901
96. *A. pulchriventer* (Giles), 1901
97. *A. saxicola* Edwards, 1922
98. *A. shortti* Barraud, 1923
99. *A. simlensis* Edwards, 1922
100. *A. sintoni* Barraud, 1924
101. *A. stevensoni* Barraud, 1923
102. *A. subsimilis* Barraud, 1927
103. *A. suffusus* (Edwards), 1922
104. *A. unicinctus* Edwards, 1922

Subgenus **Christophersiomyia** Barraud, 1923

105. *A. annulirostris* (Theobald), 1905
106. *A. ibis* Barraud, 1931
107. *A. gombakensis* Mattingly, 1959
108. *A. thomsoni* Theobald, 1905

Subgenus **Rhinoskusea** Edwards, 1929

109. *A. longirostris* (Leicester), 1908
110. *A. portnovoensis* Tewari & Hiriyan, 1991

Subgenus **Stegomyia** Theobald, 1901

111. *A. aegypti* (Linnaeus), 1762
112. *A. albolineatus* (Theobald), 1904
113. *A. albopictus* (Skuse), 1894
114. *A. annandalai* Theobald, 1910
    var. quadricinctus Barraud, 1923
115. *A. craggi* Barraud, 1923
116. *A. desmotes* Giles, 1904
117. *A. edwardsi* Barraud, 1923
118. *A. flavopictus* Yamada, 1921
119. *A. krombeini* Huang, 1975
120. *A. mediopunctatus* Theobald, 1905
    var. *perplexus* Leicester, 1908
    var. *submediopunctatus* Barraud, 1923
    var. *sureilensis* Barraud, 1934
121. *A. novalbopictus* Barraud, 1931
122. *A. pseudalbopictus* Borel, 1928
123. *A. subalbopictus* Barraud, 1931
124. *A. scutellaris* (Walker), 1859
125. *A. unilineatus* (Theobald), 1906
126. *A. vittatus* (Bigot), 1861
127. *A.* w-*albus* Theobald, 1905

Subgenus **Aedimorphus** Theobald, 1903

128. *A. alboscutellatus* (Theobald), 1905
129. *A. caecus* (Theobald), 1901
130. *A. culicinus* Edwards, 1922
131. *A. jamesi* (Edwards), 1914
132. *A. lowisi* Theobald, 1910
133. *A. nigrostriatus* Barraud, 1927
134. *A. niveoscutellum* Theobald, 1905
135. *A. nummatus* Edwards, 1923
136. *A. ostentatio* (Leicester), 1908

137. *A. pallidostriatus* (Theobald), 1907
138. *A. pipersalatus* (Giles), 1901
139. *A. punctifemore* (Ludlow), 1921
140. *A. stenoetrus* (Theobald), 1907
141. *A. syntheticus* Barraud, 1928
142. *A. taeniorhynchoides* (Christophers), 1911
143. *A. trimaculatus* (Theobald),1905
144. *A. vexans* (Meigen), 1830

Subgenus **Banksinella** Theobald, 1907

145. *A. lineatopennis* (Ludlow), 1905

Subgenus **Diceromyia** Theobald, 1911

146. *A. agastyai* Tewari & Hiriyan, 1991
147. *A. kanarensis* Edwards, 1934
148. *A. reubenae* Tewari & Hiriyan, 1991
149. *A. periskeletus* (Giles), 1902
150. *A. iyengari* Edwards, 1923
151. *A. micropterus* (Giles), 1901
152. *A. reginae* Edwards, 1922
153. *A. ramachandrai* Reuben, 1967

Subgenus **Aedes** Meigen, 1818

154. *A. abditus* Barraud, 1931
155. *A. agrestis* Barraud, 1931
156. *A. andamanensis* Edwards, 1922
157. *A. atrius* Barraud, 1928
158. *A. butleri* Theobald, 1901
159. *A. cautus* Barraud, 1928
160. *A. clavatus* Barraud, 1931
161. *A. comatus* Barraud, 1931
162. *A. hirsutipleura* Barraud, 1928
163. *A. indicus* (Theobald), 1907
164. *A. lugubris* Barraud, 1928
165. *A. pseudodiurnus* Theobald, 1910
166. *A. pseudomediofasciatus* (Theobald), 1910
167. *A. rami* Barraud, 1928
168. *A. sigmoides* Barraud, 1928
169. *A. vallistris* Barraud, 1928
170. *A. uniformis* (Theobald), 1910
171. *A. yusafi* Barraud, 1931

Subgenus **Cancraedes** Edwards, 1929

172. *A. cancricomes* Edwards, 1922

173. *A. simplex* (Theobald), 1903

Subgenus **Paraedes** Edwards, 1934

174 *P. barraudi* Edwards, 1934
175. *P. argyrurus* Edwards, 1934

12. Genus **HEIZMANNIA** Ludlow, 1905

176. *H. chandi* Edwards, 1922
177. *H. complex* (Theobald), 1910
178. *H. covelli* Barraud, 1929
179. *H. funerea* Leicester, 1908
180. *H. greeni* (Theobald), 1905
181. *H. himalayensis* Edwards, 1922
182. *H. indica* (Theobald), 1905
183. *H. metallica* (Leicester), 1908
184. *H. viridis* Barraud, 1929

13. Genus **HAEMAGOGUS** Williston, 1896

185. *H. discrepans* Edwards, 1922
186. *H. tripunctatus* (Theobald), 1908

14. Genus **ARMIGERES** Theobald, 1901

Subgenus **Armigeres** Theobald, 1901

187. *A. aureolineatus* (Leicester), 1908
188. *A. kuchingensis* Edwards, 1915
189. *A. obturbans* Walker, 1860
190 *A. subalbatus* (Coquillett), 1898
191. *A. theobaldi*, Barraud, 1934

Subgenus **Leicesteria** Theobald, 1904

192. *A. annulipalpis* (Theobald), 1910
193. *A. annulitarsis* Leicester, 1908
194. *A. dentatus* Barraud, 1927
195. *A. digitatus* Edwards, 1914
196. *A. flavus* (Leicester), 1908
197. *A. inchoatus* Barraud, 1927
198. *A. longipalpis* Leicester, 1904
199. *A. magnus* (Theobald), 1908
200. *A. omissus* Edwards, 1914

15. Genus **CULEX** Linnaeus, 1758

Subgenus **Lutzia** Theobald, 1903

201. *C. fuscanus* Wiedemann, 1820
202. *C. halifaxi* Theobald, 1903

203. *C. raptor* Edwards, 1922
204. *C. vorax* Edwards, 1921

Subgenus **Barraudius** Edwards, 1921

205. *C. modestus* Ficalbi, 1890

Subgenus **Eumelanomyia** Theobald, 1909

206. *C. brevipalpis* Giles, 1902
207. *C. castrensis* Edward, 1922
208. *C. iphis* Barraud, 1924
209. *C. Khazani* Edward, 1922
210. *C. malayi* Leicester, 1908
211. *C. mohani* Sirivanakarn,
212. *C. pluvialis* Barraud, 1924
213. *C. tenuipalpis* Barraud, 1924

Subgenus **Lophoceratomyia** Theobald, 1905

214. *C. cinctellus* Edwards, 1922
215. *C. flavicornis* Barraud, 1924
216. *C. fraudatrix* Theobald, 1905
217. *C. infantulus* Edwards, 1922
218. *C. lasiopalpis* Sirivanakarn
219. *C. mammilifer* Leicester, 1908
220. *C. minor* Leicester, 1908
    var. *bengalensis* Barraud, 1934
221. *C. minutissimus* (Theobald), 1907
222. *C. pholeter* Bram & Rattanarithikul, 1967
223. *C. plantaginis* Barraud, 1924
224. *C. seniori* Barraud, 1934
225. *C. raghavanii* Rahman, Chowdhury and Kalra, 1968
226. *C. rubithoracis* Leicester, 1908
227. *C. uniformis* Theobald, 1905

Subgenus **Culiciomyia** Theobald, 1907

228. *C. bailyi* Barraud, 1934
229. *C. fragilis* Ludlow, 1903
230. *C. pallidothorax* Theobald, 1905
231. *C. nigropunctatus* Edwards, 1926
232. *C. shebbearei* Barraud, 1924
233. *C. spathifurca* Edwards, 1915
234. *C. viridiventer* Giles, 1901
235. C. *ramakrishnii* Wattal and Kalra, 1965

Subgenus **Culex** Linnaeus, 1758

236. *C. barraudi* Edwards, 1922

237. *C. bitaeniorhynchus* Giles, 1901
     var. *ambiguus* Theobald, 1903
238. *C. cornutus* Edwards, 1922
239. *C. edwardsi* Barraud, 1923
240. *C. epidesmus* (Theobald), 1910
241. *C. fuscifurcatus* Edwards, 1934
242. *C. fuscitarsis* Barraud, 1924
243. *C. fuscocephalus* Theobald, 1907
244. *C. gelidus* Theobald, 1904
245. *C. hutchinsoni* Barraud, 1924
246. *C. mimulus* Edwards, 1915
247. *C. mimeticus* Noe, 1899
     var. *mimuloides* Barraud, 1924
248. *C. nilgiricus* Edwards, 1916
249. *C. quinquefasciatus* Say, 1823
250. *C. sinensis* Theobald, 1903
251. *C. sitiens* Wiedemann, 1928
252. *C. theileri* Theobald, 1903
253. *C. tritaeniorhynchus* Giles, 1901
254. *C. univittatus* Theobald, 1901
255. *C. vagans* Wiedemann, 1928
256. *C. vishnui* Theobald, 1901
257. *C. whitei* Barraud, 1923
258. *C. whitmorei* (Giles), 1904

III. Subfamily Anophelinae

1. Genus **ANOPHELES** Meigen, 1818

Subgenus **Anopheles** Meigen, 1818

259. *A. ahomi* Choudhury, 1929
260. *A. aitkeni* James, 1903
261. *A. annandalai* Prashad, 1918
262. *A. argyropus* (Swellengrebel, 1914)
263. *A. baileyi* Edwards, 1929
264. *A. barbirostris* van der Wulp, 1884
265. *A. barbumbrosis* Strickland and Choudhury, 1927
266. *A. barianensis* James, 1911
267. *A. bengalensis* Puri, 1930
268. *A. crawfordi* Reid, 1953
269. *A. culiciformis* Cogill, 1903
270. *A. gigas* Giles, 1901
     var. *simlensis* James, 1911
     var. *baileyi* Edwards, 1929

271. A. *insulaeflorum* (Swellengrebe & Swellengrebel de Graff), 1920
272. A. *interruptus* Puri, 1929
273. A. *lindesayi* Giles, 1900
274. A. *nilgiricus* Christophers, 1924
275. A. *nigerrimus* Giles, 1900
276. A. *nitidus* Harrison, Scanlon and Reid, 1973
277. A. *peditaeniatus* Leicester, 1908
278. A. *pinjaurensis* Barraud, 1932
279. A. *roperi* Reid, 1950
280. A. *sinensis* Wiedemann, 1828
281. A. *sintoni* Puri, 1920
282. A. *umbrosus* Theobald, 1903

Subgenus **Cellia** Theobald, 1902

283. A. *aconitus* Donitz, 1902
284. A. *annularis* Van der Wulp, 1884
285. A. *balabacensis* Baisas, 1936
286. A. culicifacies Giles, 1901
287. A. *dirus* Peyton and Harrison, 1979
288. A. *d'thali* Patton, 1905
289. A. *elegans* James, 1903
290. A. *fluviatilis* James, 1902
291. A. *jamesi* Theobald, 1901
292. A. *jeyporiensis* James, 1902
    var. *candidiensis* Koidzumi, 1924
293. A. *karwari* (James, 1903)
294. A. *kochi* Grassi, 1899
295. A *maculatus* Theobald, 1901
    var. *willmori* (James, 1903)
296. A. *majidi* Young & Majid, 1928
297. A. *minimus* Theobald, 1901
298. A. *moghulensis* Christophers, 1924
299. A. *multicolor* Camboulin, 1902
300. A. *nivipes* Theobald, 1903
301. A. *pallidus* Theobald, 1901
302. A. *philippinensis* Ludlow, 1902
303. A. *pseudojamesi* Strickland and Choudhary, 1927
304. A. *pseudowillmori* Theobald, 1910
305. A. *pulcherrimus* Theobald, 1902
306. A. *sergentii* Theobald, 1907
307. A. *splendidus* koidzumi, 1920
308. A. *stephensi* Liston, 1901

var. mysorensis Sweet and Rao, 1937

309. *A. subpictus* Grassi, 1899

var. vadakadiensis Doraisamy

310. *A. sundaicus* Rodenwaldt, 1926
311. *A. tesselatus* Theobald, 1901
312. *A. theobaldi* Giles, 1901
313. *A. turkhudi* Liston, 1901
314. *A. vagus* Donitz, 1902
315. *A. varuna* Iyengar, 1924
316. *A. willmorei* (James), 1903

(2) FAMILY DIXIDAE

1. Genus **DIXA** Meigen, 1818

Subgenus **Dixa** s. str.

317. *D. bifasciata* Brunetti, 1911
318. *D. bistriata* Brunetti,1911
319. *D. christophersi* Edwards, 1934
320. *D. maculipennis* Brunetti, 1911
321. *D. montana* Brunetti, 1911
322. *D. ochrilineata* Brunetti, 1911
323 *D. platystyla* Edwards, 1934
324. *D. trinotata* Edwards, 1934
325. *D. zeylanica* Senior White, 1924
326. *Dixa* spec. indet. 1
327. *Dixa* spec. indet. 2

Subgenus **Paradixa** Tonnoir, 1924

328. *D. kashmirensis* Edwards, 1934

(3) FAMILY CHAOBORIDAE

1. Genus **CHAOBORUS** Lichtenstein, 1800

Subgenus **Sayomyia** Coquillett, 1903.

329. C. *asiaticus* Giles, 1901
330. C. *indicus* Giles, 1905

- - - - - - - - -

INDEX

EXPLANATION TO PLATES

Plate No. 1. *Anopheles* mosquito: (a) Larva, (b) Pupa, and (c) Engorged female adult.

Plate No. 2. *Culex quinquefasciatus*: (a) Larvae, (b) Pupa, (c) Emerging adult, (d) Male, (e) Unfed female, and (f) Engorged female.

Plate No. 3. *Aedes aegypti*: (a) Pupa, (b) Engorged female, (c) Unfed female, and (d) Female at feeding job.

Plate No. 4. (a-d) *Aedes aegypti* in various stages of feeding.

Plate No. 5. *Mansonia uniformis*: Eggs laid in raft, and attached to plant, (b) Larvae attached to plant root, (c) Pupa, (d) Unfed female, and (e) Engorged female.

Plate No. 6. *Coquillettidia* mosquito: (a) Larva attached to plant tissue, and (b) Female;- *Toxorhynchites* mosquito: (c) Larva, (d) Pupa, and (e) Adult (note the downwardly curved proboscis adapted for feeding on plant juices).

Plate No. 7. (a) *Culiseta* sp., (b) *Aedes vexans*, (c) *Ochlerotatus* sp., and (d) *Aedeomyia* sp.

Plate No. 8. Different types of mosquito breeding habitats: (a) Seepage water, (b) 'tanka', an underground household/community water reservoir in desert environment, (c) Weeded canal water, (d) 'beri', another type of underground water reservoir, built in the concavity of a pond, in the desert environment, and (e) Discarded tyres.

Plate - 1

a

b

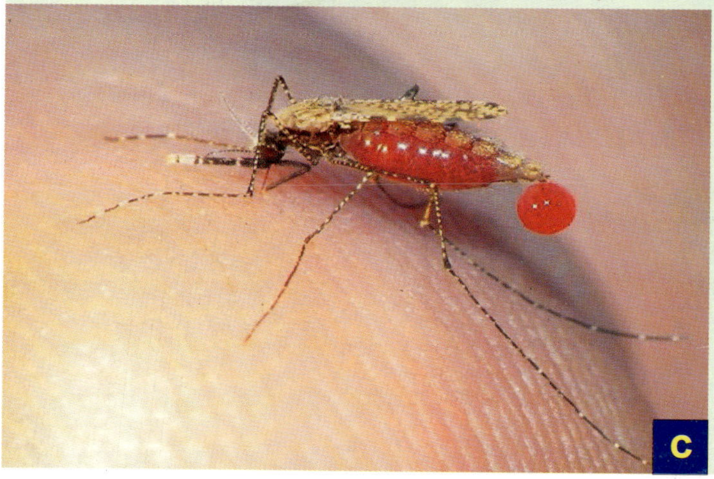

c

Plate - 2

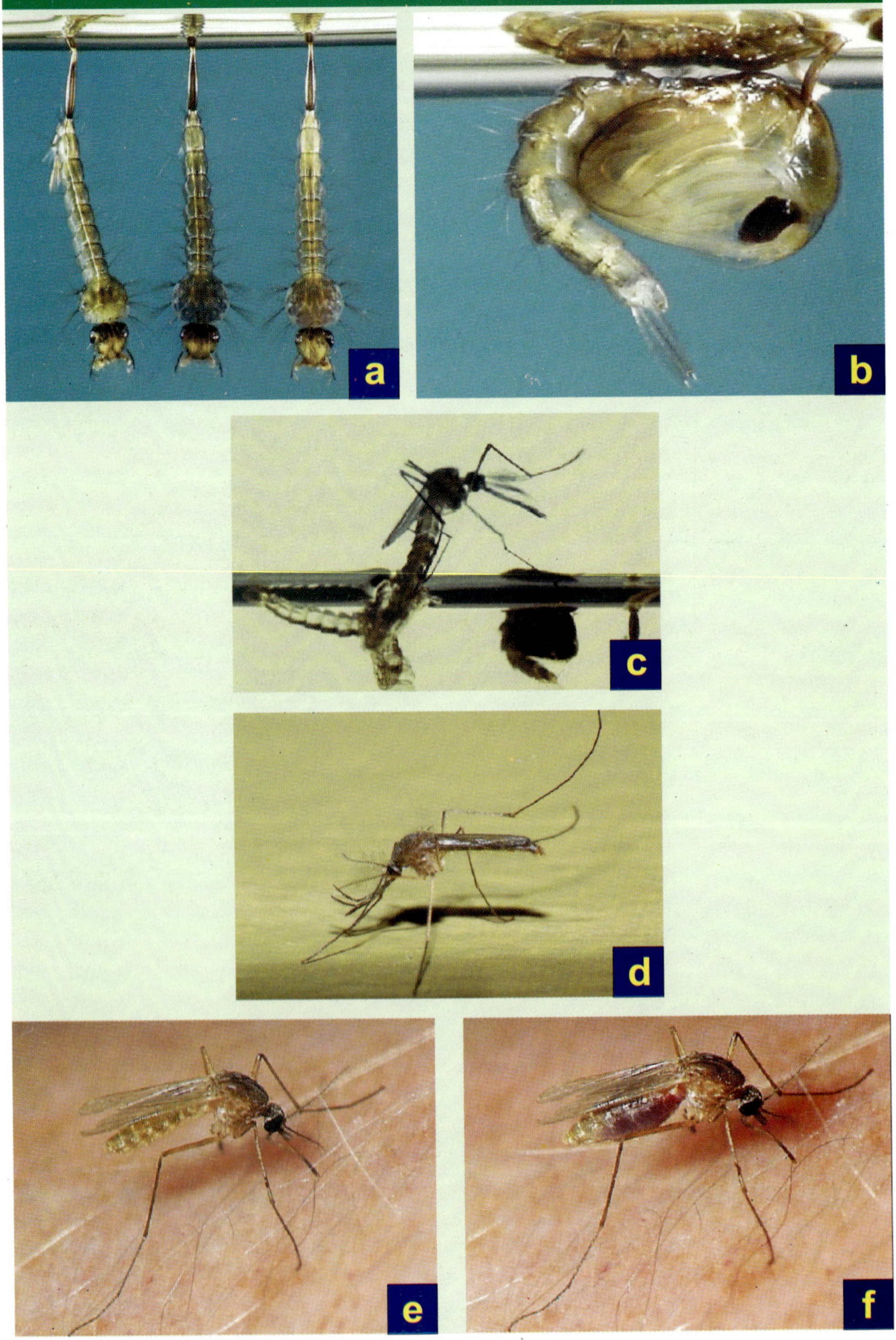

Plate - 3

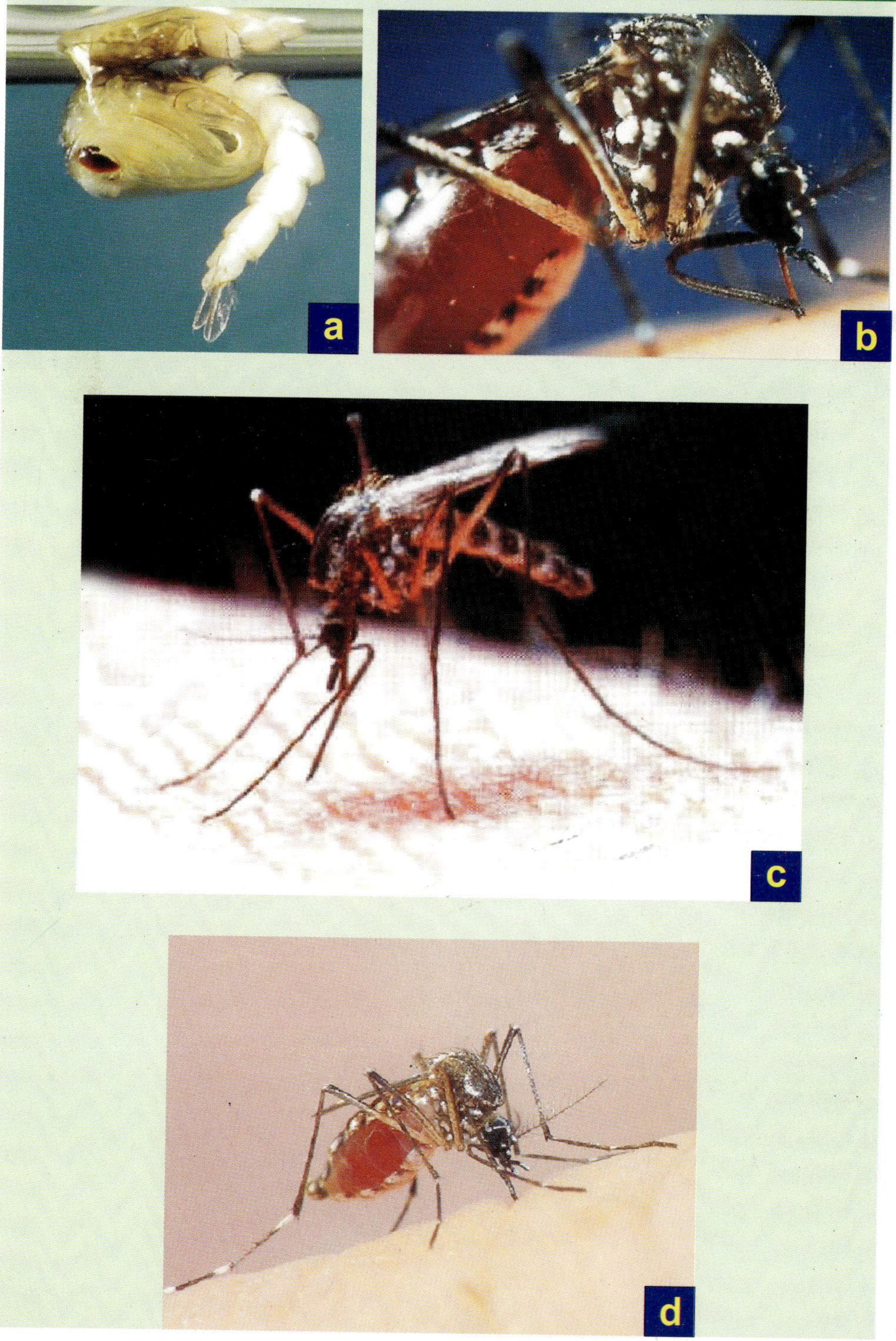

Plate - 4

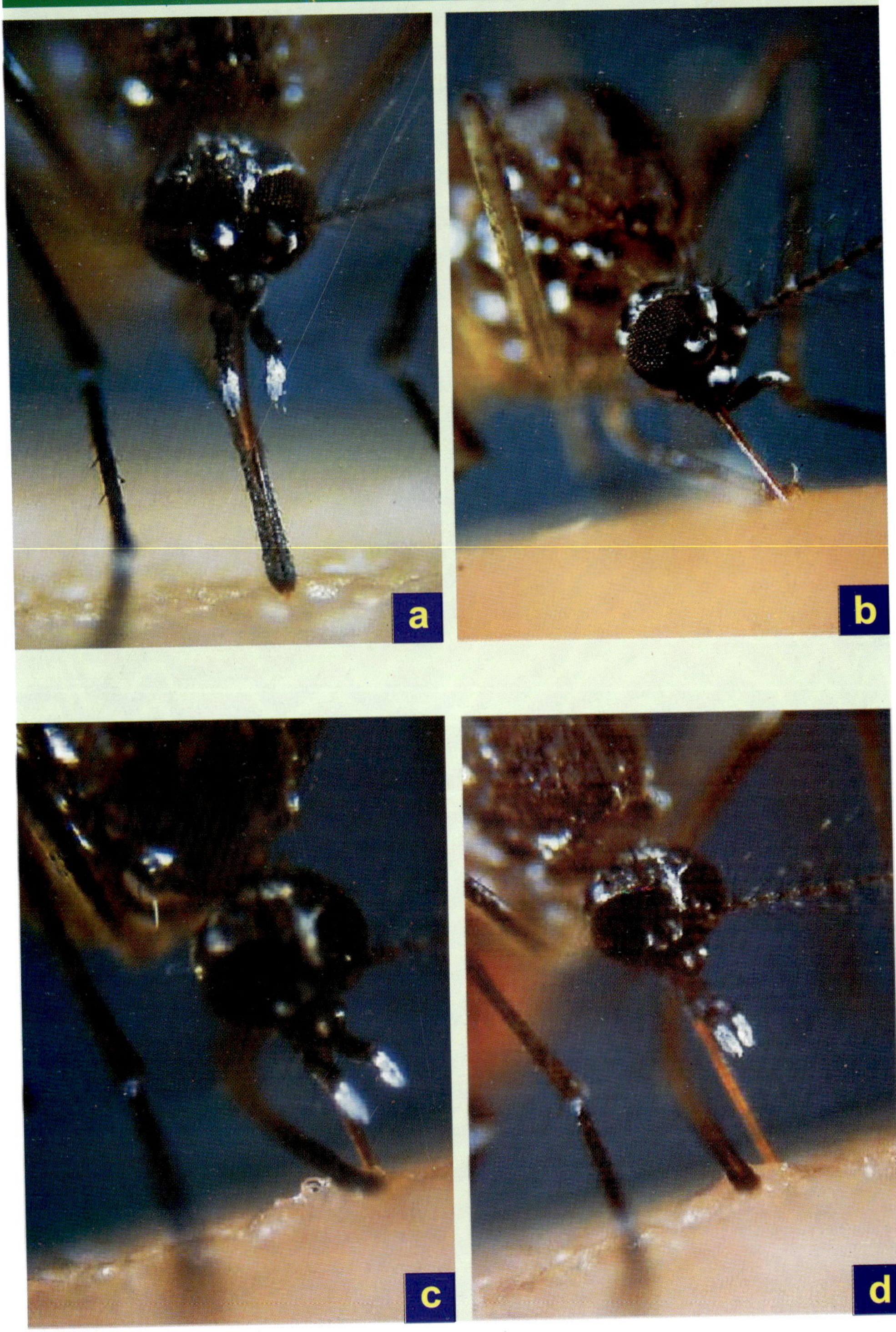

Plate - 5

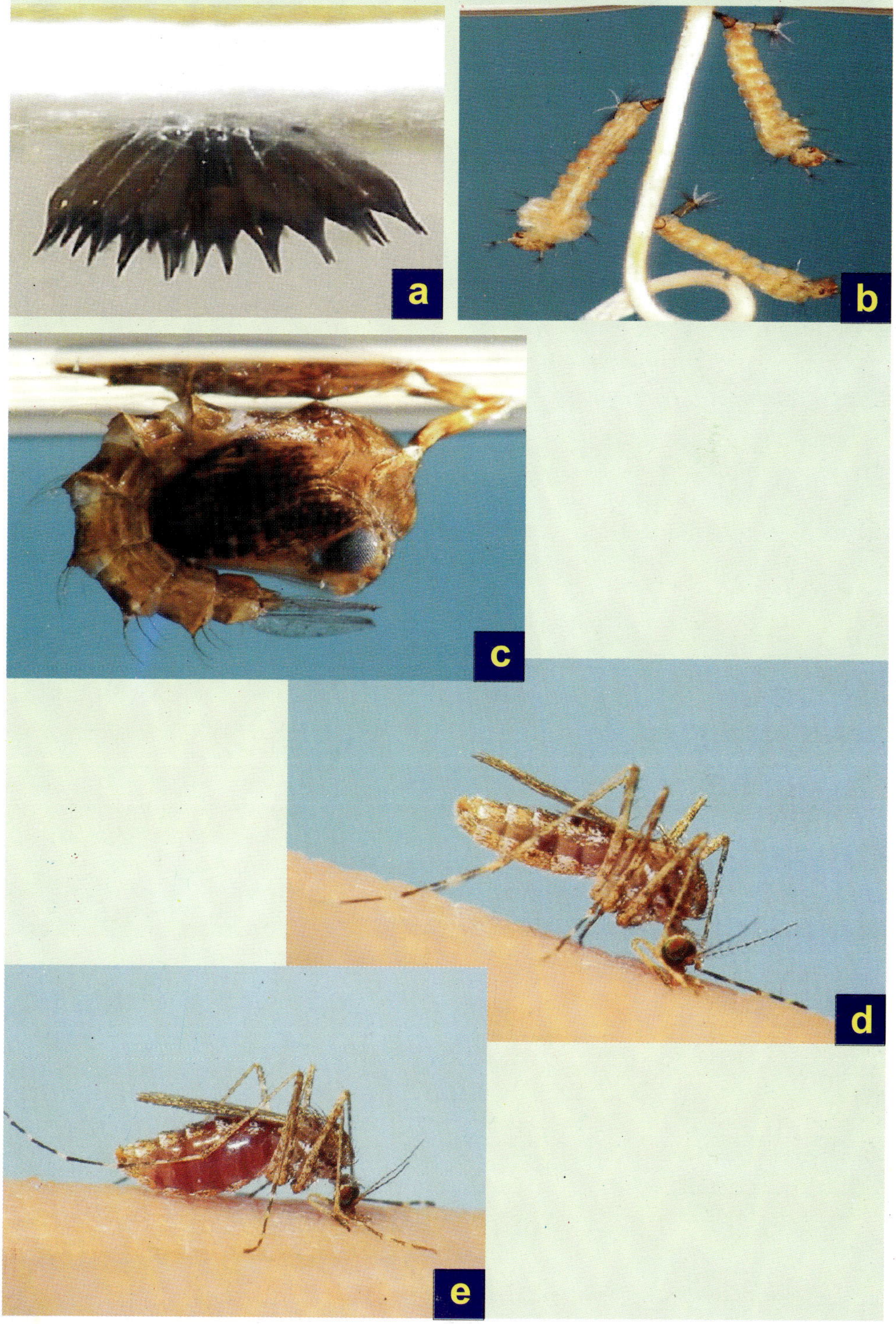

Plate - 6

Plate - 7

Plate - 8

Errata

Reference	Read	Instead
p. vi, para last, line last	anthropophagic	anthropoagic
page Contents	*Foreword*	*Foreward*
p. 2, para 2, line 4 from end	*female fatale* (or *femme fatale*)	*fenmale fatale*
p. 42, item (6), line 4	1995	1985
p. 51, last para, line 3 from end	"the Peking man"	""the Pekin man"
p. 181, para 2, line 8 from end	*Plasmodium falciparum*	*Plasmodium malariae*
p. 218, item 2, line 3	*feng, shui* and *chi*	*fang, sui* and *chi*
p. 229, Table 14.2, col. Area, last line under 1908	Punjab-Amritsar, Delhi, Palwal, Miani, Gujarat	Punjab
p. 229, Table 14.2, col. Mortality rate	Amritsar (62.5/1000 pop.), Delhi (100/1000 pop.), Palwal (420/1000 pop.), Miani (403/1000 pop.), Gujarat (247/1000 pop.).	Unknown